Sustainable Development and Biodiversity

Volume 13

Series editor

Kishan Gopal Ramawat
Botany Department, M.L. Sukhadia University, Udaipur, Rajasthan, India

This book series provides complete, comprehensive and broad subject based reviews about existing biodiversity of different habitats and conservation strategies in the framework of different technologies, ecosystem diversity, and genetic diversity. The ways by which these resources are used with sustainable management and replenishment are also dealt with. The topics of interest include but are not restricted only to sustainable development of various ecosystems and conservation of hotspots, traditional methods and role of local people, threatened and endangered species, global climate change and effect on biodiversity, invasive species, impact of various activities on biodiversity, biodiversity conservation in sustaining livelihoods and reducing poverty, and technologies available and required. The books in this series will be useful to botanists, environmentalists, marine biologists, policy makers, conservationists, and NGOs working for environment protection.

More information about this series at http://www.springer.com/series/11920

K.G. Ramawat · M.R. Ahuja
Editors

Fiber Plants

Biology, Biotechnology and Applications

Springer

Editors
K.G. Ramawat
Botany Department
M.L. Sukhadia University
Udaipur, Rajasthan
India

M.R. Ahuja
Genetics and Biotechnology
Zobel Forestry Associates
New Paltz, NY
USA

ISSN 2352-474X ISSN 2352-4758 (electronic)
Sustainable Development and Biodiversity
ISBN 978-3-319-44569-4 ISBN 978-3-319-44570-0 (eBook)
DOI 10.1007/978-3-319-44570-0

Library of Congress Control Number: 2016948625

Printed on acid-free paper

This Springer imprint is published by Springer Nature
The registered company is Springer International Publishing AG
The registered company address is: Gewerbestrasse 11, 6330 Cham, Switzerland

Preface

Fiber plants have been integral part of the human society. Fiber and subsequently fabric preparation was associated with rise and fall of various civilizations as well as considered as a parameter of living standards. Cultivation of fiber crops is as old as human civilization. Acceleration in population growth, reduction in cultivable land, and availability of freshwater for irrigation associated with climate change have profound effects for the capability of agriculture to meet the world's demands for food, feed, fiber, and fuel. Success depends on the recognition and exploit of existing molecular techniques, finding new sources as well as the increasing development of farming systems that use saline water and integrate nutrient flows.

The productivity of fiber crops, worldwide, is severely hampered by the prevalence of pests, weeds, and pathogens apart from various environmental factors. Several beneficial agronomic traits, viz. early maturity, improved fiber quality, and heat tolerance, have been successfully incorporated into fiber crops by employing conventional hybridization and mutation breeding.

Now, new advances in biotechnology are making it possible to develop plants that contain new genes which could not be introduced by sexual means. These advances in genetic engineering offer great new opportunities for improvement and sustainable use of fiber crops.

Fiber plants: Biology, biotechnology, and applications are presented with an aim to provide information about resources, their utilization, and technology available for their improvement. The purpose of this book is to assess the potential effects of biotechnological approaches particularly genetic modification on present scenario of fiber crop cultivation and improved production. The topics covered include biology, biotechnology, genomics, and applications of fiber crops such as cotton, flax, jute, and banana. The proposed book is to provide comprehensive and broad subject-based reviews, useful for students, teachers, researchers, and all others interested in the field. The field has been kept wide and general to accommodate the wide-ranging topics. How biotechnology can affect and solve the problems of fiber crops has been presented by world's leading scientists and expert of the field. This book will be useful to agriculturists, biotechnologists, botanists, industrialists, and those governments involved in planning of fiber crop cultivation.

Udaipur, India K.G. Ramawat
New Paltz, NY, USA M.R. Ahuja

Contents

Contributors

M.R. Ahuja New Paltz, NY, USA

David M. Anderson Dow AgroSciences LLC, Phytogen Seed Company, Corcoran, CA, USA

Kumar Anupam Physical Chemistry, Pulping and Bleaching Division, Central Pulp and Paper Research Institute, Saharanpur, Uttarpradesh, India

Saikat Kumar Basu Department of Biological Sciences, University of Lethbridge, Lethbridge, AB, Canada

Vimlesh Bist Physical Chemistry, Pulping and Bleaching Division, Central Pulp and Paper Research Institute, Saharanpur, Uttarpradesh, India

Julio Bruna Food Packaging Laboratory, Department of Food Science and Technology, Faculty of Technology, University of Santiago de Chile, Estación Central, Santiago, Chile; Center for the Development of Nanoscience and Nanotechnology (CEDENNA), University of Santiago de Chile, Estación Central, Santiago, Chile

William Cetzal-Ix Instituto Tecnológico de Chiná, Campeche, Mexico

Debajeet Das Division of Crop Improvement, Biotechnology Unit, ICAR-Central Research Institute for Jute and Allied Fibres (CRIJAF), Nilganj, Barrackpore, Kolkata, West Bengal, India

Rodrigo Duno de Stefano Herbario CICY, Centro de Investigación Científica de Yucatán, A. C. (CICY), Mérida, Mexico

Chandrakanth Emani Department of Biology, Western Kentucky University-Owensboro, Owensboro, KY, USA

María J. Galotto Food Packaging Laboratory, Department of Food Science and Technology, Faculty of Technology, University of Santiago de Chile, Estación Central, Santiago, Chile; Center for the Development of Nanoscience and

Nanotechnology (CEDENNA), University of Santiago de Chile, Estación Central, Santiago, Chile

Benjamin P. Graham Department of Plant and Microbial Biology, North Carolina State University, Raleigh, NC, USA

Abel Guarda Food Packaging Laboratory, Department of Food Science and Technology, Faculty of Technology, University of Santiago de Chile, Estación Central, Santiago, Chile; Center for the Development of Nanoscience and Nanotechnology (CEDENNA), University of Santiago de Chile, Estación Central, Santiago, Chile

Candace H. Haigler Department of Crop Science, North Carolina State University, Raleigh, NC, USA; Department of Plant and Microbial Biology, North Carolina State University, Raleigh, NC, USA

Pran Gobinda Karmakar Biotechnology Unit, Division of Crop Improvement, ICAR-Central Research Institute for Jute and Allied Fibres (CRIJAF), Nilganj, Barrackpore, Kolkata, West Bengal, India

Yasuo Kato Biotechnology Research Center and Department of Biotechnology, Toyama Prefectural University, Imizu, Toyama, Japan

Takao Kishimoto Biotechnology Research Center and Department of Biotechnology, Toyama Prefectural University, Imizu, Toyama, Japan

Priti Shivhare Lal Physical Chemistry, Pulping and Bleaching Division, Central Pulp and Paper Research Institute, Saharanpur, Uttarpradesh, India

Nur Alam Mandal Division of Crop Improvement, Biotechnology Unit, ICAR-Central Research Institute for Jute and Allied Fibres (CRIJAF), Nilganj, Barrackpore, Kolkata, West Bengal, India

Taiji Nomura Biotechnology Research Center and Department of Biotechnology, Toyama Prefectural University, Imizu, Toyama, Japan

Shinjiro Ogita Biotechnology Research Center and Department of Biotechnology, Toyama Prefectural University, Imizu, Toyama, Japan; Department of Life Sciences, Faculty of Life and Environmental Sciences, Prefectural University of Hiroshima, Shoubara, Japan

Arun K. Patra Department of Textile Chemistry, The Technological Institute of Textile and Sciences, Bhiwani, India

Kanniah Rajasekaran USDA-ARS Southern Regional Research Center, New Orleans, LA, USA

K.G. Ramawat Badgaon, Udaipur, India

Francisco Rodríguez Food Packaging Laboratory, Department of Food Science and Technology, Faculty of Technology, University of Santiago de Chile, Estación

Central, Santiago, Chile; Center for the Development of Nanoscience and Nanotechnology (CEDENNA), University of Santiago de Chile, Estación Central, Santiago, Chile

Hamidou F. Sakhanokho Thad Cochran Southern Horticultural Laboratory, USDA-ARS, Poplarville, MS, USA

Debabrata Sarkar Division of Crop Improvement, Biotechnology Unit, ICAR-Central Research Institute for Jute and Allied Fibres (CRIJAF), Nilganj, Barrackpore, Kolkata, West Bengal, India

Pratik Satya Division of Crop Improvement, Biotechnology Unit, ICAR-Central Research Institute for Jute and Allied Fibres (CRIJAF), Nilganj, Barrackpore, Kolkata, West Bengal, India

Arvind Kumar Sharma Physical Chemistry, Pulping and Bleaching Division, Central Pulp and Paper Research Institute, Saharanpur, Uttarpradesh, India

Nagendra Kumar Singh ICAR-National Research Centre on Plant Biotechnology (NRCPB), Pusa, New Delhi, India

Solichin Solichin Department of Mechanical Engineering, Universitas Negeri Malang, Malang, East Java, Indonesia

Michael R. Stiff Department of Crop Science, North Carolina State University, Raleigh, NC, USA

Heru Suryanto Department of Mechanical Engineering, Universitas Negeri Malang, Malang, East Java, Indonesia

J. Rich Tuttle Department of Crop Science, North Carolina State University, Raleigh, NC, USA

Uun Yanuhar Biotechnology Laboratory, Department of Fisheries and Marine Science, University of Brawijaya, Malang, Indonesia

Part I
Biology

Chapter 1
Fiber Plants: An Overview

K.G. Ramawat and M.R. Ahuja

Abstract Fiber cells are present in all plants with varied shape, sizes, and composition. Fiber crops are source of commercially exploited fibers since beginning of civilization and fiber/fabric consumption is used as indicator of civilization of society. Much of the need of rural populations for fiber is met with material harvested from wild. That is of great concern for biodiversity conservation and sustainable utilization of resources. New technologies are used to understand the fiber development and formation, diversity of plants, and consequently improvement strategies are developed using plant cell cultures and genetic engineering. New technologies are being developed to obtain fiber and fiber products from sources as diverse as agriculture waste, baggasses, vegetable and fruit processing, and other industrial waste. Plant fibers are finding new and diverse applications and usage like dietary fibers, biodegradable films in food industry, natural fiber composites, biopolymers, biofuels, and pharmaceuticals besides improved textiles. This article summarizes the scenario about new technologies and sustainable exploitation of fiber plants and their products.

Keywords Fiber plants · Biotechnology · Natural fiber composites · Biofuels · Cotton · Linen · Flax

1.1 Introduction

Increase in population, particularly in developing countries has forced us to develop new technologies to meet the demand for food and clothes; fiber comes second to food. Natural fiber is a preferred material for fabric. Textile fiber is a generic term

K.G. Ramawat (✉)
221, Landmark Treasure Town, Badgaon, Udaipur 313011, India
e-mail: kg_ramawat@yahoo.com

M.R. Ahuja
60, Shivertown Road, New Paltz, NY 12561, USA
e-mail: mrahuja@hotmail.com

K.G. Ramawat and M.R. Ahuja (eds.), *Fiber Plants*, Sustainable Development and Biodiversity 13, DOI 10.1007/978-3-319-44570-0_1

for any type of fiber (natural, synthetic, or artificial) that form the basis of a textile product (yarns, fabrics, knits, nonwovens, etc.) and having a length at least hundred times greater than its diameter. Viscose fiber is obtained by chemical modification of cotton. Cotton, flax, ramie, jute, sun hemp, hemp, and kenaf are commonly used and commercially exploited fibers. When we talk about plants, basic questions come to our mind are sustainable resource utilization, conservation of germplasm of fiber plants, efficient use of new technologies to increase quality and quantity of fiber, and to make life of farmers better.

Photosynthesis is the largest photochemical reaction on the planet Earth and ~40 % of the carbon assimilated is used to produce cellulose, one of the main components of plant cell walls. Much of this cellulose is accumulated in the form of plant fibers. Cellulose may be present in pure form in plants, but it is generally associated by hemicelluloses, lignins, and comparably small amounts of soluble compounds (Wondraczek and Heinze 2015). Cellulose is the main component of wood (40–50 wt %), baggasses (35–45 wt%), bamboo (40–55 wt%), straw (40–50 wt%), flax (70–80 wt%), hemp (75–80 wt%), jute (60–65 wt%), kapok (70–75 wt%), and ramie (70–75 wt%). As compared to this cotton is a pure cellulose fiber containing more than 90 wt% (Hon 1996). Cellulose is accumulated not only in fibers like cotton but also in tree woods (1,750,000 kilo tons, kt), bamboo (10,000 kt), cotton linters (18,450 kt), jute (2300 kt), flax (830 kt), sisal (378 kt), hemp (214 kt), and ramie (100 kt) in 1 year (Eichhorn et al. 2001). Biotechnological methods and genetic engineering used to improve the quality of fiber are described in this book in Chap. 8.

Fibers are present in all plants, and when commercially exploited, the plants become fiber crops. Cellulose is a homoglycan constituted by β (1 → 4) linked D-glucopyranose units. Cellulose is a high molecular weight polysaccharide made up of repeating cellobiose units producing a linear chain in which both intra-chain and inter-chain molecular hydrogen bonds occur to link the chains (Donato et al. 2015), which in turn produce microfibrils, matrices, and multilayered cell walls (Fig. 1.1). These molecules are arranged very systematically and symmetrically in the supramolecular structures, from small initial fibrils (with a length between 1.5 and 3.5 nm), microfibrils (between 10 and 30 nm), to macrofibrillar bands whose length can be on the order of several hundred nm (Klemm et al. 2005). The process of polymerization and structures formation varies depending on the plant source. Plant fibers from different sources vary in length, color, composition, strength, durability, and resistance to water. Cellulosic fibers are more resilient than lingo-cellulosic fibers. Wood fibers are relatively large fibers. During secondary wall formation, lignin is deposited over the primary cellulose wall making the fiber thick, hard, and strong. Cotton fiber is an elongation of seed coat cells and an example of one of the longest cells (Hill 1972). These physical, chemical, morphological, and anatomical properties of plant fibers are discussed in Chap. 12 in this book in order to assess their potential toward production of pulp and paper.

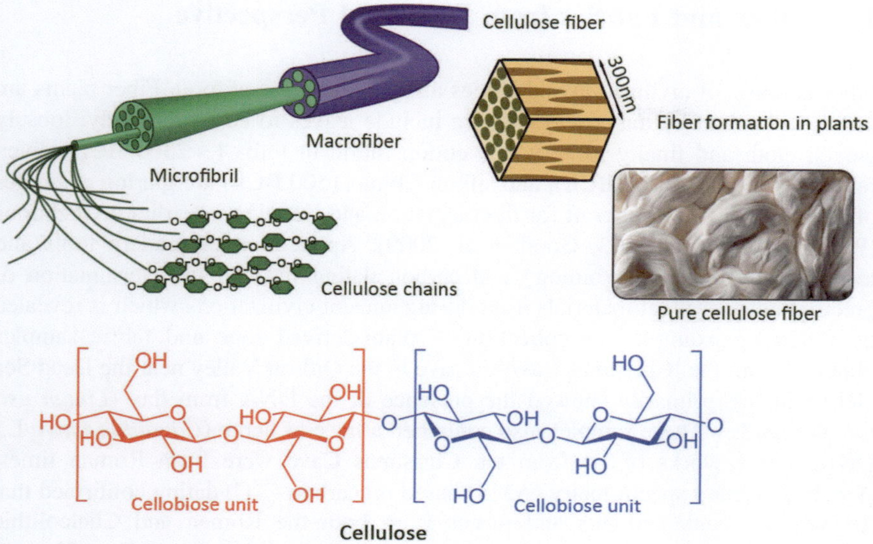

Fig. 1.1 Fiber formation from cellulose chains and structure of cellulose

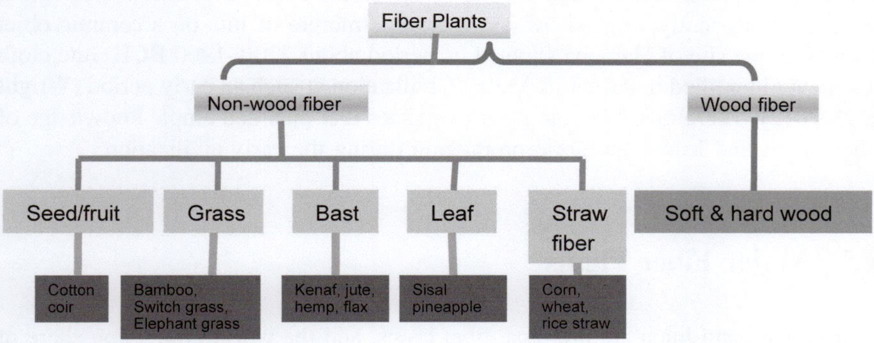

Fig. 1.2 Different sort of fibers obtained from plants

Fibers are distributed all over the plant from roots to leaf, stem and fruits and seeds. These fibers are harvested from the plant part for commercial and/or local usage. Now we know more and more about health beneficial effects of fiber present in fruits, vegetables, and grains, known as dietary fibers. Fibers are classified as soft or hard fibers, surface fibers, or endogenous fibers. Natural fibers are processed to make them suitable for different purposes. Different types of fibers obtained from plants are presented in Fig. 1.2.

1.2 Fiber and Fabrics from Historical Perspective

Fiber is fabric of civilization and comes after requirement of food. Fiber plants are useful to humans in many ways. These include leaves to cover the body; loosely woven cloth and finally to very fine cotton fabric in India (~2300 BCE); linen fabric in Egypt (~1300 BCE); and silk in China (1500 BCE) are glaring examples of rapid development of craft for fiber isolation and weaving in earlier civilizations (Pandey and Gupta 2003; Good et al. 2009). Application of modern tools and techniques of molecular biology and carbon dating resulted in determination of precise nature of plant materials used during ancient civilizations which is revealed by following examples. A collection of plant-derived rope and fabric samples obtained from the 'Christmas Cave' (a cave in the Qidron Valley near the Dead Sea and Qumran), primarily showed the presence of the DNA from flax (*Linum usitatissamum* L.). These samples also contained a trace of hemp (*Cannabis sativa* L.) DNA. These works of art from the Christmas Cave were from Roman times. Accelerator mass spectrometry (AMS) based on carbon (^{14}C) dating confirmed that the samples contained bits and pieces from both the Roman and Chalcolithic periods (Murphy et al. 2011). Similarly, excavations at the site of Kara Tepe in northwestern Uzbekistan made known evidence for the production of cotton (*Gossypium* sp.) in this region dated to ~300–500 CE (Brite and Marston 2013). These archaeobotanical remains help to understand the spread of Old World cotton production. Recently, analysis of a conserved structure of jute on a ceramic objet d'art from the site of Harappa showed its period about 2200–1900 BCE. Jute cloth was never identified in the Indus Valley Civilization at such an early period (Wright et al. 2012). Therefore, there are clear evidences that man had ample knowledge of fiber plants and learnt the fabric production during the early civilization.

1.3 Major Fiber Plants

Cotton, jute, and linen are principal fiber crops, and the world production share of cotton and jute is presented in Figs. 1.3 and 1.4. Current cotton production (~120 Million bales, MB) spreads throughout the world with China, USA and India are major producers while jute (~3.3 MT) is mainly produced in India and Bangladesh. France is major producer of hemp (65,00 Million Tons) and flax (113,000 Million Tons) (Holbery and Houston 2006; http://faostat3.fao.org/browse/Q/QC/E). Cotton has many species and cultivar varieties like Egyptian cotton (*Gossypium barbadense*), upland cotton (*G. hirsutum*), Levant cotton (*G. herbaceum*), and tree cotton (*G. arboreum*) due to cultivation in different continents since time immemorial while most of the bast fiber crops have limited geographical distribution and cultivars such flax (*Linum usitattisimum*), jute (*Corchorus*

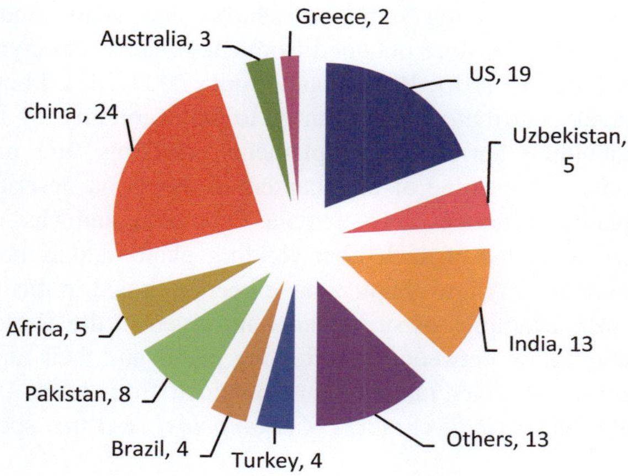

Fig. 1.3 Production share of major cotton producing countries in the world, based on FAO

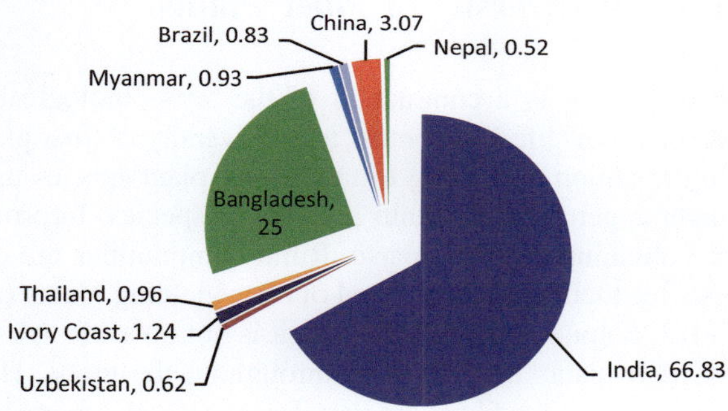

Fig. 1.4 Production share of major jute producing countries in the world, based on FAO

capsularis, *C.olitorius*), ramie (*Boehmeria nivea*), and kenaf (*Hibiscus cannabinus*). Details of varieties, cultivation practices, processing and distribution in the world are beyond scope of this paper but can be viewed in books on economic botany (Hill 1972; Levetin and McMahon 2008). Several Chaps. (2, 5–10) in this book are devoted about applications of modern tools to improve the quality and the plants (cotton, linen, jute, and bamboo) and the fiber obtained from these plants.

1.4 Lesser Known Fiber Plants

Several hundred plant species in different geographical regions of the world are used as source of various types of finished or unfinished fibers and collected from wild. Fiber-yielding weeds hold the second position after edible weeds in their economic importance. In almost all parts of the world, local agriculturists use various types of fiber plants for their daily need to tie their agricultural produce, fuel wood, packaging small households and straw, and women prepare household articles from fiber obtained from such plants (Anonymous 2010; Isaza et al. 2013; Sahu et al. 2013; Pandey and Gupta 2003). The fiber production at local/regional level contributes significantly to the economy of the region in various ways, including agricultural implements, clothing, and products for other household operations. Some of the chapters in this book describe some new promising fiber plants of Indonesia, Guatemala, Colombia, and Mexico (Chaps. 3, 4). Each region has long list of such fiber-yielding plants and is beyond the scope of this brief overview. A few chapters have been included in this book on such new plants and some examples of such plants are given in the Table 1.1. There is no taxonomic correlation between fiber-yielding plants and their families or geographical region. Similarly, fiber materials are obtained from herbs (*Agave*, *Ananas*, *Leptadenia*), shrubs (*Urtica*), climbers (*Cissampelos*), and tree species (*Bombax* spp.).

1.5 Biodiversity of Fiber Plants

Biodiversity is a contraction of the term 'biological diversity' and refers to the diversity of 'life.' In context to biodiversity of fiber plants, it implies to biodiversity in the region pertaining to number of plant species used for obtaining fiber and/or diverse germplasm within a genus or species. Information about wild plants used for obtaining fiber is sparse. Rural communities are completely dependent on forests for their day-to-day need of fiber-yielding plants (Anonymous 2010; Isaza et al. 2013; Sahu et al. 2013). Hence, it is not possible to estimate the number of species of fiber plants used by man throughout the globe. However, the number of plant species used for fibers are very large, e.g., in America over 1000, in the Philippines over 800, and in India over 450 species (representing 82 families and 273 genera). In India, presently eight species contribute as major cultivated fiber crops. Major families (with number of genera) contributing as fiber genetic resources are Malvaceae (63), Fabaceae (31), Arecaceae (25), Utricaceae (24), Tiliaceae (21), Sterculiaceae (21), and Asclepiadaceae (15) (Pandey and Gupta 2003). Continuous collection of plant material from wild without replenishing them results in loss of germplasm and biodiversity. Therefore, domestication of plants by developing appropriate techniques of cultivation, understanding their seed biology, and methods of propagation and conservation efforts are required to exploit these plants and developing sustainable agroecosystem. Effective use of biotechnological methods such as in vitro cryopreservation, DNA fingerprinting-based breeding system, micropropagation, and ecorestoration supports the in situ conservation which is need of hour for overexploited plants (Goyal et al. 2014, 2015).

Table 1.1 Lesser known fiber plant resources (based on Anonymous 2010; Sahu et al. 2013; Pandey and Gupta 2003)

Plant species	Family	Plant part	Usage/remarks
Ananas comosus (L.)Merr.	Bromeliaceae	Leaves	Ropes, fishing nets, strings, and fabrics
Agave cantula Roxb, *A. sisalana* Perrine *ex* Engelm	Agavaceae	Leaf	Making ropes, mats, twines, nets, cordage
Boehmeria nivea (L.) Gaud.	Urticaceae	Bast	Thread, cord, sacks, paper, gas mantles
Bombax ceiba L., *Ceiba pentandra* (L.) Gaertn.	Bombacaceae	Fruit inner wall	Filling pillow, mattresses
Calotropis procera (Ait.) Rox Br *Leptadenia pyrotechnica*	Asclepiadaceae	Stem bark	Ropes
Cannabis sativa L.	Cannabaceae	Stem bark	Textile industry; production of cordage; manufacturing Sailcloth; canvas goods
Cissampelos pariera L.	Menispermaceae	Stem	Ropes
Crotalaria juncea L.	Fabaceae	Stem bark	Fishing nets, gunny bags, coarse cloths and mattress
Desmodium elegans DC.	Fabaceae	Stem, bark	As rope
Erianthus arundinaceus (Retz.) Jesw.ex Heyne	Graminaceae	Leaf	Ropes, twine, paper
Erianthus munja (Roxb.) Jesw	Graminaceae	Leaf and stem	Baskets, mats, cordage
Furcaria foetida (L.) Haw.	Amaryllidaceae	Leaf	Mats, ropes, cordage
Helicteres isora L.	Sterculiaceae	Stem, bark	Ropes
Hibiscus cannabinus L.	Malvaceae	Stem bark	Ropes, mask, collar belt, carry bag
Musa textilis Nees	Musaceae	Leaf	Marine cordage, bags
Saccharum spontaneum L.	Poaceae	Leaf	Mats, ropes
Typha angustata Bory and Chaub.	Typhaceae	Leaf	Mats, ropes, basket
Urtica dioica L.	Urticaceae	Stem	Cordage, sacks

1.6 Biotechnological Approaches

Biotechnological tools and techniques applied to understand and improve the fiber and fiber plants are presented in Fig. 1.5. These include understanding the basic structure and function of genome of the fiber plants for obtaining various value-added products to uplift the socioeconomic status of the farmers. Development of technology for nanomaterials, new biomolecules, biosensors, enzymes, and biofuels

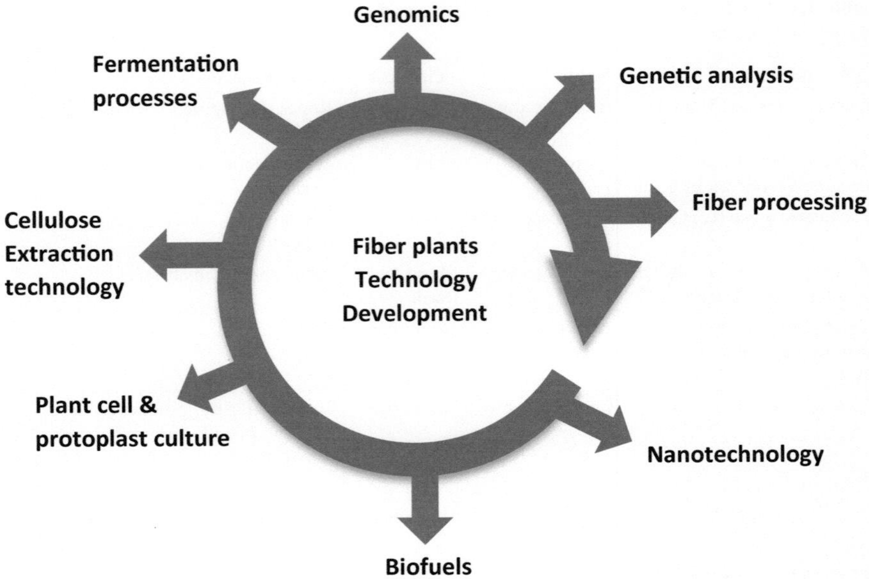

Fig. 1.5 Different technologies being developed using plant fibers and fiber plants

from various plant resources have economic and environmental consequences because of their renewable, biodegradable, and ecofriendly nature. Biotechnological inputs are explored to increase the quality and quantity of fibers in plants. Enzymes have replaced the chemicals used in the processing of fibers and preparation of fabric (see Chaps. 5 and 12). It is needless to mention that enzymes are economically sustainable and environmentally friendly reusable biomolecules (Radhakrishnan 2014). Toward the development of cheap fiber-based sensor, paper-based sensing devices do not require hardware or definite technical skill. The immobilization of biomolecules onto cellulose is a key step in the development of these sensing devices (Credou and Berthelot 2014). These are inexpensive, rapid, and user-friendly and therefore can easily used at diagnostic centers.

Biotechnological approaches like regeneration and micropropagation are gaining importance in past decades (Goyal et al. 2014, 2015). Somatic embryogenesis is not only a method of rapid multiplication but can also leads to improvement through genetic transformation. A few chapters on regeneration of cotton (Chap. 6), bamboo (Chap. 7) and transgenic cotton (Chaps. 8, 10) are included in this book where details of techniques and literature survey can be viewed. Plant cell and tissue regeneration, somatic embryogenesis, protoplast isolation, culture and fusion and cell suspension cultures, anther and microspore culture, and gene transfer and expression by genetic transformation in flax were presented in an excellent review by Millam et al. (2005). With the understanding of fiber development, the main objective of breeding fiber plants is to increase growth rate and gibberellins accumulation may lead to higher growth rate and fiber formation. In a model system, GA-2 oxidase enzyme production was blocked resulting in higher production of fiber material in tobacco plants (Dayan et al. 2010). Fiber based materials are used to develop newer technological material such as nanomaterials. Therefore, biotechnological inputs are making a dent not only in understanding process of fiber formation but also in its improvement and diversification of products produced to make the life more comfortable.

1.7 Applications of Fiber

A wide range of applications of fiber and fiber products are being developed because large quantities of agricultural by-products are available as renewable sources. Agro-based biofibers are very suitable for composite, biofuels, textile, pulp and paper manufacture because of their composition, properties, and structure. Additionally, they can also produce fuel, chemicals, enzymes, and food (Fig. 1.6). Agricultural residue from the cultivation of corn, wheat, rice, sorghum, barley, sugarcane, pineapple, banana, and coconut are the major sources of agro-based biofibers (Michelin et al. 2015; Reddy and Yang 2005). Natural fiber composites

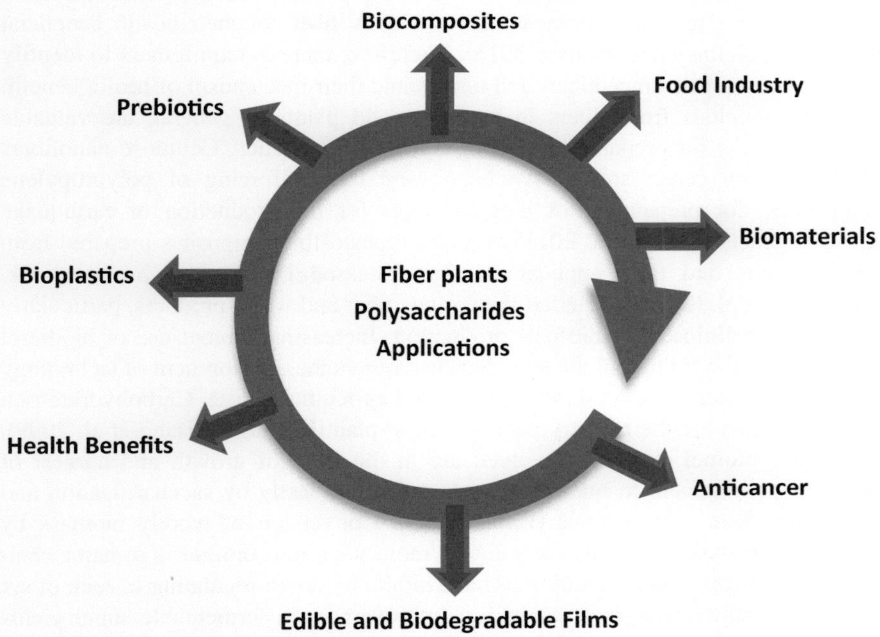

Fig. 1.6 Applications of plant fibers and their products

prepared from kenaf, hemp, flax, jute, and sisal with thermoplastic and thermoset matrices are finding way as automotive parts because of reductions in weight, cost, and CO_2 emission (Ashori 2008; Holbery and Houston 2006). Bioenergy and biomaterials obtained from fiber crops are considered as promising substitutes for conventional sources and fossil fuel, considering the increased concern about rapidly depletion of non-renewable resources and emission of carbon. As bioenergy and biomaterials raw material, fiber crops offer ecological advantages over conventional ones, such as carbon sequestration, energy savings, the reduction of greenhouse gases, and are renewable resources. However, other environmental factors such as acidifying the soil and eutrophication emissions should also be considered before advanced use of fiber crops (Fernandoa et al. 2015).

Plant residues available from industrial process like cane sugar production, paper mills, agricultural produce, seed oil production, vegetable and fruit production and processing waste, and cotton and other fiber waste are important sources of many products such as energy in the form of alcohol, fermentation products, new biomolecules for various usage including pharmaceuticals and several types of composites for industrial use (Costa et al. 2015; Donato et al. 2015; Reddy and Yang 2005). Cellulose-based biomaterials are being developed for human health and welfare as never drying wound healing membranes, various types of implants, and for tissue engineering (Wondraczek and Heinze 2015).

It is rightly stated that 'you are what you eat'(Brüssow and Parkinson 2014). Eating foods rich in plant fiber promotes health by changing the composition and metabolic products of gut bacteria. The growing public awareness about nutrition and health care results in more intake of dietary fiber for their health beneficial properties (Whitney and Manore 2015). Therefore, there is requirement to identify newer sources of neutraceuticals and understand their mechanism of health benefit.

Polysaccharides from fibers in particular and plants in general, are valuable building blocks for preparation of composites or bioplastics. Cellulose nanofibers extracted from cereal straw have been used for reinforcing of polypropylene composites, for preparation of biocomposites for the production of earthquake resistant panels (Kalia et al. 2011). Various types of biocomposites prepared from natural fibers and their applications are discussed in Chap. 3 in this book. Ragauskas et al. (2006) discussed the various fiber and wood products, particularly use of hemicelluloses for biofuels production. Increasing percent use of biodiesel and ethanol in fossil fuel in the next decade requires the development of technology for their economic production from forest and agriculture waste. Carbohydrate-rich plant cell walls are the primary energy sink in plant biomass (Hisano et al. 2009). The major biofuel expenses involved are in the form of growth and harvest of biomass, pretreatment to breakdown cell walls, and lastly by saccharification and their conversion into biofuels (Rubin 2008). Conversion of woody biomass by enzymatic breakdown, particularly lignin molecules, into biofuel is a major challenge. Modification in lignin in transgenic alfalfa by down-regulating in each of six lignin biosynthetic enzymes resulted in improvement in fermentable sugar yields for biofuel production (Poovaiah et al. 2014; Chen and Dixon 2007). It is evident from the above account that applications of fiber and its products are as diverse as the fiber plants itself. These applications are at various stages of development and industrial processes based on these technologies will be available soon for the benefit of human health and environment.

1.8 Conclusions

In this chapter, we have presented a brief scenario of fiber crops from ancient civilization to modern technology of molecular biology and genomics. Newer applications of fibers in diverse fields such as health, composites, biodegradable films, biofuels, biopolymers, and pharmaceuticals are presented in brief. The major emphasis of present-day research is to find out alternatives for non-renewable resources and plants offer both, renewable resources and are environmental friendly too.

Efforts are required to understand the mechanism of fiber formation and elongation, control of genes involved in fiber elongation, conservation by in situ and ex situ techniques of germplasm resources of vast number of plants used by local communities, value-added products formation to uplift economic situation of local people, creation of inventory of such plants, and breeding of crops based on their evaluation on the basis of techniques of molecular biology. Diversification of products from fiber plants such as healthy biodegradation and edible biofilms, automotive components, new bioactive molecules for pharmaceutical industries, biopolymers and biofuels offers new avenues for the crops plants containing fiber and holds promise for better economic strengthening of agriculture-based communities, which is a major share of population of the world. This has also beneficial environmental consequences and less dependence on fossil fuel.

References

Anonymous (2010) Promising fibre-yielding plants of the Indian Himalayan region. G.B. Pant Institute of Himalayan Environment and Development, Kosi-Katarmal, Almora-263 643, Uttarakhand (India). pp 1–56

Ashori A (2008) Wood–plastic composites as promising green-composites for automotive industries. Bioresour Technol 99(11):4661–4667

Brite EB, Marston JM (2013) Environmental change, agricultural innovation, and the spread of cotton agriculture in the Old World. J Anthropol Archaeol 32:39–53

Brüssow H, Parkinson SJ (2014) You are what you eat. Nat Biotechnol 32:243–245

Chen F, Dixon RA (2007) Lignin modification improves fermentable sugar yields for biofuel production. Nat Biotechnol 25:759–761. doi:10.1038/nbt1316

Costa SM, Aguiar A, Luz SM, Pessoa A Jr, Costa SA (2015) Sugarcane straw and its cellulosic fraction as raw materials for obtainment of textile fibers and other bioproducts. In: Ramawat KG, Mérillon J-M (eds) Polysaccharides: biotechnology and bioactivity. Springer, Switzerland

Credou J, Berthelot T (2014) Cellulose: from biocompatible to bioactive material. J Mater Chem B 2:4767–4788

Dayan J, Schwarzkopf M, Avni A, Aloni R (2010) Enhancing plant growth and fiber production by silencing GA 2-oxidase. Plant Biotechnol J 8(4):425–435

Donato PD, Poli A, Taurisano V, Nicolaus B (2015) Polysaccharides from bioagro-waste new biomolecules-life. In: Ramawat KG, Mérillon J-M (eds) Polysaccharides: biotechnology and bioactivity. Springer, Switzerland

Eichhorn SJ, Baillie CA, Zafeiropoulos N, Mwaikambo LY, Ansell MP, Dufresne A, Entwistle KM, Herrera-Franco PJ, Escamilla GC, Groom L, Hughes M, Hill C, Rials TG, Wild PM (2001) Current international research into cellulosic fibers and composites. J Mater Sci 36:2107–2131

Fernandoa AL, Duartea MP, Vatsanidoub A, Alexopoulouc E (2015) Environmental aspects of fiber crops cultivation and use. Ind Crops Prod 68:105–115

Good I, Kenoyer JM, Meadow R (2009) New evidence for early silk in the Indus civilization. Archaeometry 50:1–10

Goyal S, Arora J, Ramawat KG (2014) Biotechnological approaches to medicinal plants of aravalli hills: conservation and scientific validation of biological activities. In: Ahuja MR, Ramawat KG (eds) Biotechnology and biodiversity, sustainable development and biodiversity, vol 4. Springer, Switzerland, pp 203–245

Goyal S, Sahrma V, Ramawat KG (2015) A review of biotechnological approaches to conservation and sustainable utilization of medicinal lianas in India. In: Parthasarathy N (ed) Biodiversity of lianas, sustainable development and biodiversity, vol 5. Springer, Switzerland, pp 179–210

Hill AF (1972) Economic botany: a textbook of useful plants and plant products, 2nd edn. Tata McGraw-Hill Publishing Company Ltd, New Delhi, pp 18–51

Hisano H, Nandakuma R, Wang ZY (2009) Genetic modification of lignin biosynthesis for improved biofuel production. In Vitro Cell Dev Biol Plant 45:306–313

Holbery J, Houston D (2006) Natural-fiber-reinforced polymer composites in automotive applications. JOM 58:80–86

Hon DN-S (1996) Cellulose and its derivatives: structures, reactions, and medical uses. In: Dumitriu S (ed) Polysaccharides in medical applications. Marcel Dekker, New York, pp 87–105

Isaza C, Bernal R, Howard P (2013) Use, production and conservation of palm fiber in South America: a review. J Hum Ecol 42(1):69–93

Kalia S, Dufresne A, Cherian BM, Kaith BS, Averous L, Njuguna J, Nassiopoulos E (2011) Cellulose-based bio- and nanocomposites: a review. Int J Polymer Sci 2011:35, art id 837875. doi:10.1155/2011/837875

Klemm D, Heublein B, Fink H, Bohn A (2005) Cellulose: fascinating biopolymer and sustainable raw material. Angew Chem Int Ed 44:3358–3393

Levetin E, McMahon K (2008) Plants and society, science egineering, 5th edn. McGraw Hill, New York, pp 299–322

Michelin M, Ruiz HA, Silva DP, Ruzene DS, Teixeira JA, Polizeli MLTM (2015) Cellulose from lignocellulosic wastes: a biorefinery processing perspective. In: Ramawat KG, Mérillon J-M (eds) Polysaccharides: biotechnology and bioactivity. Springer, Switzerland

Millam S, Obert B, Petrova A (2005) Plant cell and biotechnology studies in *Linum usitatissimum*—a review. Plant Cell, Tissue Organ Cult 82(1):93–103

Murphy TM, Ben-Yehuda N, Taylor RE, Southon JR (2011) Hemp in ancient rope and fabric from the Christmas Cave in Israel: talmudic background and DNA sequence identification. J Archaeol Sci 38(10):2579–2588

Pandey A, Gupta R (2003) Fibre yielding plants of India, genetic resources, perspectives for collection and utilization. Nat Prod Radiance 2(4):194–204

Poovaiah CR, Rao NM, Soneji JR, Baxter HL, Stewart CN Jr (2014) Altered lignin biosynthesis using biotechnology to improve lignocellulosic biofuel feedstocks. Plant Biotechnol J 12 (9):1163–1173

Radhakrishnan S (2014) Application of biotechnology in the processing of textile fabrics. In: Textile science and clothing technology, roadmap to sustainable textiles and clothing. pp 277–325
Ragauskas AJ, Nagy M, Kim DH, Eckert CA, Hallett JP, Liotta CL (2006) From wood to fuels: integrating biofuels and pulp production. Ind Biotechnol 2(1):55–65. doi:10.1089/ind.2006.2.55
Reddy N, Yang Y (2005) Biofibers from agricultural byproducts for industrial applications. Trends Biotechnol 23:22–27
Rubin EM (2008) Genomics of cellulosic biofuels. Nature 454:841–845
Sahu SC, Pattnaik SK, Dash SS, Dhal NK (2013) Fibre-yielding resources of Odisha and traditional fibre preparation knowledge: an overview. Indian J Nat Prod Resour 4(4):339–347
Whitney S, Manore MM (2015) Dietary fiber: simple steps for managing weight and improving health. ACSM'S Health Fit J 19(1):9–16
Wondraczek H, Heinze T (2015) Cellulose biomaterials. In: Ramawat KG, Mérillon J-M (eds) Polysaccharides: biotechnology and bioactivity. Springer, Switzerland, pp 289–328
Wright RP, Lenz DL, Beubien HF, Kimbrough CK (2012) New evidence for jute (*Corchorus capsularis* L.) in the Indus civilization. Archaeol Anthropol Sci 4:137–143

Chapter 2
The Global Importance of Transgenic Cotton

David M. Anderson and Kanniah Rajasekaran

Abstract The origins of transgenic cotton are reviewed including the original objectives, early efforts to establish the technical capabilities, selection of initial traits for development, market place benefits, and global acceptance of the technology. Further consideration is given to cotton's place in the effort to meet the projected demands for food and fiber over the next 50-year horizon, traits and technologies under development, and the need for close public and private research collaboration in order to address the issues facing the world's farmers as they work to meet those demands. Impact of transgenic cotton on global economy, environment, genetic diversity, and safety is also highlighted.

Keywords Cotton · Transgenic cotton · Cotton traits · Global economy · Genetic diversity

2.1 Historical Perspective

In a book published in 1957 (Brown et al. 1957), James Bonner emphasized that significant problems would face the world's agricultural producers as they sought to keep pace with the needs of the growing population. First, James envisioned ongoing pressure on agricultural productivity and an elevation of the costs of production as a consequence of industrialization attracting more and more of the world's labor force at the expense of farm labor. Science and technology were posited as the most likely means of increasing overall farm output in order to produce the food (calories) and fiber required to feed and clothe the world's

---

D.M. Anderson
Dow AgroSciences LLC, Phytogen Seed Company, P.O. Box 787, 850 Plymouth Avenue, Corcoran, CA 93212-0787, USA

K. Rajasekaran (✉)
USDA-ARS Southern Regional Research Center, 1100 Robert E. Lee Blvd., New Orleans, LA 70124-4305, USA
e-mail: Kanniah.Rajasekaran@ars.usda.gov

K.G. Ramawat and M.R. Ahuja (eds.), *Fiber Plants*, Sustainable Development and Biodiversity 13, DOI 10.1007/978-3-319-44570-0_2

increasing population. Bonner also foresaw a second challenge to agricultural output in the form of public resistance to the adoption of those very technologies that would be essential to achieving the necessary gains in per acre productivity to feed the world's growing population. On this point, James cited a commonly encountered public sentiment of his day that chemical fertilizers would be poisonous to plants if used to replace manure. The more the things change, the more they stay the same? Cotton biotechnology is nothing more, or less, than developing and applying technology to improve farm output in the face of decreased labor availability and increasing costs of same, reducing costs associated with controlling yield-reducing pests, and improving the value of the fiber cotton farmers produce. In many respects, the problems foreseen by James Bonner 58 years ago are where our present writing begins and ends.

It is appropriate that 23 years after the publication of "The Next Hundred Years," James, along with the J.G. Boswell Company, became a co-founder of the plant biotechnology company Phytogen. Phytogen was an early entrant in the field of plant biotechnology with a primary focus on developing and deploying biotechnologies in the improvement of cotton through increasing output per acre, reducing costs of production, and in improving the overall value received by farmers for the seed and fiber produced on a per acre basis. Phytogen was very successful both in the development of many of those technologies and in the commercialization of transgenic cotton, mirroring the advancements made by many researchers across other major row crops in elevating productivity per acre and delivering on-farm economic improvement through the development and deployment of various biotechnologies.

In our discussion of transgenic cotton and its global impact, it is useful for us to establish what we do and do not mean by the phrase "transgenic." The Dictionary of Botany published in 1980 (Little and Jones 1980), which sat on the corner of the laboratory bench at the time Phytogen began laboratory operations, did not define the term. Clearly, when this dictionary was published, the same year in which Phytogen was founded, the science underlying transgenic plants was yet to be created. It is worth noting the remarkable advancements that have been made in the very short intervening period since those very first days in the laboratory. The online Oxford Advanced Lerner's Dictionary (2015) defines the term as "having genetic material introduced from another type of plant or animal." This definition underscores the advances made in the development and application of biotechnologies and the general public awareness of same as the referenced dictionary is a Web-based resource used by the general public. This definition does have its limitations for our purposes here, as transgenic plants in general, cotton included, can and do carry genetic material from bacteria, fungi, animals, or other plants. Transgenic cotton can also carry genetic material from other cotton plants which has been removed, altered in some particular way, and then reintroduced. One such example would be cotton plants resistant to sulfonylurea herbicides created by isolating a cotton gene for acetohydroxyacid synthase (AHAS), introducing point mutations at particular serine (653) or tryptophan (574) codons, and then reintroducing these same mutated AHAS genes back into cotton to create cotton plants resistant to certain sulfonylurea and imidazolinone herbicides (Grula et al. 1995; Rajasekaran et al. 1996a, b). For the purposes of this discussion, we will consider a transgenic cotton plant as one which carries a heritable foreign gene construct irrespective of the source of the foreign DNA. Excluded will be cotton plants carrying heritable traits resulting from the use of other means of altering the genome such as somaclonal variation, EMS mutation, or site-directed in situ modification via technologies such as zinc finger nucleases (Rajasekaran et al. 1996a; Cai et al. 2008; Shukla et al. 2009). In each of these latter examples, heritable alterations in the genome can be introduced and/or useful phenotypes selected, but no foreign DNA per se remains in the resulting plant. Hence, they will not be considered transgenic.

The development of transgenic cotton began in earnest in 1980 in parallel with the broader focus by many private and public organizations in development of transgenic plants per se. It is also of note that this is the year in which the United States Supreme Court ruled that living organisms could be patented so long as it could be demonstrated that they were the products of man and not naturally existing in nature (Diamond v Chakrabarty 1980). This ruling opened the way for infusions of research dollars from private industry as patent protection was necessary to ensure the commercial success of transgenic plants. Protection of technology is essential to business success as it enables shareholders and the investing public the opportunity to recover and then profit from the large dollar investments that are required to develop and bring biotechnologies to market. Such traits must be delivered in the best germplasm (seeds) available in order to be useful to farmers, and most seed companies are privately held. Private industry was thus best positioned to carry the weight of both trait development, introgression of those traits into leading cultivars, and subsequent delivery to farmers in the form of seeds. Partnerships between public and private institutions were viewed as essential to the entire endeavor. Creative scientists armed with unbounded optimism took on the challenge of solving some of the globes' most difficult crop production problems by developing and then applying plant biotechnology.

Those of us working in the field of plant biotechnology in general, and cotton biotechnology in particular, recognized the need to (a) develop gene constructs that would function in cotton plants; (b) establish methods for introducing heritable foreign gene constructs, including selectable markers, into cotton cells; (c) select cells that had been genetically altered and which were expressing the introduced genes ("transformed"); (d) establish the ability to regenerate fertile cotton plants from cells; and (e) build useful libraries of regulatory sequences and genes, thereby enabling the development of multiple generations of cotton plants with combinations of useful traits. While there were relatively few laboratories focused on cotton in the early 1980s, competition to establish these capabilities for plants in general from both private and public sector laboratories was spirited. All things considered, progress came remarkably quickly, both for plants in general and for cotton in particular. Davidonis and Hamilton (1983) reported the important finding that at least one Coker cultivar was amenable to somatic embryogenesis, while parallel work extended regeneration success to other commercial cottons including high fiber quality Acala and Pima varieties and a broad range of additional Midsouth upland cotton varieties (Rangan and Zavala 1984; Trolinder and Goodin 1987; Shoemaker et al. 1986; Rangan and Rajasekaran 1997; Sakhanokho et al. 2001; Sakhanokho and Rajasekaran 2016). The development of vectors, selectable markers, and transformation in plants emerged from a period of intense, creative public and private research on a global scale in the early 1980s, both for *Agrobacterium*-mediated (Fraley et al. 1983; Hoekema et al. 1983; De Block et al. 1984; Horsch et al. 1985; An et al. 1985) and for transformation with "naked" DNA per se (Anderson 1985; Potrykus 1991). Successful cotton transformation was achieved during this same period (Firoozabady et al. 1987; Umbeck et al. 1987; Chlan et al. 1999; Rajasekaran 2004; Rangan et al. 2004). Selectable markers initially included genes conferring resistance to antibiotics such as kanamycin (Fraley et al. 1983) and hygromycin (Waldron et al. 1985). For many practical reasons, public researchers still make wide use of antibiotic resistance markers for selection, while private companies now typically avoid antibiotic resistance markers and use genes for resistance to various herbicide tolerance traits such as glufosinate (Thompson et al. 2005), AHAS inhibitors (Rajasekaran et al. 1996b) and on occasion glyphosate (Rathore et al. 2008).

2.2 Initial Traits and Trait Development

For the farmer, viability of his or her farming operation is driven by three simple factors including (a) total productivity per acre, (b) costs of production on a per acre basis, and (c) the market value for what is produced. Put simply, and this is a global reality, the grower needs higher yield, lower costs of production and a good price for what he produces. From the very beginning, the objective of cotton biotechnology across programs (public and private) has been to deliver against those three primary grower needs. Our own approach has been to view yield and stability of yield as best addressed initially through genetics, capturing of native traits from race stocks and diploid species, and the development of marker-based breeding tools and deployment of genome-wide selection capabilities to enable efficient and cost-effective introgression and stacking of those traits into high-yielding cultivars. The per unit value for what is sold depends largely on the quality of the fiber. The spinning performance of the fiber produced from a cotton crop, driven by its individual fiber properties such as length, strength, fineness, and maturity, will determine the extent to which farmers receive discounts or premiums for the harvested crop and hence drive output trait value. Given the complexity of the genetics conditioning overall spinning performance, and the relatively untapped reservoir of genetic resources for fiber improvement in cotton breeding populations, accessions, and race stocks, we likewise viewed fiber quality and output trait value as being best addressed initially by the aforementioned genetic tools. It is our perspective that single-gene (transgenic) constructs will not likely add sufficient economic value to the complex genetic systems undergirding spinning performance within the foreseeable future.

For our program, it seemed apparent that the largest near-term opportunities for improving on-farm economic return utilizing biotechnology would come from developing in-plant resistance to lepidopteran insects and tolerance to broad-spectrum herbicides. Clearly, researchers in other companies reached the same conclusions. Namely, that insect resistance would reduce costly spraying to control insects, reduce the labor costs associated with said applications, and have the benefit of driving yield improvement in situations where weather conditions precluded field application or insect attack was below the treatment threshold but still negatively impacting yield. Concerns regarding increasing costs of production driven in large part by increasing labor costs and a shrinking pool of agricultural labor echo the theme voiced by Bonner in 1957 referenced earlier. With respect to herbicide tolerance, weeds compete for sunlight, water, and nutrients, harbor insect pests, host pathogens, and create trash that can end up in the lint, thereby reducing crop output trait (fiber) value. We, along with others, saw that resistance to environmentally safe, broad-spectrum herbicides such as glyphosate would simplify control practices and reduce overall labor costs. Personal conversations with cotton growers in 1981 reinforced these conclusions, both for insect resistance and for glyphosate tolerance.

The first glyphosate-tolerant plant was developed at Phytogen in the early 1980s resulting in the patent now owned by Dow AgroSciences which issued covering same (Rangan et al. 2004), while the first commercial glyphosate-tolerant ("Round-up Ready™") cotton variety was developed by Monsanto and introduced in 1997 (Rathore et al. 2008). US acres planted to glyphosate-tolerant cotton reached 65 % by 2006 and 93 % by 2009, and at present, approximately 98 % of US cotton acres are glyphosate tolerant (Roundup-Ready Flex and Glytol from Bayer Crop Sciences) (Fig. 2.1). Glyphosate tolerance has been a compelling tool in the hands of US cotton growers. It is important to note that the first transgenic cotton commercialized anywhere was developed by Calgene and was resistant to the herbicide Bromoxynil (BXN™ cotton; Stalker et al. 1988). The BXN™ cotton system, introduced in 1995, was excellent technology but ultimately lacked extensive uptake by growers due to its relatively narrow weed control spectrum versus glyphosate along with uncertainties for the future of the technology regarding a potential ban on BXN cotton due to certain environmental concerns (Kamalick 1997).

While herbicide tolerance in cotton is seen to be of keen importance to US cotton growers, the development of in-planta lepidopteran insect resistance has been even more crucial not only to US growers, but has been breakthrough technology globally. Monsanto lead the way in the USA with the introduction of Bollgard™ cotton in 1996, which comprised a gene for a single active component, the delta-endotoxin Cry1Ac from *Bacillus thuringiensis* (Perlak et al. 1990). Since the introduction of Bollgard™ cotton, the number of US trait providers has increased

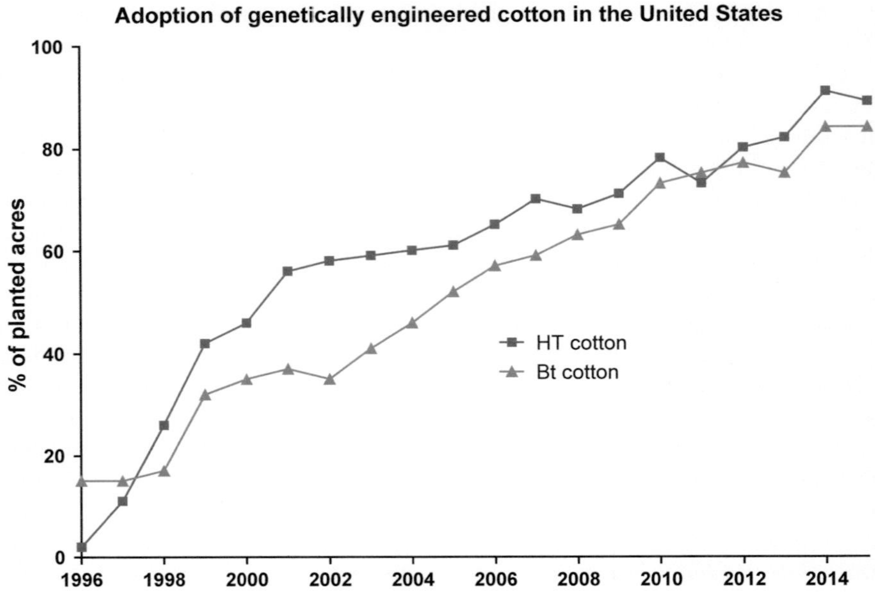

Fig. 2.1 Adoption of GM cotton in the USA from 1996 to 2015. Redrawn using data from USDA, National Agricultural Statistics Service

with a concomitant expansion of the spectrum of insects controlled, more effective resistance management, as well as giving farmers freedom of choice when determining trait and germplasm combinations best for their individual production needs. US growers now select between trait providers including Monsanto (Bollgard™ and Bollgard II™) and Dow AgroSciences (WideStrike™ and WideStrike 3™, and Bayer Crop Science (TwinLink™). Globally, Monsanto's Bollgard™ cotton has been of keen importance in Australia, India, and China. Dow AgroSciences' WideStrike™ trait is the preferred insect resistance technology in Brazil. Early assumptions that insect resistance and herbicide tolerance would deliver the most near-term value to cotton growers been borne out. In the USA, and Australia, where these technologies were first introduced in combination, varieties "stacked" with both insect resistance ("IR") and herbicide tolerance ("HT") dominate the market (Fig. 2.2). Clearly, growers have spoken very clearly as to the importance of these technologies in improving productivity while managing costs. For them, it is not an intellectual debate. In many cases, access has been the difference between prosperity and loss of one's farm and livelihood. Regulatory restraints have slowed the pace at which growers have had access to these technologies outside the USA, but IR traits have enjoyed rapid uptake when made available and have delivered excellent value to those growers fortunate enough to have access.

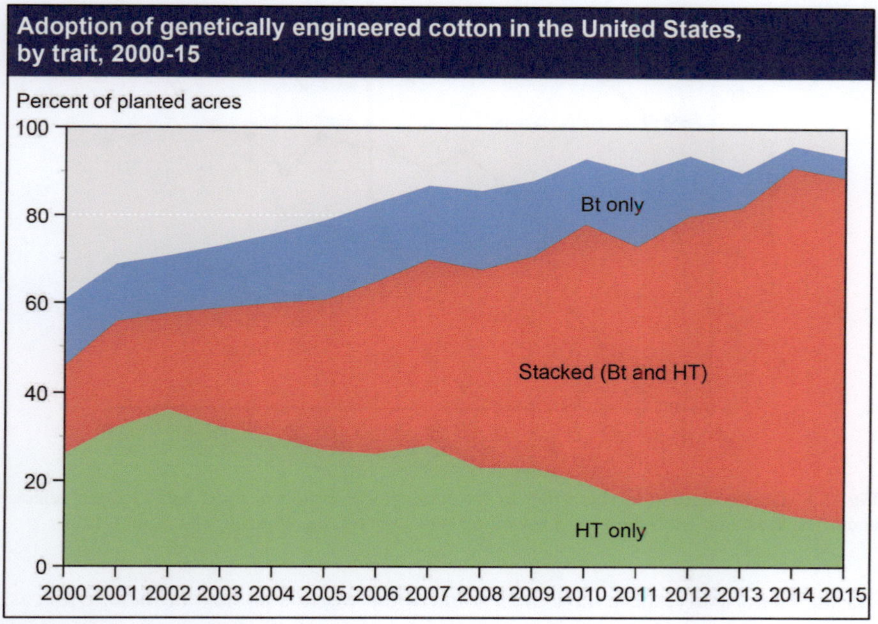

Fig. 2.2 Adoption of GM cotton in the USA, by trait, 2000–2015. *Source* USDA-Economic Research Service using data from USDA, National Agricultural Statistics Service, June Agricultural Survey

2.3 Global Adoption Is Driven by Benefits

In 1995, zero percent of global cotton production was planted to transgenic cotton. Twenty years later, it is estimated that roughly 25 million of the world's 37 million hectares of cotton production is planted to varieties carrying one or more biotech traits (Fernandez-Cornejo et al. 2014; James 2014). If accurate, then cotton (68 %) stands second only to soybeans (82 %) in terms of global hectares planted to transgenic varieties, and the same studies indicate that the adoption of transgenic cotton is increasing at about 5 % per year. Clearly, adoption of transgenic cotton varieties has proceeded quickly whenever transgenic cotton varieties have become available. US cotton growers went from 0 % planted to transgenic cotton to 85 % in 4 years. China went from 0 to 65 % in 4 years. India went from 0 to 90 % in 8 years (Fig. 2.3). Bt cotton has resulted in a virtual doubling of the national production of cotton fiber in India with almost no new hectares added under cultivation. India is now a major exporter of cotton fiber rather than a net importer, driving significant value back to the individual cotton farmer. There are approximately 6.3 million cotton farmers in India with an average cotton farm size of about

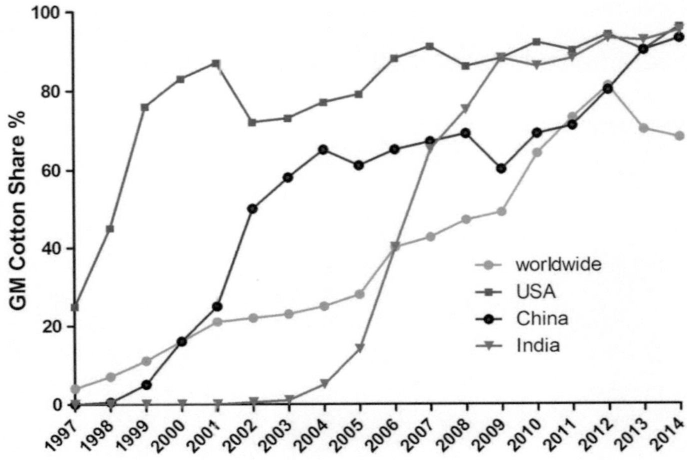

Fig. 2.3 Transgenic cotton share (%) in the total global cotton acreage and in three selected countries. Redrawn from data provided by GMO-Compass.org, ISAAA, and USDA-ERS

1.5 ha. Historically, these have been some of the world's poorest farmers, with cash from operations rarely meeting yet alone exceeding costs of production. For them, Bt cotton has been a revolution in productivity and economic opportunity. Cotton is an important cash crop for them, and one estimate places the economic benefit back to the India's cotton farmers at $5.1 billion (Choudhary and Gaur 2010). Furthermore, Bt cotton significantly reduces pesticide spraying, thereby increasing the safety of their farming activities. A similar story has unfolded in Burkina Faso, where Bt cotton was introduced in 2008 and now represents more than 70 % of the nation's production. Yields have increased an estimated 20 % over conventional cotton, and profitability to growers has increased by 50 %. Several authors have analyzed the benefits of GM crops, and they concluded that on average, GM technology adoption has reduced chemical pesticide use by 37 %, increased crop yields by 22 %, and increased farmer profits by 68 %. Yield gains and pesticide reductions are larger for insect-resistant crops than for herbicide-tolerant crops. Yield and profit gains are higher in developing countries than in developed countries (Brookes and Barfoot 2014; Klümper and Qaim 2014). As is the case in India, Burkina Faso's cotton farmers are smallholder farmers with an average of 3 ha or less under production. Bt cotton has meant lower costs of production, greater crop safety, higher yields, and improved economic returns to Burkina Faso's smallholder cotton farmers. With the recent passage of the biosafety law in Nigeria (All Africa 2015), it is anticipated that Bt cotton will reach growers there, exhibit a similar rate and extent of adoption, and deliver the same benefits to Nigeria's smallholder cotton farmers that are being experienced by growers of transgenic cotton the world over. One of the important aspects of transgenic cotton globally is that it is one of the first, if not the first, transgenic crops to gain wide acceptance. Because of the demonstrated value delivered, transgenic cotton paves the way for similar technologies in other crops to enter these geographies. China, India, Brazil, Burkina Faso, and Australia are good examples. Nigeria will soon be another.

Those of us involved in the early days of plant biotechnology were intrigued by what we perceived to be an opportunity to increase global farm productivity, reduce costs of production, increase the safety of farming practices, and enable the production of food and fiber to keep pace with the world's growing population. Yes, we thought we could eliminate world hunger. Clearly, the industry is on its way toward delivering against that promise far beyond our wildest imaginations. When cotton farmers have been allowed to vote, and they do so with their livelihoods, they have voted overwhelmingly in favor of the benefits delivered by biotechnology. As James Bonner so aptly perceived in 1957, technology is absolutely required to deliver the on-farm productivity that will certainly be essential over the next 50 years in order to feed and clothe the world's population. We are well on our way toward achieving that objective, and the alternative is unthinkable.

2.4 Trait Pipeline

As of this writing, transgenic events comprising 16 traits and trait combinations have achieved non-regulated status for cotton in the USA and one additional trait is pending final assessment and decision (USDA APHIS data). Of these, six are insect resistance (IR) only, eight are herbicide tolerance (HT) only, and three are IRHT molecular stacks. The years ahead will see an array of new transgenic traits working their way into cotton. Some will serve to broaden the efficacy of insect resistance actives as well as fortify pest resistance management strategies. Multiple Cry and Cry-like Bt endotoxin genes encoding proteins with alternative binding sites to those presently in use will be stacked to forestall development of resistance in targeted insect pests. Additional actives will target other non-lepidopteran insects including plant bugs. New transgenic cottons developed in the USA will have the opportunity to move into other markets globally, which dictates that new IR actives in new molecular stacks will need to address potential resistance management questions on a global as opposed to a local level. The new HT traits such as 2,4-D (Enlist Duo™) from Dow AgroSciences and Dicamba (Extend™) from Monsanto are designed to provide broader efficacy, broaden the application window, and deliver better control for glyphosate-tolerant species such as pigweed.

Longer term, the industry will have the opportunity to move its focus from input (production) traits and seek the manipulation of output (fiber, oil, and meal) traits with the intention of opening up new end uses by exploring ways to alter cotton fiber to enable new end uses on a global scale as well as expanding the use of cotton meal for wider human and animal (non-ruminant) nutrition (Sunilkumar et al. 2006). In the end, the path to higher prices and better economic return to the world's cotton growers will be to create new end uses for the oil and meal and by ensuring that a higher share of the world's spinning system is devoted to cotton. This would likely have to be at the expense of man-made fibers, principally polyester. Clearly, this will require altering fiber attributes to enable new functionalities, likely via rationally designed alterations in fiber structure in order to enable new end uses. But what are those new end uses, what are the attributes that will enable them, and what alterations in fiber structure might be required? Consumers consistently identify significant unmet needs in cotton fabrics and articulate desired new functionalities including modified permeability, improved durability, shrink and wrinkle resistance, shape retention in fabric, and fire retardation. Each of these individual categories has an estimated potential market impact of $5 billion or more and, if achieved in cotton, could significantly elevate cotton's share of the global spinning system, thereby creating more value for the world's cotton growers. Given the incredible molecular complexity of a cotton fiber (see, e.g., Rapp et al. 2010), there may become apparent practical ways to rationally design and then manipulate fiber structure to either deliver the desired attributes directly, or lend themselves to post-spinning treatments that will achieve the desired performance. We will need to define what needs to be done prior to designing alterations to accomplish same. In our view, these are all improvements that can be enabled in cotton, but developing fibers that address these consumer demands will require a proactive, coordinated public and private cooperation in order to make them a reality.

2.5 Preserving Genetic Diversity

Genetic diversity is clearly desirable for long-term crop improvement, and simple sequence repeat analysis presents a picture of a cotton germplasm pool that is relatively narrow (de Magalhães Bertini et al. 2006; Liu et al. 2000). This situation will not improve in the near term due to the nature of the transgenic event regulatory approval process. The exact numbers are difficult to come by, but present estimates are that it costs somewhere between $75 million and $100 million to register a single event in cotton and bring it to market. Accordingly, the only economically viable approach to trait development is to transform a single cotton variety, examine the structure and complexity (single versus multiple copies) of the "events," measure expression of the transgene, characterize the surrounding chromosomal environment, and then pick the best event to use as the initial trait donor. Backcrossing and forward-crossing are used to move deregulated traits into new genetic backgrounds and develop cultivars with suitable agronomic and fiber properties. This approach works well to a point, but there will always be some flanking DNA from the original transformed variety (typically a Coker variety) traveling along linked with the trait of interest. A "construct-" versus "event"-based approach to deregulation would be helpful in broadening the genetic diversity of the world's cultivated gene pool. Greater than 95 % of the US cotton crop is transgenic, as much of as 68 % of global production is transgenic, and virtually all transgenic cottons cultivated globally share in part a common Coker genetic background. This is due in part to the recalcitrance of most cotton cultivars except Coker to regeneration with the methods being employed (Chlan et al. 1999; Mishra et al. 2003; Wilkins et al. 2000). Two major exceptions are the Dow AgroSciences' WideStrike™ Cry1Ac and Cry1F events referenced earlier which were generated in the Acala cotton variety GC510. This variety was selected because acceptable regeneration protocols existed and GC510 was an Acala cotton variety with superior fiber quality (Rangan and Rajasekaran 1997). As anticipated, these GC510 events carry linkages to improved fiber quality and *Verticillium* tolerance far superior to those found in Coker backgrounds. Nevertheless, the cost of registering new events mandates that the single background most amenable to transformation and regeneration must be used for initial event generation. With the high cost of new event registration, trait providers are engaging in trait cross-licensing agreements which are having the unfortunate consequence of spreading the Coker background broadly. A construct-based registration policy would allow any registered and deregulated construct to be used widely across multiple genetic backgrounds and help alleviate some of the negatives associated with event-based deregulation. Better yet, we should view transformation as one more natural breeding process and embrace new technologies that are making real strides in helping to achieve the rate of productivity required to feed and clothe the world over the next 50-year period.

2.6 An Inconvenient Truth

It appears that we have all been consuming transgenic food for thousands of years. One could say that what we did not know was not hurting us. Quoting from Kyndt et al. (2015), "This finding draws attention to the importance of plant-microbe interactions, and given that this crop has been eaten for millennia, it may change the paradigm governing the 'unnatural' status of transgenic crops," so reports Kyndt et al's fascinating paper which details the observation that domesticated sweet potato (*Ipomoea batata* L.) is transgenic, carrying expressed bacterial genes, the product of horizontal gene transfer involving *Agrobacterium* which may have occurred 8000 to 10,000 years ago. The persistence of transgenes in the selected, cultivated clones versus non-cultivated wild unselected clones is at least partly suggestive of a potential selective advantage for transgenic sweet potato, at least in the eyes of the early consuming public. Chilton et al. (1977) demonstrated that *Agrobacterium* employs a natural system for genetically transforming host plants cells. Gladyshev et al. (2008) demonstrated extensive horizontal gene transfer in rotifers. Moran and Jarvik (2010) documented horizontal gene transfer between fungi and aphids. Li et al. (2014) showed the horizontal transfer of a functional receptor from a bryophyte to ferns. McClintock (1953) described the existence of mobile genetic elements in the maize genome playing a significant role in chromosome rearrangement as well as gene expression, and there are now studies describing same across a broad range of dicots and monocots (Bureau and Wessler 1994). Rob Schilperoort's laboratory (Hooykaas van Slogteren et al. 1984) described the ability of *Agrobacterium* to infect monocots, and

Agrobacterium-mediated transformation is now a preferred method for the transformation of cereals. All this to say that we have every reason to believe what Kyndt et al. have described will prove to be a general occurrence across all plants. Some will consider it to be inconvenient, but we deem it likely that the next 10 years will witness a plethora of studies finding similar results across the breadth of crop species driving us to the inescapable conclusion that we have all been consuming a host of transgenic foods in addition to sweet potato for millennia. Indeed, based on Kyndt, it may well be determined that these natural transgenes persist because they confer a selective advantage. Certainly, that argument might be made for *Ipomoea*. "What has been will be again, what has been done will be done again; there is nothing new under the sun" (Ecclesiastes 1:9).

Organisms have been exchanging DNA since there was DNA in organisms to exchange. Biotechnologists have not invented genetic recombination, and we have not invented the recognition signals, the enzymes that cut, splice, and rejoin DNA, the regulatory elements associated with regulated transcription of genes, the ability for elements to hop in and out of genomes horizontally as well as vertically, nor the mechanisms that allow genes to rearrange or those that allow one organism to move functional genes into another. Nor are biotechnologists the inventors of horizontal gene transfer taking place in plants intra- and interspecifically around the globe 24/7. What we do is to study natural systems and then determine ways to take advantage of, improve on, or otherwise speed the pace of developments that would likely be achievable by other means but only over a much greater time frame. It might be plant molecular breeding, but it is plant breeding nonetheless built on a foundation of mechanisms established thousands of years ago. Mules can still pull plows, but tractors make it possible to feed the world. Over time, mankind might possibly develop immunity to smallpox, but we are better off having the vaccine. Cotton biotechnologists use natural systems to assemble genes for resistance to pests, transfer them into cotton plants, and in so doing increase productivity per hectare, enable safer production practices, and improve the economics of farming, particularly for the smallholder farmers that have been proven to be those that benefit most from access. These efforts should be embraced if for no other reason than the humanitarian good served by so doing.

Approximately 175 million hectares of transgenic crop production is underway globally in 2015, 25 million hectares of which are cotton, and the industry has had an incredible record for increased safety and increased farm productivity. We have already seen that transgenic cotton raised productivity in Burkina Faso by 20 % and in the country of India by 100 % (James 2014). These are the kinds of gains that need to be made to keep pace with the world's increasing population and accordant need for food and fiber. It is indeed an inconvenient truth that transgenic plants including cotton, in spite of their record of safety and productivity gains, have been subject to much public scrutiny, debate, and incredibly costly regulation (borne primarily by US farmers and US consumers). This even while other approaches to food production make claims that go untested scientifically, stand virtually unregulated, and have resulted in thousands of foodborne illnesses and several deaths on a global scale (Hanola and Pauly 2011; Popoff 2011). Biodynamic farming and organic production practices might feed the affluent, but cannot provide for the 9.5 billion people expected to live on earth by the year 2050. The relatively good times we presently enjoy with respect to food availability on a global scale may in fact also be feeding the complacency which drives our dalliance with regulatory processes that restrict genetic diversity and slow the deployment of new technologies that will be essential to expanding the production of food and fiber at a pace required by the increasing population (Hoisington et al. 1999). Feeding and clothing the additional 2.5 billion people that will arrive over the next 35 years will require an increase in food and fiber production of 35.7 %. The realistic options are to (a) increase productivity of existing farmland by 35.7 %, or (b) bring a minimum of some 3.6 million km^2 of new land under production. That number would surely be a minimum because the most fertile land is already being tilled. Considering the fact that 70–80 % of the new farmland brought under production takes place by deforestation (Kendall and Pimental 1994), the selection of option b would have devastating environmental effects. Increasing overall agricultural productivity will not be solved by any single approach. The best existing farming methods must be coupled with the best new developments in farming practice including technology. Rather than embracing technology, there has been created an enormous regulatory bureaucracy, spending billions of dollars annually to regulate that which is going on naturally across the plant kingdom, which restricts access to technology and results in a narrowing of the germplasm base, and which is slowing the rate at which gains in per hectare productivity could and should be achieved. And for what real purpose? Perhaps Andersen (1909) said it as well as anyone. Given the fact that we have been consuming transgenic plants for as long as man has been consuming plants, we can hope that at some point in the near future, molecular breeding including the creation of transgenic plants will be looked at as any other breeding process and no longer be subject to the kinds of regulation being faced today.

2.7 Back to the Future

We began this writing referencing projections made by James Bonner some 59 years ago with respect to what one might expect the future to hold for agriculture as the world becomes industrialized and more populated. His observations remain spot on. Increasing agricultural productivity on a per hectare basis is critical, reducing costs of production is essential to maintaining the economic incentive to keep fertile land under production, the quality and value of what is produced must be elevated in order to keep pace with rising costs and drive that same economic return, and we must continue our development of technologies resulting in safer farming practices. This was an appropriate foundation on which to build our discussion of the global importance of transgenic cotton. It has been just 36 years since we and others began in earnest to invent, refine, and deploy cotton biotechnologies. We have made tremendous strides in understanding and utilizing what plants already understand and utilize with respect to optimizing gene expression, shuffling coding regions, exchanging regulatory elements, and effecting advances in overall environmental fitness. In these relatively short years, we have seen transgenic cotton go from 0 to 68 % of the world's cotton acres and the safety and performance record is remarkable. No one is forcing smallholder cotton farmers to plant transgenic cotton, but there are many using misdirected law and regulatory processes to forcibly prevent many from doing so. We have achieved productivity gains as measured in yield per hectare between 12 and 100 % depending on the country. Safer production practices have been enabled by in-plant insect resistance and the poorest of the world's smallholder cotton farmers have benefitted the most when they have been allowed access to the technology. These are accomplishments which the entire industry and public institutions are and should be proud of. We do not live in an either/or world. It will take far more than biotechnology to enable cotton to keep pace with the increasing need for natural fiber, vegetable oil, and meal, but biotechnology is surely needed. Marker-based technologies and genome-wide selection, for example, are enabling the capture and movement of native traits from race stocks and diploid species as well as allowing us to do a far better job of selecting parents in our crossing programs. Advances in how we break negative linkages are allowing extremely high-quality fiber to be carried in high-yielding cultivars. A significant portion of yield improvement in cotton production comes from the development of better systems for planting, cultivating, and GPS guidance for the purpose of "surgically" applying fertilizers, nutrients, and controlling pests. All of that notwithstanding, there remain many traits that can and will be brought into cotton via transgenic methods much more efficiently than by any other of the means just described. All of these cross-functional approaches must be embraced and deployed if we are to meet the demands of the world's population in the year 2050.

References

All Africa.com, 23 April 2015. Jonathan Signs Biosafety Bill Into Law, Nigeria

An G, Watson BD, Stachel S, Gordon MP, Nester EW (1985) New cloning vehicles for transformation of higher plants. EMBO J 4(2):277–284

Andersen HC (1909) The emperor's new clothes tales. In: Eliot CW (ed) The harvard classics. P.F. Collier & Son, New York

Anderson D (1985) Plant vector. US patent #432,842

Brookes G, Barfoot P (2014) Economic impact of GM crops: the global income and production effects 1996–2012. GM Crops Food 5:65–75

Brown H, Bonner J, Weir J (1957) The next hundred years. The Viking Press, New York

Bureau TE, Wessler SR (1994) Stowaway: a new family of inverted repeat elements associated with the genes of both monocotyledonous and dicotyledonous plants. Plant Cell 6:907–916

Cai CQ, Doyon Y, Ainley WM, Miller JC, Dekelver RC, Moehle EA, Rock JM, Lee YL, Garrison R, Schulenberg L, Blue R, Worden A, Baker L, Faraji F, Zhang L, Holmes MC, Rebar EJ, Collingwood TN, Rubin-Wilson B, Gregory PD, Urnov FD, Petolino JF (2008)

Targeted transgene integration in plant cells using designed zinc finger nucleases. Plant Mol Biol 69(6):699–709

Chilton M-D, Drummond MH, Merlo DJ, Sciaky D, Montoya AL, Gordon MP, Nester EW (1977) Stable incorporation of plasmid DNA into higher plant cells: the molecular basis of crown gall tumorigenesis. Cell 11:263–271

Chlan CA, Rajasekaran K, Cleveland TE (1999) Transgenic cotton (*Gossypium hirsutum* L.). In: Bajaj YPS (ed) Biotechnology in agriculture and forestry, vol 46. Transgenic Crops ISpringer, Berlin, pp 283–301

Choudhary B, Gaur K (2010) Bt cotton in India: a country profile. ISAAA Series of Biotech Crop Profiles. ISAAA, Ithaca

Davidonis GH, Hamilton RH (1983) Plant regeneration from callus tissue of *Gossypium hirsutum* L. Plant Sci Lett 32:89–93

De Block M, Herrera-Estrella L, Van Montagu M, Schell J, Zambryski P (1984) Expression of foreign genes in regenerated plants and in their progeny. EMBO J 3(8):1681–1689

de Magalhães Bertini CHC, Schuster I, Sediyam T, Gonçalves de Barros E, Moreira MA (2006) Characterization and genetic diversity analysis of cotton cultivars using microsatellites Cândida. Gen Mol Biol 29(2):321–329

Diamond v Chakrabarty (1980) 447 US 303. US Supreme Court Decision

Ecclesiastes 1:9. Holy Bible, New International Version®, NIV® Copyright ©1973, 1978, 1984, 2011 by Biblica, Inc.®

Fernandez-Cornejo J, Wechsler S, Livingston M, Mitchell L (2014) Genetically engineered crops in the United States. USDA-economic research report no. 162

Firoozabady E, DeBoer DL, Merlo DJ, Halk EL, Amerson LN, Rashka KE, Murray EE (1987) Transformation of cotton (*Gossypium hirsutum* L.) by *Agrobacterium tumefaciens* and regeneration of transgenic plants. Plant Mol Biol 10:105–116

Fraley R, Rogers S, Horsch R (1983) Use of a chimeric gene to confer antibiotic resistance to plant cells. In advances in gene technology: molecular genetics of plants and animals. Plenum Press, New York

Gladyshev EA, Meselson M, Arkhipova IR (2008) Massive horizontal gene transfer in bdelloid rotifers. Science 320:1753–1756

Grula JW, Hudspeth RL, Hobbs SL, Anderson DM (1995) Organization, inheritance and expression of acetohydroxyacid synthase genes in the cotton allotetraploid *Gossypium hirsutum*. Plant Mol Biol 28:837–846

Hanola V, Pauly S (2011) New York times. June 11, p A5

Hoekema A, Hirsch PR, Hooykaas PJJ, Schilperoort RA (1983) A binary plant vector strategy based on separation of *vir*-and T-region of the *Agrobacterium tumefaciens* Ti-plasmid. Nature 303:179–180

Hoisington D, Khairallah M, Reeves T, Ribaut J-M, Skovmand B, Taba S, Warburton M (1999) Plant genetic resources: what can they contribute toward increased crop productivity? Proc Natl Acad Sci USA 96:5937–5943

Hooykaas van Slogteren GMS, Hooykaas PJJ, Schilperoort RA (1984) Expression of Ti-plasmid genes in monocotyledonous plants infected with *Agrobacterium tumefaciens*. Nature 311: 763–764

Horsch R, Fry J, Hoffmann N, Eichholtz D, Rogers S, Fraley R (1985) A simple and general method for transferring genes into plants. Science 227(4691):1229–1231

James C (2014) Global status of commercialized biotech/GM crops. ISAAA briefs no. 49. ISAAA, Ithaca, NY

Kamalick J (1997) US EPA herbicide ban cuts Calgene market. 29 Dec 1997 ICIS News 19:39

Kendall HW, Pimental D (1994) Constraints on the expansion of the global food supply. Ambio 23:198–205

Klümper W, Qaim M (2014) A meta-analysis of the impacts of genetically modified crops. PLoS ONE 9:e111629

Kyndt T, Quispe D, Zhai H, Jarret R, Ghislain M, Liu Q, Gheysen G, Kreuze JF (2015) The genome of cultivated sweet potato contains *Agrobacterium* T-DNAs with expressed genes: An example of a naturally transgenic food crop. Proc Natl Acad Sci USA 112(18):5844–5849

Li F-W, Villarreal JC, Kelly S, Rothfels CJ, Melkonian M, Frangedakis E, Ruhsam E, Sigel EM, Der JP, Pittermanni J, Burge DO, Pokorny L, Larsson A, Chen T, Weststrand S, Thomas P, Carpenter E, Zhang Y, Tian Z, Chen L, Yan Z, Zhu Y, Sun X, Wang J, Stevenson DW, Crandall-Stotler BJ, Shaw AJ, Deyholos MK, Soltis DE, Graham SW, Windham MD, Langdale JA, Wong GK-S, Mathews S, Pryer KM (2014) Horizontal transfer of an adaptive chimeric photoreceptor from bryophytes to ferns. Proc Natl Acad Sci USA 111:6672–6677

Little RJ, Jones CE (1980) A dictionary of botany. Van Norstrand Reinhold Company, Inc, Hoboken

Liu S, Cantrell RG, McCarty JC Jr, Stewart JMcD (2000) Simple sequence repeat based assessment of genetic diversity in cotton race stock accessions. Crop Sci 40:1459–1469

McClintock B (1953) Induction of instability at selected loci in maize. Genetics 38:579–599

Mishra R, Wang HY, Yadav NR, Wilkins TA (2003) Development of a highly regenerable elite Acala cotton (*Gossypium hirsutum* cv. Maxxa)—a step towards genotype independent regeneration. Plant Cell, Tissue Organ Cult 73:21–35

Moran NA, Jarvik T (2010) Lateral transfer of genes from fungi underlies carotenoid production in aphids. Science 328:624–627

Oxford Advanced Lerner's Dictionary (2015) Oxford University Press, New York, NY

Perlak FJ, Deaton RW, Armstrong TA, Fuchs RL, Sims SR, Greenplate JT, Fischoff DA (1990) Insect resistant cotton plants. Nat Biotechnol 8:939–943

Popoff M (2011) Blame organic food industry for *E. coli* outbreak. Real Clear Science June 29, 2011. http://www.realclearscience.com/articles/2011/06/29/blame_organic_industry_for_e_coli_outbreak_106245.html Accessed July 2015

Potrykus I (1991) Gene transfer to plants: assessment of published approaches and results. Ann Rev Plant Physiol Plant Mol Biol 42:205–225

Rajasekaran K (2004) *Agrobacterium*-mediated genetic transformation of cotton. In: Curtis IS (ed) Transgenic crops of the world—essential protocols. Springer, Berlin, pp 243–254

Rajasekaran K, Grula JW, Anderson DM (1996a) Selection and characterization of mutant cotton (*Gossypium hirsutum* L.) cell lines resistant to sulfonylurea and imidazolinone herbicides. Plant Sci 119:115–124

Rajasekaran K, Grula JW, Hudspeth RL, Pofelis S, Anderson DM (1996b) Herbicide-resistant Acala and Coker cottons transformed with a native gene encoding mutant forms of acetohydroxyacid synthase. Mol Breed 2:307–319

Rangan TS, Rajasekaran K (1997) Regeneration of cotton plant in suspension culture. US patent #5,695,999

Rangan TS, Zavala T (1984) Somatic embryogenesis in tissue culture of *Gossypium hirsutum* L.). In Vitro 20:256

Rangan TS, Anderson DM, Rajasekaran K, Grula JW, Hudspeth RL, Yenofsky RL (2004) Transformed cotton plants. US patent #6,753,463

Rapp RA, Haigler CH, Flagel L, Hovav RH, Udall JA, Wendel JF (2010) Gene expression in developing fibres of upland cotton (*Gossypium hirsutum* L.) was massively altered by domestication. BMC Biol 15(8):139

Rathore K, Sunilkumar G, Cantrell R, Hague S, Reding H (2008) Cotton. In: Kole C, Hall TC (eds) Compendium of transgenic crop plants. Transgenic sugar, tuber and fiber crops, vol 7. Wiley-Blackwell, Chichester, pp 199–238

Stalker DM, McBride KE, Malyj LD (1988) Herbicide resistance in transgenic plants expressing a bacterial detoxification gene. Science 242:419–423

Shukla VK, Doyon Y, Miller JC, KeKelver RC, Moehle EA, Worden SE, Mithcell JC, Arnold NL, Gopalan S, Meng X, Choi VM, Rock JM, Wu Y, Katibah GE, Zhifang G, McCaskill D, Simpson MA, Blakeslee B, Greenwalt SA, Butler HJ, Hinkley SJ, Zhang L, Rebar EJ, Gregory PD, Urnov FD (2009) Precise genome modification in the crop species *Zea mays* using zinc-finger nucleases. Nature 459:437–441

Sunilkumar G, Campbell LM, Puckhaber L, Stipanovic RD, Rathore KS (2006) Engineering cottonseed for use in human nutrition by tissue-specific reduction of toxic gossypol. Proc Natl Acad Sci USA 103:18054–18059

Thompson GD, Pellow JW, Braxton LB, Haygood RA, Richburg JS, Lassiter RB, Haile FJ, Huckaba RM, Willrich MM, Langston VB, Richardson JM, Mueller JP (2005) WideStrike: a new stacked insect resistant trait for cotton. In: Proceedings of the 2005 Beltwide cotton conference, National Cotton Council, New Orleans, LA

Trolinder NL, Goodin JR (1987) Somatic embryogenesis and plant regeneration in cotton (*Gossypium hirsutum* L.). Plant Cell Rep 6:231–234

Sakhanokho HF, Zipf A, Rajasekaran K, Saha S, Sharma GC (2001) Induction of highly embryogenic calli and plant regeneration in upland (*Gossypium hirsutum* L.) and Pima (*Gossypium barbadense* L.) cottons. Crop Sci 41:1235–1240

Sakhanokho HF, Rajasekaran K (2016) Cotton regeneration in vitro (Chapter 6). In: Ramawat KG, Ahuja MR (eds) Fiber plants. Sustainable development and biodiversity, vol 13. Springer, pp xxx-xxx

Shoemaker RC, Couche LJ, Galbraith DW (1986) Characterization of somatic embryogenesis and plant regeneration in cotton (*Gossypium hirsutum* L.). Plant Cell Rep 5:178–181

Umbeck P, Johnson G, Barton K, Swain W (1987) Genetically transformed cotton (*Gossypium hirsutum* L.) plants. Nat Biotechnol 5:263–266

Waldron C, Murphy E, Roberts J, Gustafson G, Armour S, Malcolm S (1985) Resistance to hygromycin B: a new marker for plant transformation studies. Plant Mol Biol 5:103

Wilkins TA, Rajasekaran K, Anderson DM (2000) Cotton biotechnology. Crit Rev Plant Sci 19:511–550

Chapter 3
Natural Cellulose Fiber from Mendong Grass (*Fimbristylis globulosa*)

Heru Suryanto, Solichin Solichin and Uun Yanuhar

Abstract The global waste problems resulting from the use of synthetic fiber are becoming increasing environmental concerns. It would be better if the synthetic fibers give way to the natural fibers as renewable resources for environmental sustainability. New sources of natural fibers are being developed in recent years as natural fibers offer many advantages over synthetic fibers. Mendong grass is one of the natural sources of fiber. It is easy to grow and cultivate, and it offers several harvests from one plantation. The fiber has found many applications for small-scale industries and helps in economic welfare of small farmers. This chapter provides a general overview of mendong grass cultivation and obtaining fiber. The chemical, physical, mechanical, and thermal properties and prospective application of the mendong fiber are also presented.

Keywords Mendong · Fiber structure · Mechanical properties · Thermal properties

3.1 Introduction

Agricultural crops, forest trees, and other plant species have many uses for the farming community. Plant-based materials have been used traditionally for food and feed. Biobased polymeric products based on green materials such as plant and agricultural stocks are the basis for forming a portfolio of sustainable, eco-efficient

H. Suryanto (✉) · S. Solichin
Department of Mechanical Engineering, Universitas Negeri Malang,
Jl. Semarang 6, Malang, East Java, Indonesia
e-mail: heru.suryanto.ft.@um.ac.id

U. Yanuhar
Biotechnology Laboratory, Department of Fisheries and Marine Science,
University of Brawijaya, Jl. Veteran, Malang, Indonesia

K.G. Ramawat and M.R. Ahuja (eds.), *Fiber Plants*, Sustainable Development and Biodiversity 13, DOI 10.1007/978-3-319-44570-0_3

products which compete with synthetic products in market. The production of chemicals and materials from biobased feedstocks is expected to increase from today's 5 % level to about 12 % in 2010, about 18 % in 2020, and about 25 % in 2030 (Mohanty et al. 2005). Expectations are that the production of bulk chemicals from renewable resources could reach 113 million tons by 2050. It represents 38 % of all organic chemical production (de Jong et al. 2012).

Environmental sustainability-based technology is a global issue to move away from synthetic material to renewable resources. The synthetic fiber benefited human in various ways. Synthetic fibers are very durable and non-degradable, depending on their composition and the particular application. The disposal of parts made of synthetic fiber, such as composite for packaging containers and trash bags, also creates an environmental problem. It requires alternative ways to secure sustainable world development. Renewable biomaterials can be used as an alternative to replace the synthetic products.

Natural fibers have been offering many advantages over the synthetic fibers in recent years. The advantages of natural fiber as reinforcement composite are low price, low density, easy to be separated, abundantly available, renewable, biodegradable, and no health hazard (Li et al. 2007; Mu et al. 2009). Several alternatives of fiber sources, especially agricultural by-products such as ramie (Marsyahyo et al. 2008), banana (Venkateshwaran and Elayaperumal 2010), kenaf (Akil et al. 2011), hemp (Beckermann and Pickering 2008), Sisal (Li et al. 2000), Indian grass (Liu et al. 2004), Napier grass (Reddy et al. 2009), and Pineapple leaves (Mishra et al. 2004), have been used to produce cellulose fibers.

Traditionally, mendong grass has been used for a long time by the community for mats, rope fibers, and other product such as handbags, baskets, and furniture mats. In Indonesia, the grass is grown as a crop cultivated in some regions of Java, Sumatra, and Nusa Tenggara. Estimated production of mendong in Java, Indonesia, was 14,000 tons/year (Suryanto et al. 2014b). Since it has an economic potency, mendong needs more intensive cultivation.

3.2 Biology of Mendong Grass (*Fimbristylis globulosa*)

3.2.1 Taxonomy

Fimbristylis is a genus of sedges that known commonly as a fimbristyle, fimbry, or fringe-rush. Several continents have native species, but many species have been introduced to regions where they are not native. Mendong grass (*Fimbristylis globulosa*) was categorized as cyperaceae family and genus of *Fimbristylis* Vahl. This species is a synonym of *Fimbristylis umbellaris* (Table 3.1).

Table 3.1 Taxonomy of mendong grass (USDA 2015)

Kingdom	*Plantae*—Plants
Subkingdom	*Tracheobionta*—Vascular plants
Superdivision	*Spermatophyta*—Seed plants
Division	*Magnoliophyta*—Flowering plants
Class	*Liliopsida*—Monocotyledons
Subclass	*Commelinidae*
Order	*Cyperales*
Family	*Cyperaceae*—Sedge family
Genus	*Fimbristylis* Vahl—fimbry
Species	*Fimbristylis globulosa* (Retz.) Kunth—globe fimbry

3.2.2 Ecology

Mendong grass is originated in the Southeast Asia. This plant requires a watery environment for better growth. Therefore, mendong grass is easily found in the technically irrigated rice farm or swamps where there is always standing water year-round (Fig. 3.1). Mendong grass can grow well in the area that has an altitude of 300–700 m above the sea level, provided there is enough water, and exposed to full sunlight. These plants do not require particular soil types, but it would be better if planted in the sandy soils. In the marshy soils, mendong plants can also grow well. Mendong plants require plenty of water similar to the rice plants. Therefore, the mendong plants should not face water shortage, especially in the dry season. The mendong plants that lack water will turn yellow producing trunk of inferior quality. Well-maintained mendong plants flourish and produce good quality stalk mendong for long term, which are very strong.

Before the harvest is conducted, the water that inundated the plant area is removed in advance so that the surface of the land is visible and harvesting of mendong can be done easily. Mendong harvest is done by cutting the stalks mendong 3 cm above the surface of the ground using the sharp sickle leaving the clump of roots in the soil. After 1 month, clumps will sprout again and can be harvested after 3.5–4.5 months. This cycle is repeated up to 5 times in 2 years. After that, the plant was dismantled for the processing of land for the next planting. Farmers can save costs for soil tillage by harvesting five times the grass from one sowing. Managing the harvest and post-harvest should be done adequately and correctly to maintain the quality of mendong straw.

3.2.3 Morphology and Structure

Mendong grass is an annual plant with morphological characteristics such as stalks green shiny, rhizomes short, fibrous roots, and grooved (Fig. 3.2). Mendong leaves

Fig. 3.1 Mendong grass in land (**a**), harvest of mendong grass (**b**), and dried mendong grass (**c**)

are often reduced to sessile, hairy on the edges and have a small fibula. Mendong leaves grow on the top of the stem with some strands. Mendong straw is actually a flower stalk. The straw is compact, slender, hollow, 0.2–0.4 cm in diameter, and fast becoming stiff and looks like a cylinder but almost flattened beneath the flower stalk. Straw length can reach 1.5–1.7 m. This straw is harvested and used in the manufacture of various goods for human needs.

Mendong straw contains fiber bundles, vascular bundles, xylem, phloem, and aerenchyma (Fig. 3.3a). The most mendong fibers are located under the epidermis. Some fibers present near the vascular bundles in the middle of the straw. Fibers are a bit flat shaped with varied length, and pores can be seen on the fiber wall. In the transverse sections, the straw consists of 5–12 vascular bundles which mostly located in the center of the straw (Fig. 3.3a, b). The fiber bundle consists of some individual fibers (Fig. 3.3f, g). Each fiber has a lumen, middle lamella, primary wall, and a secondary wall (Fig. 3.3d). The primary wall is usually very thin (<1 μm), but the secondary wall is thick. It is composed of three layers, consists of microfibrils with a different orientation that contains larger quantities of cellulose molecules ($\sim$80 %). This wall is the main contributor to the overall properties of fiber. The microfibrils present parallel to each other forming a steep helix around the cell (Akil et al. 2011).

Fig. 3.2 Mendong grass: **a** single mendong grass, **b** flower, and **c** root

3.3 Mendong Fiber Properties

3.3.1 Chemical Composition

The plant contains large amounts of water due to its semiaquatic habitat. Based on the dry weight of the plant, all plant-based polymers were composed of sugars (carbohydrates) in combination with lignin and with lower amounts extractable proteins, starch, and inorganic materials. These chemicals are present in outer cell wall layer consists of primary and secondary wall. The chemical composition varies in each plant, even in the different parts of the same plant and in different plants depending upon geographic location, age, climate, and soil conditions (Rowell et al. 2000).

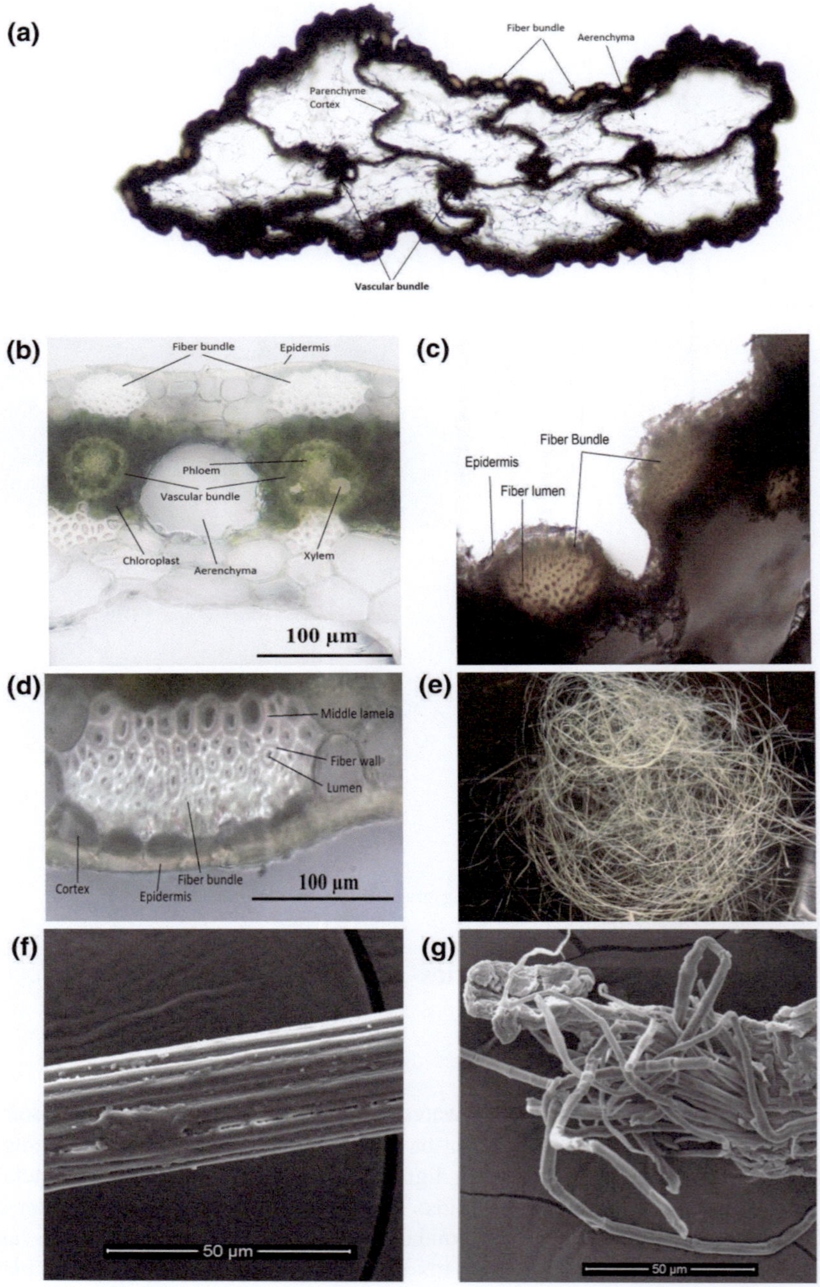

Fig. 3.3 Structure of mendong fiber: **a** dry mendong straw cutoff, **b** component of fresh mendong straw, **c** fiber bundle in dry mendong straw, **d** fiber bundle in wet mendong straw (observed by optic microscope), **e** extracted mendong fiber, **f** single-fiber bundle, and **g** single-cell fiber (SEM ×2000)

The properties of fiber are influenced by the chemical composition, particularly cellulose. Cellulose determines the strength of fibers because the cellulose has a high modulus of 45 GPa in the plant (Mwaikambo and Ansell 2006). Hemicellulose is a polysaccharide with low molecular weight. It often forms copolymers with glucose, glucuronic acid, mannose, arabinose, and xylose. It may take the form of random, amorphous branched, or nonlinear structure with low strength. Hemicellulose easily hydrolyzed by dilute acid or alkali, or enzyme hydrolysis (Summerscales et al. 2010). At the plant fiber level, hemicellulose serves as a matrix for cellulose (Bergander and Salmen 2002) and responsible for moisture absorption, both bio- and thermal degradation of the fibers.

Lignin provides rigidity to the plants. It is present localized to the luminal surface and around porous wall area to maintain the strength of the wall and helps transport water. Lignin is resistant to microorganisms attack due to the presence of aromatic rings, which provides resistance to the anaerobic processes (Bismarck et al. 2005). Lignin is thermally stable but responsible for the UV degradation of the fibers (Yi et al. 2010; Akil et al. 2011). Lignin strength is 100 times higher compared with hemicellulose at 70 % moisture level (Cousins 1976); thus, lignin can influence the fiber structure, properties, and morphology.

The mendong fiber is composed of cellulose of 72.14 %, hemicellulose 20.2 %, lignin 3.44 %, extractive matter 4.2 %, and moisture of 4.2–5.2 %. Table 3.2 shows a comparison of chemical content of others fibers with the mendong fiber. It is clear from these data that the mendong fiber has high cellulose content but lower than established fiber such as cotton and flax.

Table 3.2 Chemical content of mendong fiber as compared to other natural fibers

Fiber	Cellulose (%)	Hemicellulose (%)	Lignin (%)	Others (%)	Moisture content (%)	References
Mendong	72.14	20.2	3.44	4.2	4.2–5.2	Suryanto et al. (2014b)
Cotton	85–90	1–3	0.7–1.6	5.4–13.3	8–10	Foulk et al. (2011)
Flax	85	9	4	2	8.76–10	Foulk et al. (2011)
Jute	58–63	20–24	12–15	–	10.99	Wang et al. (2009)
Rice straw	64	–	8	28	9.8	Reddy and Yang (2006)
Sea grass	57	28	5	10	–	Davies et al. (2007)

3.3.2 Physical Properties

The mendong fiber bundle is consist of some single-cell fiber having 9.16 and 923 μm diameter and length, respectively (Suryanto et al. 2014b). The mendong fiber varies in shape and diameter. The average diameter of the fiber is 33.4 μm with the aspect ratio and density of 101 and 0.892 g/cm^3, respectively (Table 3.3). The physical properties of mendong are dependent on the species, maturity, and fertilization and site of growth. Comparison of physical properties of other fibers is shown in Table 3.3. The mendong fiber has low density compared with cotton, flax, rice straw, jute, and sea grass fiber.

Biofiber can be regarded as a composite of cellulose fibrils, formed in a matrix of lignin and hemicelluloses (Jayaraman 2003). The structure and the properties of the fibers are influenced by both dimension and arrangement in fiber bundle. High aspect ratio of fibers will improve the modulus and strength by optimizing stress transfer between the matrix and the cellulose.

The total content of cellulose and non-cellulose fiber constituents determines the structure, properties, and affect to the crystallinity (Reddy and Yang 2005). The mendong fiber was arranged by the crystalline structure of cellulose. The semicrystalline cellulose structure of mendong produced three peaks at 2θ of 16.5°, 22.5°, and 34.5°. The third peak at 34.5° corresponds to 1/4 of the length of one cellobiose unit and arises from ordering along the fiber direction. It is sensitive to the alignment of the chains into fibrils (Cheng et al. 2011). The amorphous component showed the little-diffracted intensity around 18 (Fig. 3.4). The peaks showed reflections at crystal planes of (011), (002), and (400). Widening at the 16.31° refer to non-cellulose materials such as hemicellulose and lignin in the fibers. The major intensity at an angle $2\theta = 22.5°$ has the same angle relative to the structure of cellulose Iβ($2\theta = 22.3°$). Thus, the structure of the cellulose fibers is cellulose Iβ mendong in which the unit Iβ cellulose structure is monoclinic (Bismarck et al. 2005). Both crystallinity and crystalline index of the mendong fiber were 70.7 and 58.6 %, respectively (Table 3.4), and the cellulose fibers extracted from the

Table 3.3 Physical properties of mendong fiber as compared to other natural fibers

Fiber	Density (g/cm^3)	Diameter (μm)	Fiber aspect ratio (average)	Reference
Mendong	0.892	33.8 ± 5.6	101	Suryanto et al. (2014b)
Cotton	1.5–1.6	12–38	1919	Gassan and Bledzki (1999) and Rouison et al. (2004)
Flax	1.5	40–600	1000	Gassan and Bledzki (1999), Foulk et al. (2011) and Rouison et al. (2004)
Rice straw	1.36	4–16	74	Reddy and Yang (2006), Abe and Yano (2009) and Rowell et al. (2000)
Jute	1.3	26.0	100	Gassan and Bledzki (1999), Park et al. (2006) and Rowell et al. (2000)
Sea grass	1–1.5	5	–	Davies et al. (2007)

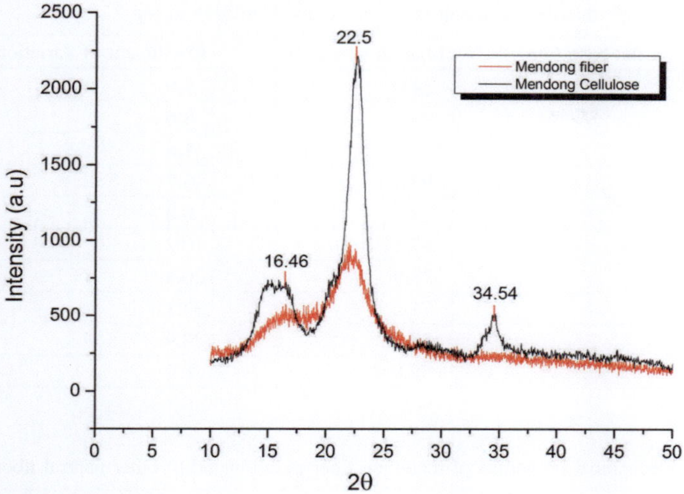

Fig. 3.4 Diffractogram of both fiber and cellulose of mendong

Table 3.4 Structure of mendong fiber as compared to other natural fibers

Fiber	Crystallinity (%)	Crystalline index (%)	Crystalline size (nm)	Microfibril angle (deg)	References
Mendong	70.7	58.6	14.3	22.2	Suryanto et al. (2014a)
Cotton	78.7	68	5–7	–	Ioelovich and Leykin (2008)
Flax	77	70	5.4	5–10	Kaith and Kalia (2008) and Bismarck et al. (2005)
Rice straw	62.8	57	3.75	19.4	Reddy and Yang (2006)
Jute	68.89	65.8	29.25	16.9	Wang et al. (2009), Mwaikambo (2009) and Sinha and Rout (2009)

mendong have crystallinity and crystalline index for 85.8 and 83.5 %, respectively. It indicates that the fiber mendong contains non-crystalline materials such as hemicellulose, lignin, and pectin which should be cleaned to make the fibers strong.

3.3.3 Mechanical Properties

The mechanical properties of natural fibers were affected by the fiber structure, chemical composition, and numbers of defects in a fiber. Mendong straw has enough strength homogeneous up to a length of 60 cm from the base of the stem with a coefficient of variation of <15 %. After 60 cm, the strength of the straw has a variation that is too high (>20 %), as shown in Table 3.5.

Table 3.5 Strength distribution along mendong straw from base to top

Distance from the base (cm)	Load at break (N)	Coefficient of variation (%)
0–10	74.0	9.6
10–20	98.7	8.6
20–30	100.1	1.6
30–40	94.8	13.7
40–50	89.1	8.0
50–60	87.3	10.2
60–70	78.9	20.4
70–80	71.2	32.1
80–90	59.1	22.9
90–100	52.8	20.9

Table 3.6 Mechanical properties of mendong fiber as compared to other natural fibers

Fiber	Tensile strength (MPa)	E-modulus (GPa)	Specific strength (kN m/kg)	References
Mendong	452 ± 47	17.4 ± 3.9	507	Suryanto et al. (2014a, b)
Cotton	287–597	5.5–12.6	179–398	Gassan and Bledzki (1999) and Rouison et al. (2004)
Flax	345–1035	27.6	230–690	Gassan and Bledzki (1999), Foulk et al. (2011) and Rouison et al. (2004)
Rice straw	450	26	331	Reddy and Yang (2006), Abe and Yano (2009) and Rowell et al. (2000)
Jute	1316	91.9	1012	Gassan and Bledzki (1999), Park et al. (2006) and Rowell et al. (2000)
Sea grass	573 ± 120	1	458	Davies et al. (2007)

The mendong fiber had tensile strength, elastic modulus, and the specific strength of 452 MPa, 17.4 GPa, and 507 kN m/kg, respectively (Table 3.5). The mendong fiber has a relatively high tensile strength, and fiber mendong has a lower density so that the specific strength of the mendong fiber is over cotton, rice straw, and sea grass fiber, but lower than jute and flax fiber (Table 3.6).

3.3.4 *Thermal Properties*

Thermal properties of the mendong fiber were observed by thermogravimetric test. The heat resistance of the fiber can be seen from the decomposition process. Curves' loss of mass and the mass loss were obtained using a sample of approximately 20 mg of sample (powdered mendong fiber), with an inert gas (Argon), the

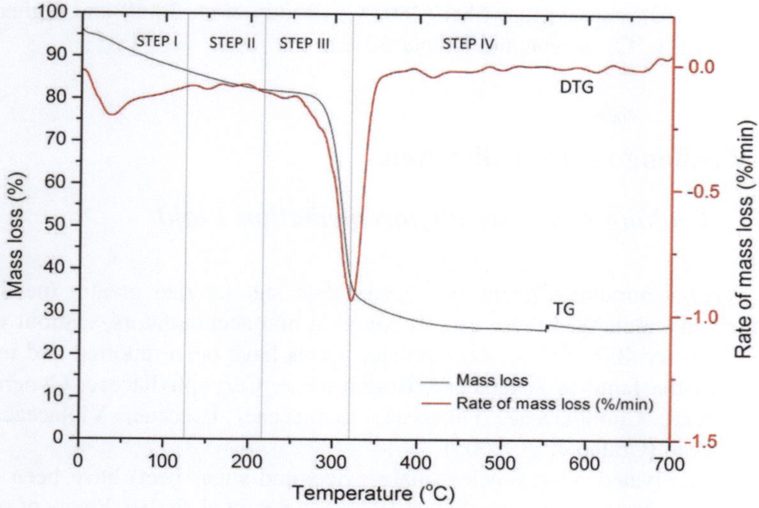

Fig. 3.5 Decomposition curve of the mendong fiber in an inert atmosphere (Argon) with a heating rate of 10 °C/min

heating rate of 10 °C/min. The mendong fiber decomposition test results are shown in Fig. 3.5.

Based on Fig. 3.5, it is observed that the decomposition of the samples is a exothermic process of chemical reaction that releases a significant amount of heat and shows the break down of organic material (Sonibare et al. 2005). The decomposition by thermal degradation of the whole sample shows four main stages associated with degradation of the mendong fiber. The first step is the initial devolatilization, characterized by the first basin in the reduction rate curve. This stage is related to the release of water content, and volatile compounds are very light (Chen et al. 2011). Devolatilization at the mendong fiber occurs at temperatures up to 164 °C. The second step is a transition period, which is indicated by the rate of mass loss. This is relatively stable and shows the decrease in release of volatile compounds and start of degradation of the fiber. This stage occurs until the temperature reaches 250 °C. In third step, the fiber decomposes rapidly, and the decomposition of complete biomass occurs at 321 °C temperature, which further decomposes until temperature reaches exact 350 °C. The fourth step is the slow combustion reaction. Residual mass shows a very slow decomposition which is characterized by low mass loss and the amount of mass that is relatively stable up to 700 °C temperature.

From Fig. 5, it is observed that the fiber mendong is less resistance to heat degradation as the mass is lost at a constant rate until the temperature reaches 250 °C. When it is compared to other fibers, this temperature is lower than bagasse (273 °C)

(Han et al. 2010), napier grass fiber (280 °C) (Reddy et al. 2009) and higher than maize fiber (211 °C) (Bavan and Kumar 2012).

3.4 Mendong Grass Utilization

3.4.1 Mendong Grass as Phytoremediation Plant

Metal hyper-accumulator plants can accumulate and tolerate greater metal concentrations in shoots than those usually found in non-accumulators, without visible symptoms. Over 400 of hyper-accumulator plants have been reported and include members of the families Asteraceae, Brassicaceae, Caryophyllaceae, Cyperaceae, Flacourtiaceae, Cunoniaceae, Fabaceae, Lamiaceae, Poaceae, Violaceae, and Euphorbiaceae (Gratão et al. 2005).

Several cultivated plant species (maize, rice, and sugar beet) have been established to use as metal phytoremediation (Poniedzialek et al. 2010). Plants of several grass families are also used for phytoremediation (Żurek et al. 2013). Vetiver grass (*Vetiveria zizanioides*) can absorb and promote biodegradation of organic wastes (2,4,6-trinitroluene, phenol, ethidium bromide, benzo[α]pyrene, atrazine, and heavy metals (Danh et al. 2009; Chen et al. 2004). Cyperaceae plants are capable of improving soil and water contaminated by heavy metals and toxic materials. Some species of *Fimbristylis* were applied as phytoremediation plants, which are *Fimbristylis globulosa* (Kurnia et al. 2004; Sa'ad et al. 2011), *Fimbristylis cymosa* (Paquin et al. 2006), *Fimbristylis dichotoma* (Muhammad et al. 2013), *Fimbistylis miliacea* L. (Akutam et al. 2014), and *Fimbristylis littoralis* (Nwaichi et al. 2015).

3.4.2 Mendong Straw as Craft Material

Mendong straw is used for several craft items such as woven handicrafts and wicker crafting mats, hats, ropes, bags, wallets, fancy paper, and others. A good mendong straw has good length and flexibility. Once harvested, the mendong straw is dried in the sun, for 4–6 h in dry season or for several days in rainy season. Drying twice produces good quality mendong straw. The first drying is performed immediately after the harvest, while the second drying is conducted after the first drying and soaking in water overnight. After coloring, mendong is used to make into various forms of handicrafts as shown in Fig. 3.6.

Fig. 3.6 Craft product from mendong straw: **a** rope, **b** mat, **c** bag, and **d** fitting basket

3.4.3 Mendong Fiber as Reinforcement in the Polymer Composite

The natural fiber has several advantages if applied in polymer composites because they are low price, low density, can be easily separated, abundantly available, renewable, biodegradable, and have no health hazard (Li et al. 2007; Mu et al. 2009). Fiber from crop plants such as ramie, jute, and hemp have already been established as reinforce fiber for the composite. Some of these fibers from crop have applied as reinforcement in the polymer composite such as rice straw (Reddy and Yang 2006), wheat straw (Reddy and Yang 2007a), Indian grass (Liu et al. 2004), switch grass (Reddy and Yang 2007b), and napier grass (Reddy et al. 2009).

High specific strength is the characteristic of the mendong fiber which is worthy to explore as reinforcement material in the polymer composite. The high specific strength is making it suitable for lightweight composites with applications in the field of road transport as a complementary component. Before its use for composite reinforcement, mending fiber should be soaked in sodium hydroxide solution with a concentration of 5 % for 2 h to increase the strength of the fiber as well as cleaning of fiber surface. Such treatment increases fiber strength to about 10 % (Suryanto et al. 2014a). As reinforcement composite, the mendong fiber has a critical length of 630 μm in matrix epoxy and interface shear strength of 11.1 MPa (Suryanto et al. 2015). This value is lower than hemp fiber in polypropylene matrix (Beckermann and Pickering 2009) and ramie fiber in polypropylene matrix (Awal et al. 2011). This low value of critical fiber length indicates the better stress transfer of the mendong fiber as reinforcement in the polymer composite. With the low (0.63 mm) critical length of fiber and the convenience in the extraction process, the processing to make composite is easier with these fibers.

3.4.4 Mendong Fiber as Source of Microcrystalline Cellulose

Microcrystal cellulose (MCC) is cellulose with fine size. Microcrystal cellulose had been used in different fields such as both binder and filler in medical tablets, fat replacer and stabilizer in the food industry, and a composite material in the plastic industry (Terinte et al. 2011). It was characterized by the size (diameter in micrometers) of the fibers. These fibers consist of crystalline cellulose that has a width of about 5 nm and a length of about 20–30 nm (Leppänen et al. 2009).

Usually, MCC is obtained from woody pulp. It means that it is produced from the trees following deforestation. There is a need for environment-friendly process with slowdown of the fast global deforestation. The use of plants having short life cycle, such as mendong grass, needs to be encouraged. The initial research was conducted with by extracting MCC from cellulose fibers of mendong through a chemical extraction sequence (Fig. 3.7). The results obtained show MCC with a crystallinity of 83 %. This value is lower by 3 % compared with the commercial MCC (Suryanto et al. 2013).

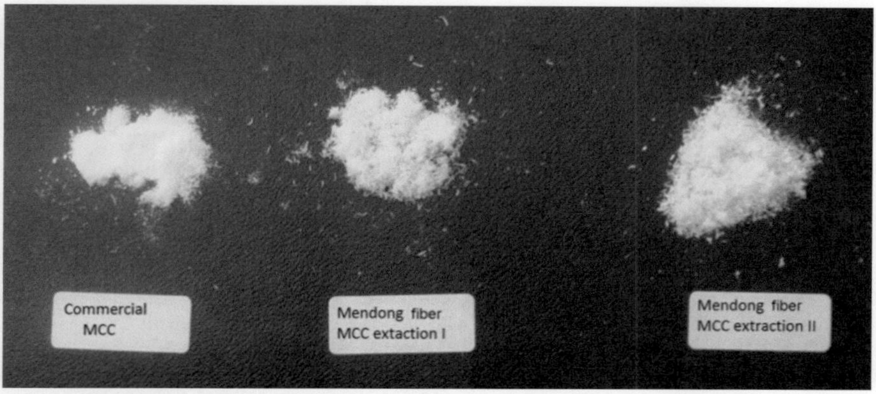

Fig. 3.7 Extraction of MCC from mendong fiber

3.5 Conclusion

Mendong grass is the plant with a potential future that has a variety of applications for the needs of community. This plant has been successfully characterized in the biological structure, its properties such as physical, mechanical, chemical content, and thermal degradation in comparison with other natural fibers. Because of high cellulose content and specific strength, mendong fiber is an excellent material to be used in the fields of biocomposites and handicrafts, as well as a source of cellulose microcrystal. Exploitation of this plant needs to be done so that this plant can be applied to other fields.

Acknowledgments Gratefulness to the Ministry of Research, Technology and Higher Education, Indonesia, through the competitive research Grant 2013 and the fundamental research Grant with Contract No. 9.4.3/UN32.14/LT/2015.

References

Abe K, Yano H (2009) Comparison of the characteristics of cellulose microfibril aggregates of wood, rice straw and potato tuber. Cellulose 16:1017–1023. doi:10.1007/s10570-009-9334-9

Akil HM, Omar MF, Mazuki AAM et al (2011) Kenaf fiber reinforced composites: a review. Mater Des 32:4107–4121. doi:10.1016/j.matdes.2011.04.008

Akutam A, Pappoe ANM, Armah FA, Enu-Kwesi L (2014) Phytoremediation potential of indigenous Ghanaian grass and grass-like species grown on used motor oil contaminated soils. J Ecol Environ 37:41–51. doi:10.5141/ecoenv.2014.006

Awal A, Cescutti G, Ghosh SB, Müssig J (2011) Interfacial studies of natural fibre/polypropylene composites using single fibre fragmentation test (SFFT). Compos Part A 42:50–56. doi:10.1016/j.compositesa.2010.10.007

Bavan S, Kumar M (2012) Morphological and thermal properties of maize fiber composites. Fibers Polym 13:887–893. doi:10.1007/s12221-012-0887-0

Beckermann GW, Pickering KL (2008) Engineering and evaluation of hemp fibre reinforced polypropylene composites: fibre treatment and matrix modification. Compos Part A Appl Sci Manuf 39:979–988. doi:10.1016/j.compositesa.2008.03.010

Beckermann GW, Pickering KL (2009) Engineering and evaluation of hemp fibre reinforced polypropylene composites: micro-mechanics and strength prediction modelling. Compos Part A 40:210–217. doi:10.1016/j.compositesa.2008.11.005

Bergander A, Salmen L (2002) Cell wall properties and their effects on the mechanical properties of fibers. J Mater Sci 37:151–156. doi:10.1023/A:1013115925679

Bismarck A, Mishra S, Lampke T (2005) Plant fibers as reinforcement for green composites. In: Mohanty AK, Misra M, Drzal LT (eds) Natural fibers, biopolymer, and biocomposites. CRC Press Tailor and Francis Group, Boca Raton

Chen Y, Shen Z, Li X (2004) The use of vetiver grass (*Vetiveria zizanioides*) in the phytoremediation of soils contaminated with heavy metals. Appl Geochem 19:1553–1565. doi:10.1016/j.apgeochem.2004.02.003

Chen C, Ma X, Liu K (2011) Thermogravimetric analysis of microalgae combustion under different oxygen supply concentrations. Appl Energy 88:3189–3196. doi:10.1016/j.apenergy.2011.03.003

Cheng G, Varanasi P, Li C et al (2011) Transition of cellulose crystalline structure and surface morphology of biomass as a function of ionic liquid pretreatment and its relation to enzymatic hydrolysis. Biomacromolecules 12:933–941. doi:10.1021/bm101240z

Cousins WJ (1976) Elastic modulus of lignin as related to moisture content. Wood Sci Technol 10:9–17. doi:10.1007/BF00376380

Danh LT, Truong P, Mammucari R et al (2009) Vetiver grass, *Vetiveria zizanioides*: a choice plant for phytoremediation of heavy metals and organic wastes. Int J Phytoremediat 11:664–691. doi:10.1080/15226510902787302

Davies P, Morvan C, Sire O, Baley C (2007) Structure and properties of fibres from sea-grass (*Zostera marina*). J Mater Sci 42:4850–4857. doi:10.1007/s10853-006-0546-1

De Jong E, Higson A, Walsh P, Wellisch M (2012) Product developments in the bio-based chemicals arena. Biofuels Bioprod Biorefin 6:606–624. doi:10.1002/bbb.1360

Foulk J, Akin D, Dodd R, Ulven C (2011) Production of flax fiber for biocomposite. In: Kalia S, Kaith BS, Kaur I (eds) Cellulose fibers: bio- and nanopolymer composites. Springer, Berlin

Gassan J, Bledzki AK (1999) Possibilities for improving the mechanical properties of jute/epoxy composites by alkali treatment of fibres. Compos Sci Technol 59:1303–1309. doi:10.1016/S0266-3538(98)00169-9

Gratão PL, Prasad MNV, Cardoso PF et al (2005) Phytoremediation: green technology for the clean up of toxic metals in the environment. Braz J Plant Physiol 17:53–64. doi:10.1590/S1677-04202005000100005

Han W, Chen K, Yang R-D et al (2010) Utilization of bagasse fiber for preparation of biodegradable flame retarding composites (BFRCS). BioResources 5:1605–1617. doi:10.15376/biores.5.3.1605-1617

Ioelovich M, Leykin A (2008) Structural investigations of various cotton fibers and cotton celluloses. BioResources 3:170–177. doi:10.15376/biores.3.1.170-177

Jayaraman K (2003) Manufacturing sisal–polypropylene composites with minimum fibre degradation. Compos Sci Technol 63:367–374. doi:10.1016/S0266-3538(02)00217-8

Kaith BS, Kalia S (2008) Graft copolymerization of MMA onto flax under different reaction conditions: a comparative study. Express Polym Lett 2:93–100. doi:10.3144/expresspolymlett.2008.13

Kurnia U, Suganda H, Saraswati R (2004) Pollution control technology in paddy fields. In: Fahmudin A (ed) Paddy fields and Processing Technology (in Indonesian Language). Center of Research and Development for Agroclimate Land, Bogor Indonesia, pp 249–283

Leppänen K, Andersson S, Torkkeli M et al (2009) Structure of cellulose and microcrystalline cellulose from various wood species, cotton and flax studied by X-ray scattering. Cellulose 16:999–1015. doi:10.1007/s10570-009-9298-9

Li X, Tabil LG, Panigrahi S (2007) Chemical treatments of natural fiber for use in natural fiber-reinforced composites: a review. J Polym Environ 15:25–33. doi:10.1007/s10924-006-0042-3

Li Y, Mai Y, Ye L (2000) Sisal fibre and its composites: a review of recent developments. Compos Sci Technol 60:2037–2055. doi:10.1016/S0266-3538(00)00101-9

Liu W, Mohanty AK, Askeland P et al (2004) Influence of fiber surface treatment on properties of Indian grass fiber reinforced soy protein based biocomposites. Polymer (Guildf) 45:7589–7596. doi:10.1016/j.polymer.2004.09.009

Marsyahyo E, Soekrisno S, Rohardjo HSB, Jamasri J (2008) Identification of ramie single fiber surface topography influenced by solvent-based treatment. J Ind Text 38:127–137. doi:10.1177/1528083707087835

Mishra S, Mohanty AK, Drzal LT et al (2004) A review on pineapple leaf fibers, sisal fibers and their biocomposites. Macromol Mater Eng. doi:10.1002/mame.200400132

Mohanty AK, Misra M, Drzal LT, Selke SE, Harte BR, Hinrichsen G (2005) Natural fibers, biopolymers, and biocomposites: an introduction. In: Mohanty AK, Misra M, Drzal LT (eds) Natural fibers, biopolymers, and biocomposites. CRC Press: Boca Raton, pp 1–36

Mu Q, Wei C, Feng S (2009) Studies on mechanical properties of sisal fiber/phenol formaldehyde resin in-situ composites. Polym Compos 30:131–137. doi:10.1002/pc.20529

Muhammad S, Shah MT, Khan S et al (2013) Wild plant assessment for heavy metal phytoremediation potential along the mafic and ultramafic terrain in northern Pakistan. Biomed Res Int 2013:1–10. doi:10.1155/2013/194765

Mwaikambo LY (2009) Tensile properties of alkalised jute fibres. BioResources 4:566–588. doi:10.15376/biores.4.2.566-588

Mwaikambo LY, Ansell MP (2006) Mechanical properties of alkali treated plant fibres and their potential as reinforcement materials. I. Hemp fibres. J Mater Sci. doi:10.1007/s10853-006-5098-x

Nwaichi EO, Frac M, Nwoha PA, Eragbor P (2015) Enhanced phytoremediation of crude oil-polluted soil by four plant species: effect of inorganic and organic bioaugumentation. Int J Phytoremediat. doi:10.1080/15226514.2015.1058324

Paquin DG, Sun WH, Tang C-S, Li QX (2006) A phytoremediation study: selection of tropical and other vascular plants for decolorization of Poly R-478 dye. Remediat J 16:97–107. doi:10.1002/rem.20104

Park J, Tran S, Hwang B, Devries KL (2006) Interfacial evaluation of modified Jute and Hemp fibers/polypropylene (PP)-maleic anhydride polypropylene copolymers (PP-MAPP) composites using micromechanical technique and nondestructive acoustic emission. Compos Sci Technol 66:2686–2699. doi:10.1016/j.compscitech.2006.03.014

Poniedzialek M, Sekara A, Jedrszczyk E, Ciura J (2010) Phytoremediation efficiency of crop plants in removing cadmium, lead and zinc from soil. Folia Hortic 22:25–31. doi:10.2478/fhort-2013-0155

Reddy N, Yang Y (2005) Biofibers from agricultural byproducts for industrial applications. Trends Biotechnol 23:22–27. doi:10.1016/j.tibtech.2004.11.002

Reddy N, Yang Y (2006) Properties of high-quality long natural cellulose fibers from rice straw. J Agric Food Chem 54:8077–8081. doi:10.1021/jf0617723

Reddy N, Yang Y (2007a) Preparation and characterization of long natural cellulose fibers from wheat straw. J Agric Food Chem. doi:10.1021/jf071470g

Reddy N, Yang Y (2007b) Natural cellulose fibers from switchgrass with tensile properties similar to cotton and linen. Biotechnol Bioeng 97:1021–1027. doi:10.1002/bit.21330

Reddy K, Maheswari CU, Reddy DJP, Rajulu AV (2009) Thermal properties of Napier grass fibers. Mater Lett 63:2390–2392. doi:10.1016/j.matlet.2009.08.035

Rouison D, Sain M, Couturier M (2004) Resin transfer molding of natural fiber reinforced composites: cure simulation. Compos Sci Technol 64:629–644. doi:10.1016/j.compscitech.2003.06.001

Rowell RM, Han JS, Rowell JS (2000) Characterization and factors effecting fiber properties. In: Frollini E, Leao A, Mattoso LHC (eds) Natural polymer and agrofibre based composites. Embrapa Instrumentação Agropecuária, Sao Carlos, pp 115–134

Sa'ad NS, Artanti R, Dewi T (2011) Phyto-remediation for rehabilitation of agricultural land contaminated by cadmium and copper. Indones J Agric 4:17–21

Sinha E, Rout SK (2009) Influence of fibre-surface treatment on structural, thermal and mechanical properties of jute fibre and its composite. Bull Mater Sci 32:65–76. doi:10.1007/s12034-009-0010-3

Sonibare OO, Ehinola OA, Egashira R, KeanGiap L (2005) An investigation into the thermal decomposition of Nigerian Coal. J Appl Sci 5:104–107. doi:10.3923/jas.2005.104.107

Summerscales J, Dissanayake N, Virk AS, Hall W (2010) A review of bast fibres and their composites. Part 1—fibres as reinforcements. Compos Part A Appl Sci Manuf 41:1329–1335. doi:10.1016/j.compositesa.2010.06.001

Suryanto H, Zakia N, Marsyahyo E (2013) The exploration of cellulose nanocrystal from Mendong Fibers using Pulsed Electric Field (PEF), and the utilization for biopackaging applications (in Indonesian language). Research report, Universitas Negeri Malang

Suryanto H, Irawan YS, Marsyahyo E, Soenoko R (2014a) Effect of alkali treatment on crystalline structure of cellulose fiber from mendong (*Fimbristylis globulosa*) straw. Key Eng Mater 594–595:720–724. doi:10.4028/www.scientific.net/KEM.594-595.720

Suryanto H, Marsyahyo E, Irawan YS, Soenoko R (2014b) Morphology, structure, and mechanical properties of natural cellulose fiber from mendong grass (*Fimbristylis globulosa*). J Nat Fibers 11:333–351. doi:10.1080/15440478.2013.879087

Suryanto H, Marsyahyo E, Irawan YS et al (2015) Improvement of interfacial shear strength of mendong fiber (*Fimbristylis globulosa*) reinforced epoxy composite using the AC electric fields. Int J Poly Sci 2015, Article ID 542376, 10 pages. http://dx.doi.org/10.1155/2015/542376

Terinte N, Ibbett R, Schuster KC (2011) Overview on native cellulose and microcrystalline cellulose I structure studied by X-ray diffraction (WAXD): comparison between measurement techniques. Lenzing Ber 89:118–131

USDA (2015) Plant Profile. http://plants.usda.gov/java/profile?symbol=FIGL#. Accessed 11 Jan 2015

Venkateshwaran N, Elayaperumal A (2010) Banana fiber reinforced polymer composites—a review. J Reinf Plast Compos 29:2387–2396. doi:10.1177/0731684409360578

Wang W, Cai Z, Yu J, Xia Z (2009) Changes in composition, structure, and properties of jute fibers after chemical treatments. Fibers Polym 10:776–780. doi:10.1007/s12221-009-0776-3

Yi C, Tian L, Tang F et al (2010) Crystalline transition behavior of sisal in cycle process. Polym Compos 31:933–938. doi:10.1002/pc.20885

Żurek G, Pogrzeba M, Rybka K, Prokopiuk K (2013) Suitability of grass species for phytoremediation of soils polluted with heavy-metals. In: Barth S, Milbourne D (eds) Breeding strategies for sustainable forage and turf grass improvement. Springer, Dordrecht

Chapter 4
Sanseviera zeylanica (L.) Willd and Its Potential as a New Natural Source Fiber: A Case Study from the Yucatan Peninsula, Mexico

Rodrigo Duno de Stefano, William Cetzal-Ix and Saikat Kumar Basu

Abstract *Sansevieria zeylanica* (L.) Willd. (Asparagaceae) is a plant native to Sri Lanka, but is currently distributed in the tropics and subtropics across the world as a well-known ornamental species. In the Yucatan Peninsula, Mexico, this species grows in disturbed areas adjoining the towns and villages and has been used as a supplement to local cattle feed. However, it can yield a good quality, natural fiber from which hammocks are locally made. In the Yucatan Peninsula, particularly in the community of Euan, hammocks are being produced for almost 25 years and can be considered as an important locality for innovative use for this fiber-yielding plant. These hammocks are certainly higher in quality compared to others made with the henequen fibers, palms, cotton, and synthetic fibers. Furthermore, these hammocks are stained with natural dyes. Hence, this artistic activity that involves the novel use of an exotic plant as a source of fiber, traditional application of the native dyeing plants, and indigenous knowledge of the Mayan communities of Yucatan Peninsula should be promoted as a minor (cottage) industry that can bring economic opportunities for the people of this region.

Keywords Conservation · Biodiversity · Fiber · Yucatan peninsula · Mexico

R.D. de Stefano
Herbario CICY, Centro de Investigación Científica de Yucatán, A. C. (CICY), Calle 43. No. 130. Col. Chuburná de Hidalgo, Mérida 97200, Mexico

W. Cetzal-Ix (✉)
Instituto Tecnológico de Chiná, Calle 11 entre 22 y 28, Colonia Centro Chiná, Campeche 24050, Mexico
e-mail: rolito22@hotmail.com

S.K. Basu
Department of Biological Sciences, University of Lethbridge, Lethbridge, AB T1K 3M4, Canada

K.G. Ramawat and M.R. Ahuja (eds.), *Fiber Plants*, Sustainable Development and Biodiversity 13, DOI 10.1007/978-3-319-44570-0_4

4.1 Introduction

At present, we have become accustomed to processed plastics products and several other derivatives of different hydrocarbons. We cannot deny that this product despite the detrimental environmental impact brings some benefits to humans. Also, over time, we have seen that many natural products have been replaced by plastic or other synthetics. In Mexico, such modernization has historically impacted the socio-economic pattern of the traditional life and practices in the Yucatan Peninsula; for example, the decline in the use of henequen (*Agave furcroydes Lem.*), a principal export from the region since the middle of twentieth century. The hammock is an essential component of the Yucatec culture; however, every day it is becoming more common to see nylon hammocks compared to those made from natural fibers such as henequen and cotton.

Hammock is not a native product of Mexico but rather native to the Caribbean islands where Arakuak people used it before the arrival of the Spanish. The production of hammockson the YucatanPeninsula began possibly in the sixteenth century but its use became widespread in the Mayan population once the warp technique was adopted, a process possibly facilitated by the local cultivation of henequen (*Agave fourcroydes*) for fibers and for the preparation of ropes (Baños-Ramírez 2010). Today, it is very difficult to get hammocks of henequen or other natural fibers (except cotton), because they were replaced by plastic (nylon and polypropylene). Synthetic hammocks can cost up to three times cheaper than the most economical realized with natural fibers.

In this sense, the use of *Sansevieria zeylanica* (L.) Willd. (Asparagaceae) "*Lengua de vaca*" is particularly valuable in Yucatan State, Mexico. Another species, *S. liberica* Gérôme & Labroy (bowstring hemp)fiber is reported as filler in natural rubber compounds (Osabohien and Egboh 2007). The genus has been reported for a number of important physiological, biochemical, and medicinal properties in animals and humans (Botting 2006; Katzung et al. 2009; Ikewuchi and Ikewuchi 2009), the most prominent being its role investigated as anti-diarrheal effect (Adeyemi et al. 2009) and anti-inflammatory properties (Chinasa et al. 2011). The reasons for the importance of *S. zeylanica*in the Yucatan Peninsulaare that it is used locally and traditionally used as a natural fiber and has a long tradition in handling anddyeingof the fiber and its warp. Finally, it represents an ethnic, rural, artistic, and economic activity that would be worth to promote and stimulate as an alternative source of employment for local communities in the form of a minor or cottage industry.

4.2 Taxonomy and Systematics

The plant is a robust and upright growing plant, often occurring in dry, narrow patches and cluster with compact, fibrous root system that penetrates deep into the soil. The leaves of *S. zeylanica* are characteristically organized as a rosette. The leaf blade or lamina is thick, smooth, rubbery, succulent, and fibrous as important adaptive features to enable the plant survive successfully under hot and dry climatic regimes for storing water. The flowers are in greenish white, dull brownish, or lilac color varying between different cultivars and are usually arranged in simple or compound raceme inflorescences (Mbugua and Moore 1996). Different species of *Sansevieria* are known to have horticultural and ornamental values (Mbugua and Moore 1996), important medicinal properties (Adeyemi et al. 2009; Ikewuchi and Ikewuchi 2009; Chinasa et al. 2011), and as natural fiber-yielding plants (Osabohien and Egboh 2007).

The genus *Sansevieria* was described by Thunberg in 1794 with two species: *S. thyrsiflora* and *S. aethiopica*. Typification and nomenclature of *S. zeylanica* were resolved by Wijnands in 1973. This species was originally described by Linneaus in 1753 as *Aloe hyacinthoides* L. var. *zeylanica* L. from Ceylon ["*habitatZeylon*"] in Sri Lanka but later transferred to *Sansevieria* by Willdenow in 1799. The genus *Sansevieria* is represented by more than 50–70 species worldwide and is a native of the continents of Asia (south), Africa including Madagascar (Mbugua and Moore 1996).

Sansevieria zeylanica (L.) Willd. Sp. Pl. 2: 159. 1799
Type: *Aloe zeylanica pumila, foliis variegatis* C. Commelin, Hort. Med. Amstelod. Rar. Pl. 2: 41. T. 21. 1701
Aloe hyacinthoides L. α *zeylanica* L. Sp. Pl. 321. 1753.
Aloe zeylanica (L.) N. J. Jacq. Enum. Stirp. Vindobon. 310. 1762.
Aletris hyacinthoides (L.) L. α *zeylanica* (L.) L. Sp. Pl. ed. 2. 456. 1762.
Aletris zeylanica (L.) Miller, Gard. Dict. ed. 8. 1768.
Sansevieria indica Herter, Estud. Bot. Reg. Urug. 24: 218. 1956, nom illegit.

4.3 Hammock Production

The elaboration of *Sansevieria* hammocks is much more appreciated than henequen hammocks for its freshness, softness, strength, and durability (Arroyo-Irigoyen and Terán-Contreras 2010). The production of the hammocks with *S. zeylanica* is performed in several towns near Merida, but particularly in the town of Euan, located 30 km from the city of Merida and only 6 km from Tixkokob (Fig. 4.1). This town is well known for its long tradition in the production of nylon or cotton hammocks. The yarn obtained from *S. zeylanica* fiber is much thinner than that from henequen, but on the other hand it is much thicker than silk thread. However, the yarns of *S. zeylanica* do not cause itching or scrape as is common with hammocks derived from the

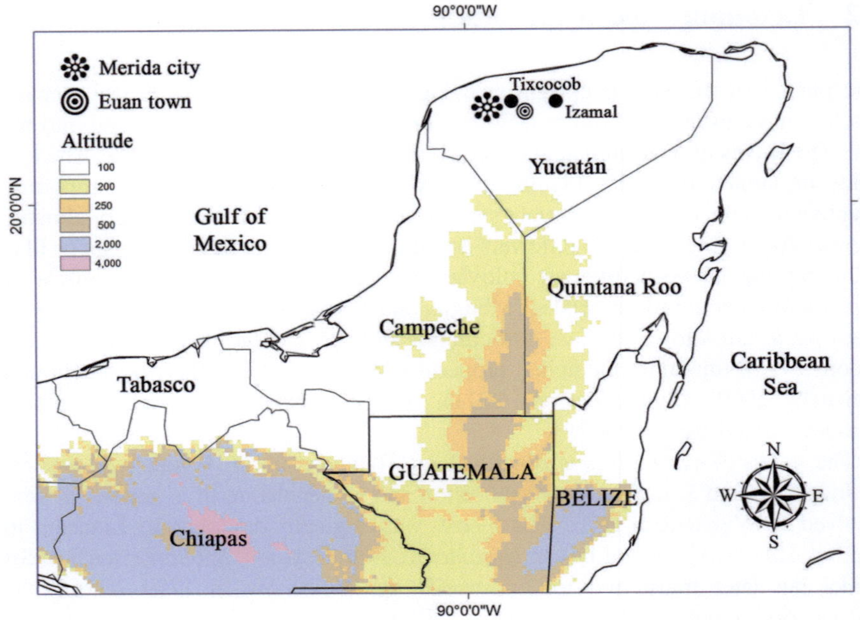

Fig. 4.1 Locality of Euan where hammock production of *Sansevieria zeylanica* is done in the Yucatan Peninsula, Mexico

henequen yarn. The prices of hammocks range between US$ 76–120 (~600–1200 Mexican pesos). The elaboration under this scheme may take up to 3 weeks, but can make two to three hammocks simultaneously.

The manufacturing of hammocksinvolves five distinct steps: (1) harvesting of leaves, (2) extraction of the raw fiber, (3) corked or braid of the fiber, (4) dyeing, and (5) warping of the hammock. Steps 1 and 2 are usually supervised by a male member of the family, whereas steps 3–5 are supervised mostly by female members.

4.3.1 Harvesting of Leaves

Sansevieria zeylanica is not cultivated in the Yucatan region. It grows as a ruderal plant and the leaves are obtained from naturalized populations in disturbed areas near some populations around the city of Merida, Yucatán, Mexico (Fig. 4.2). It was originally introduced in the region for use in the campuses of henequen as rows and protect it from fire (Arroyo-Irigoyen and Terán-Contreras 2010). The harvest of leaves is made with a long-handled narrow spade "*coa*" or a machete. The harvest is nondestructive, because only the leaves are cut and subterranean rhizomes remain intact.

Fig. 4.2 *Sansevieria zeylanica*. **a**, **b** Plants in secondary vegetation (*ruderal*) and gardens. **c** Inflorescence. **d** Single leaf. **e** Cross (*transverse*) section of the leaf blade. **f** Longitudinal section of the leaf blade. **g** Yarn sample

4.3.2 Extraction of the Raw Fiber

After obtaining the leaves, these are defibered with a shredder machine specially designed for *S.zeylanica* that is very similar to the shredder henequen but smaller. Henequen industrial shredders cannot be used in *S.zeylanica* because the fiber is thinner and can be easily damaged (Arroyo-Irigoyen and Terán-Contreras 2010). Defibration can also be done manually, but it is a very laborious mechanical process that involves hitting the leaf against a hard surface, (usually a stone) to get the fiber and is then dried for a short period of time (3 days).

4.3.3 Corked or Braid of the Fiber

After obtaining the raw fiber, it is dried thoroughly for proper processing. The next step is the preparation of yarn, a process known as corked or braid. This process involves a simple device made from a bicycle wheel that is fixed on a wooden framework and then moves a needle with a hook-shaped tip. The *Sansevieria* fiber is fixed to this hook and then spinning starts; when the single yarn reaches the desired size, it is corked once again with the help of the hook to make a double thread (Fig. 4.3).

4.3.4 Dyeing

For dyeing, the hammocks (red, green, blue, brown, etc.; Fig. 4.3g–j) up to 10 species of flowering plants are commonly used representing members of the family Fabaceae: *Caesalpinia gaumeri* Greenm. (kitim che' o kitam che'), *Haematoxylum campechianum* L. (palo de Campeche, palo tinto, tinto), *Havardia albicans* (Kunth) Britton & Rose (chukum), *Indigofera suffruticosa* Mill. (añil, platanillo), and *Lysiloma latisiliquum* (L.) Benth. (tsalam). In addition, a common colorant is used as a food additive throughout tropical America; *Bixa orellana* L. (achiote) (Bixaceae) is also used for dyeing. Among other dyes *Maclura tinctoria* (L.) Steud. (palo moral, mora) (Moraceae) or the famous cochineal insect *Dactylopius coccus* (Hemiptera) are also used (Table 4.1; Fig.4.4). It is also important to mention that addition colors are obtained by mixing the above plant extracts with synthetic compounds such as baking soda or other alkaline substances.

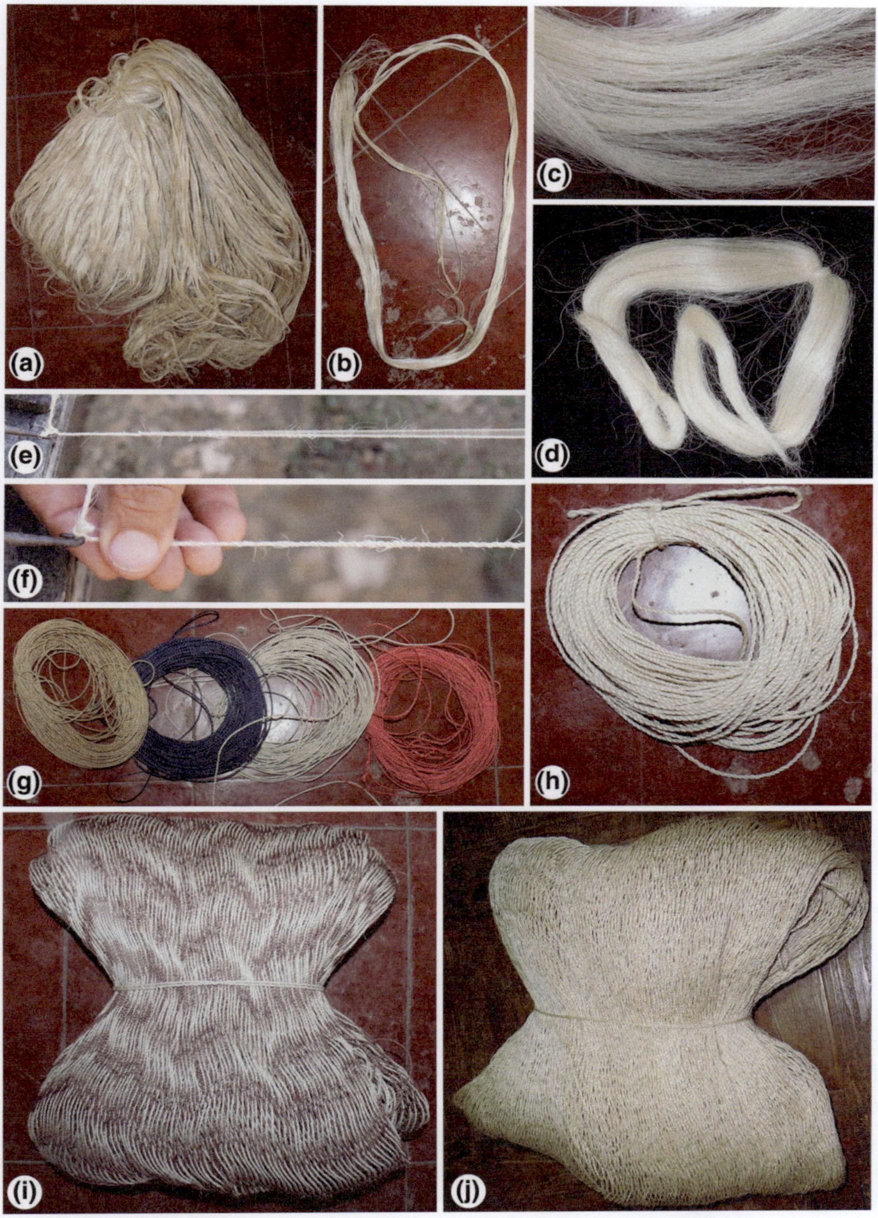

Fig. 4.3 **a** Raw fiber. **b** Dried fiber. **c**, **d** Yarn. **e**, **f** Corked or braid technique. **g** Sample of variously dyed fibers. **h** Double yarn. **i**, **j** Hammocks

Table 4.1 Plant species used for dyeing hammocks in the Yucatan Peninsula, Mexico

Species	Family	Habit	Distribution	Staining	Part
Caesalpinia gaumeri	Fabaceae	T	MEX, CA, ANT (Cuba)	Dark brown	Wood
Haematoxylum campechianum	Fabaceae	T	MEX, CA	Red	Wood
Havardia albicans[a]	Fabaceae	T	MEX		Stem
Indigofera suffruticosa	Fabaceae		MEX, EUA, AT		Stem
Lysiloma latisiliquum	Fabaceae	T	MEX, EUA (Florida), CA, ANT		Stem
Bixa orellana	Bixaceae	S	AT, ANT	Brown, Red	Fruits
Maclura tinctoria	Moraceae	T	MEX, AT	Yellow	Bark

Habit: *T* tree, *S* shrub. Distribution: *MEX* Mexico, *CA* Central America, *ANT* Antilles. *AT* Tropical America
[a]Endemic to the Yucatan Peninsula, Mexico

4.3.5 *Warping of the Hammock*

Warping technique requires a traditional horizontal or vertical frame and a single operator, usually a female member of the family. This process is performed with the help of needles of wood, plastic, or metal. The technical details of warping include the amount of fiber to be used depending on the size of the hammock (small for children, singles, and doubles) and color design, which basically is carried out at the request of the buyer (Fig. 4.5).

4.4 Conclusions

The "*lengua de vaca*" and its fiber could be used to generate handicraft activity and economic value added for many people in the rural communities of Yucatan. We have also seen its use as a raw material for manufacture of papers that could help the growth of relevant local industries. The technique of fiber extraction and production of the yarn is very artisan, but could be improved and combined with a good promotion program, to generate demand (key in this modern world) that converts to introduced species, in a productive plant, which enriches a traditional activity in the region, such as tinctorial activities and the elaboration of the hammocks. However, Arroyo-Irigoyen and Terán-Contreras (2010) mentioned that *Sansevieria* fiber has great potential but is limited by the defibration process, the lack of adequate, and close to the area of elaboration of the hammocks equipment.

Fig. 4.4 Plants used for dyeing the hammocks. **a–d** *Caesalpinia gaumeri*. **e**, **f** *Haematoxylum campechianum*. **g**, **h** *Havardia albicans*. **i**, **j** *Lysiloma latisiliquum*. **k** *Maclura tinctoria*. (*Photos* All pictures of W. Cetzal-Ix, except D. Germán Carnevali)

Fig. 4.5 Hammock production. **a**, **b**. Traditional frame with single operator. **c** Fibers for warping. **d** Differently colored fibers. **e** Finished hammocks

This traditional practice of the local Mayan communities of Yucatan Peninsula in using an exotic fiber-yielding plant simultaneously with locally available, natural, plant-based dyes should be promoted as a minor (cottage) industry to bring economic opportunities for the people of this region.

Acknowledgments The authors are grateful to the Villajuana Madariaga family for sharing their traditional knowledge in producing the hammocks. Silvia Hernández Aguilar (CICY) is also acknowledged for her assistance with the Fiber Library. Second author is grateful to CONACYT for the postdoctoral scholarship.

References

Adeyemi OO, Akindele AJ, Ogunleye EA (2009) Evaluation of the anti-diarrhoeal effect of *Sansevieria liberica*. J Ethnopharmacol 123:459–463

Arroyo-Irigoyen LE, Terán-Contreras S (2010) Artesanía y recursos naturales. In: Durán-García R, Méndez-González M (eds) Biodiversidad y Desarrollo Humano en Yucatán. CICY, PPD-FMMAM, CONABIO/SEDUMA, Mérida, pp 365–367

Baños-Ramírez O (2010) El henequén, la hamaca de henequén y el hábitat maya. In: Durán-García R, Méndez-González M (eds) Biodiversidad y Desarrollo Humano en Yucatán. CICY, PPD-FMMAM, CONABIO/SEDUMA, Mérida, pp 45–49

Botting RM (2006) Inhibitors of cycloxygenases: mechanisms, selectivity and uses. J Physiol Pharmacol 5:113–124

Chinasa EC, Stella Ifeoma IA, Obodoike EC, Chhukwuemeka ES (2011) Evaluation of anti-inflammatory property of the leaves of *Sansevieria liberica* Ger. and Labr. (Fam: Dracaenaceae). Asian Pac J Trop Med 4:791–795

Ikewuchi CC, Ikewuchi JC (2009) Amino acid, mineral and vitamin composition of *Sansevieria liberica* Ger. and Labr. Pac J Sci Tech 10:477–482

Katzung BG, Masters SB, Trevor AJ (2009) Basic and clinical pharmacology, 11th edn. The McGraw Hill Companies Inc, New York

Mbugua PK, Moore DM (1996) Taxonomic studies of the genus *Sansevieria* (Dracaenaceae). In: van der Maesen LJG, van der Burgt M, van Medenbach de Rooy JM (eds) The biodiversity of African plants. Kluwer, Dordrecht, pp 489–492

Osabohien E, Egboh SH (2007) Utilization of bowstring hemp (*Sansevieria liberica*) fibre as a filler in natural rubber compounds. J Appl Poly Sci 107:210–214

Chapter 5
Linen and Its Wet Processing

Arun K. Patra

Abstract This chapter addresses the concept of linen being used for apparels and the ways to process it for the same. The background of the fiber and a brief discussion on its cultivation and harvesting are included. An insight into chemical constitution of the fiber shows some of the stubborn natural impurities present in it. The physical properties have greatly influenced its usage as a textile material. A detailed chemical processing for linen is given in the text. Biotechnology finds a wide application in most technical fields, and in this context, the various applications of enzymes in wet processing of linen range right from retting to biofinishing.

Keywords Enzyme · Flax · Lignin · Pretreatment · Retting

5.1 Introduction

Linen had a virtual revival in the last few years. There is an upward trend in its popularity among consumers, and more so in the apparel segment. The demand for large quantities of fiber is seen in the recent years as it is a preferred material now with many fashion designers. In fact, 'flax' is the term used to describe the fiber while it is termed linen when spun into yarn or in turn woven to make a fabric. Linen which had lost its prominence for many years is now making a fashion statement in the apparel segment with its specialty of freshness, comfort, and grace.

Like a few other natural fibers, linen too suffered from lack of serious interest among researchers as well as users since mid of twentieth century. The reason could be mainly due to the fast expansion of man-made fiber industry. So during this period, many of the natural fibers had limited use only, as a component fiber in blends with synthetics. However, since early 1980s, casual and unstructured look became fashionable and that brought linen back to spotlight. The linen industry has

A.K. Patra (✉)
Department of Textile Chemistry, The Technological Institute of Textile and Sciences, Bhiwani 127021, India
e-mail: arunkpatra@rediffmail.com

K.G. Ramawat and M.R. Ahuja (eds.), *Fiber Plants*, Sustainable Development and Biodiversity 13, DOI 10.1007/978-3-319-44570-0_5

witnessed some significant changes in the recent times starting from flax cultivation to mechanical and chemical processing of linen (Sloan 1997a).

The chemical processing of this textile fiber to quite an extent resembles that of cotton, as both are cellulosic. However, depending on the typical impurities in flax, the process route is slightly modified. The machines used for cotton processing are also applicable to linen. The use of certain ancient methods such as lime boil followed by souring with hydrochloric acid is of course phased out (Tailfer 1998). Besides this, the use of sodium hypochlorite in the preparatory process is getting out of date and is replaced by hydrogen peroxide as a bleaching agent. Moreover, other eco-friendly processes are also being tried so as to restrict the use of harsh and toxic chemicals.

5.2 Historical Background

Flax is one of the oldest textile fibers. It is claimed to be used in Egypt as early as 3400 BC for coarse linens (Lewin and Pearce 1998). Evidence of its use has also been found in prehistoric lake dwellings of Switzerland. Flax seeds found in Syria and Turkey indicate that the plant might have been grown as early as 7000 BC. The origin of the word 'Linen' predates history and appears in the language of almost every European country (Cook 1968). It is widely believed that the use of the fiber had spread from Egypt to the area encompassing the present-day Israel, Jordan, Iraq, etc., during the Biblical period, and to England and France during the Roman period. Slowly, it got spread to whole of Europe. Consequently, this became one of the very widely used fibers. In the seventeenth century, Western Europe became home for linen fabric manufacture. In Ireland, although linen industry started flourishing, but in the eighteenth century, due to rise in cotton, the linen industry became recessive. The linen manufacture in Ireland got confined to Northern Ireland only, where it survives even today. However, the erstwhile Soviet Union became a major producer of flax fiber, and by 1939, it was growing nearly three-quarters of the total world output. Since the beginning of this century, the major areas of flax fiber cultivation have been the Russian Federation, Belarus, China, and various EU countries such as Germany, France, Belgium, and Holland. It is believed that the world's best flax fiber comes from Belgium and adjoining countries (Cook 1968).

5.3 Flax Cultivation and Harvest

Flax is a multicellular vegetable fiber and is termed 'bast fiber' because it is located in the bark of the single stem of the flax plant. The fiber is extracted from the bark of a dicotyledonous plant belonging to natural order Linaceae, and the species is '*Linum usitatissimum*'. As a matter of fact, the plant species, *Linum usitatissimum*, are of two types, one primarily suited as a source of fibers (flax) and the other as a source of linseed oil. Cultivation of flax oilseed is done in some of the countries such as Canada, Argentina, China, and India. The flax plant grows to a height of about 0.5–1.25 m and has a stem diameter of 1.6–3.2 mm. Seed from the plant is contained in small spherical bolls, or pods, at the top of the stalk (shown in Fig. 5.1). This is the linseed of commercial value from which linseed oil is produced. However, from textile usage point of view, flax plants that produce long and smooth stems, with minimum branching, are of significance. To avoid branching, a high planting density of 1800–2000 plants/m^2 is maintained. About one-fourth of the stem consists of fiber (Hann 2005).

Flax plants are annual plants, and moist temperate climates are best suited for growing them. Flax being a heavy feeder quickly depletes the nutrients from the land where it is planted. Hence, it should not be grown on the same land year after year and should preferably be rotated with other crops. The seed is usually sown in March/April and takes about one hundred days from sowing to pulling (harvesting). The flax is therefore harvested by mid-July/mid-August.

In general, flax takes 3–4 months time to mature. The ripening of flax is of three degrees marked by colors green, yellow, and brown. The yellow one is of course the most suitable for fiber production (Fig. 5.2). If the flax is harvested too early, i.e., when the stem is green, very fine fibers are produced but with poor strength. Contrary to this, overripe flax is marked by brown color where the stems are strong and brittle in nature, producing a large proportion of undesirable short fibers. When the flax is yellow, the fibers are long and supple and therefore are ideal for further processing. A simple diagram of stem with fiber inside is given in Fig. 5.3. The mature yellow stems are to be harvested by careful pulling. Pulling is better than stem cutting at ground surface, as cutting could lead to some loss of available fiber. Moreover, as the flax fiber in the plant tapers to point at both the upper and the lower ends, cutting the plant would result in removing the lower taper, which may prevent the fiber from being used in fine threads. Machine pulling is preferred to hand pulling, as the latter is a costly and labor-intensive operation. Modern self-propelled pullers can harvest a hectare of flax per hour (http://www.oldandsold.com; http://www.rootsweb.com).

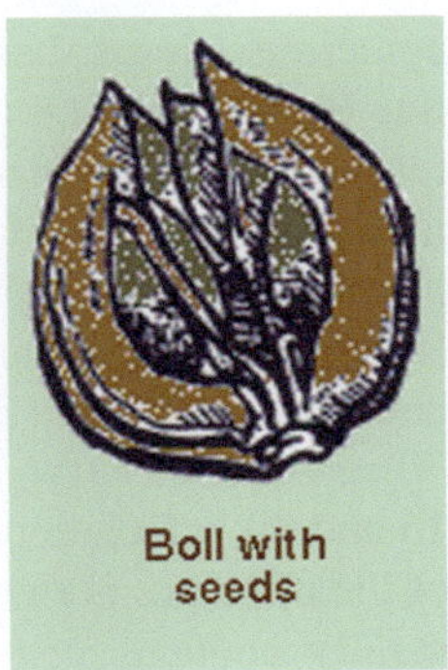

Fig. 5.1 Location of Linen seeds

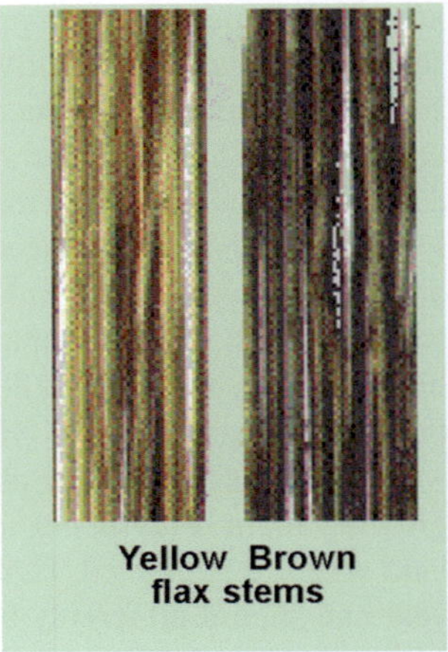

Fig. 5.2 Ripening stages of Flax stem

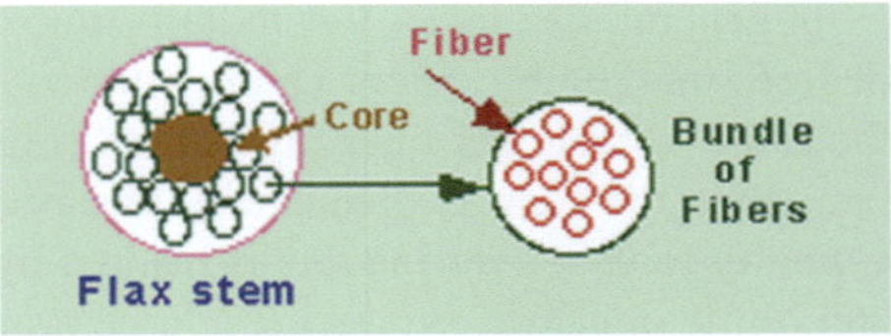

Fig. 5.3 Stem with fibre

5.4 Morphological Aspects of Plant

Flax plant has small narrow leaves not even an inch long. The branches of the plant are very slim and flexible, dividing at their tips into inflorescences bearing attractive flowers. As regards pollination of flowers, they are mostly self-pollinated, with some cross-pollination by insects. New flowers emerge for a few weeks, each developing into a round seed capsule or boll about one-third inch in diameter. Each capsule contains 4–10 seeds (http://www.rootsweb.com).

About the stem of the flax plant, its cross-section may be identified to five distinct regions. These five layers of the stem starting from the innermost are epidermis, cortex, bast layer, cambium layer, and interior woody tissue. There is a fine layer of wax on the surface of epidermis to prevent undesirable evaporation of moisture from the plant thus giving it adequate protection. Cortex, which is the next layer, is made up of nonlignified cortical cells containing pectic substances and coloring matter. In the third layer, i.e., the bast layer, fiber bundles run the full length of the stem and are surrounded by parenchyma. Each fiber bundle usually consists of between 10 and 40 individual flax fibers of 2–3 cm length and 14–30 μm diameter (typically 14–17 μm). The number of bundles may vary between 15 and 40, which depends on variety of plant and the cultivation. The primary wall of the ultimates is rich in pectin with traces of lignin in it, while the secondary wall mainly contains cellulose. The next layer, i.e., the cambium layer, has tender

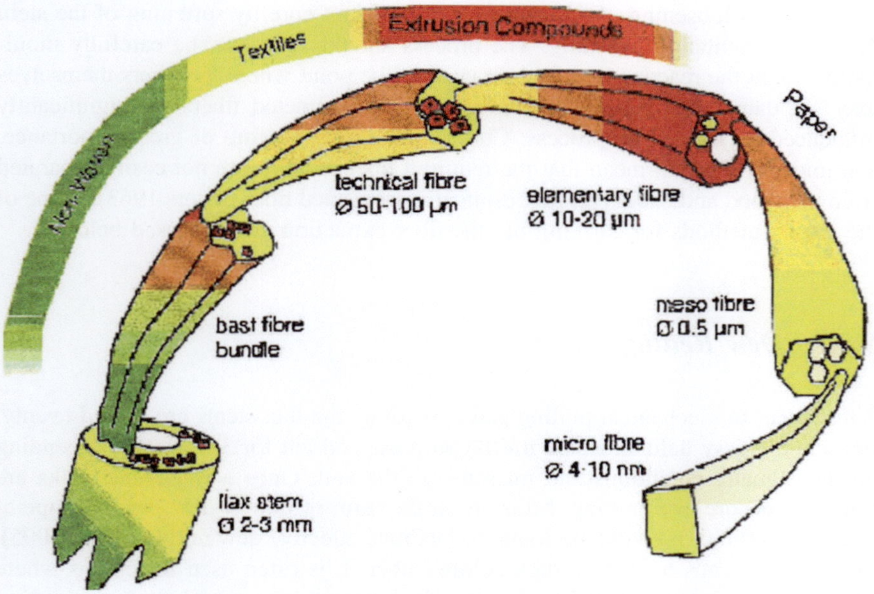

Fig. 5.4 Forms of Flax fibre

growth tissues, composed of thin-walled cells, separating the fiber from the next layer of woody tissues. This consists of thick-walled cells first, then thin-walled cells, surrounding the pith cavity which is essentially an air chamber extending through the length of the stalk (Batra 1995).

The morphology of mature bast fiber varies with climate. In a mature flax plant, fibers occupy about 25 % of total weight of plant. The occurrence of various forms of flax fibers is schematically shown in Fig. 5.4 (Van den Oever and Bos 1999). In microscopic examination, flax fibers look like small lengths of bamboo. The physical structure of the fiber ultimates is characterized by 'dislocations' or 'nodes'. The cell walls of the flax fiber are thick and have polygonal cross-section. On the contrary, an immature fiber has oval cross-section, much larger lumen, and thinner cell wall. The flax fibers are usually composed of closely packed cells of fibrillar nature that are cemented together with lignin and holocellulose (http://www.fabrics.net).

5.5 Fiber Extraction

The fiber extraction process, termed as 'retting', is in fact controlled rotting. As discussed, the flax fibers are held together in the stems by woody matter and cellular tissues. The fibers are therefore freed from these materials by a fermentation process. This loosening of flax fibers from the plant core by softening of the stem through fermentation is retting. The process should, however, be carefully monitored so that the reaction does not proceed to the point where the fibers themselves may get damaged. The quality and quantity of extracted fibers are significantly influenced by the retting process. Correct retting is therefore of great importance, and underretting will mean that the required fiber bundles are not easily separated from the wood and excess retting results in a weakened fiber (Peters 1963). Some of the retting methods for carrying out the fiber extraction are discussed below.

5.5.1 Dew Retting

Subsequent to mechanical pulling and deseeding, the flax stems are spread evenly, often in a grassy field allocated for the purpose, and left for 3–7 weeks, depending on the climatic conditions and microbes in the soil. Once a week, the stalks are turned to ensure even retting. Moisture in the form of rain or dew, warm temperatures, and little or no wind are found to facilitate effective dew retting (Hann 2005). Dew retting tends to yield a dark-colored fiber. It is often used in regions where water is in short supply like Russia. The advantage of this method over other retting methods is that it is easily mechanized and is less energy-intensive as it utilizes natural air-drying. In fact, dew retting is the most widely practiced form of retting and is the source of the bulk of good-quality linen textiles worldwide. However, dew retting can be practiced only in certain geographical locations as appropriate temperature and moisture are of importance. Other demerits are the process produces relatively coarse fiber compared to that by water retting and involves use of land, suspending its use for other agricultural purposes till retting is complete.

5.5.2 Water Retting

The lack of control over conditions is a major lacuna with dew retting, although a large amount of available linen is treated in this way only. The fiber quality can be improved if the retting conditions and bacterial attack of the straw are controlled. In fact, in the process, the impurities such as pectin and lignin are removed efficiently, leaving the ultimate fibers soft and clean. Water retting is one such technique which produces finer fibers, but is quite labor-intensive and costly. The process often referred as 'dam retting' used to be done traditionally, by keeping flax stalks immersed for about ten days in water in dams, ditches, or slow-running streams and rivers. These have been now replaced with tank retting. Tanks may be open or closed, in which cold water retting requires 8–14 days for completion. Increasing the temperature from 30 to 40 °C can decrease retting time to 3–4 days. The tanks have a cascade design and generally use three to four interconnected tanks with continuous flowing water. Warm water retting is practiced in Belgium, Holland,

Poland, and Romania. To decrease energy cost and effluent, chemical additions are often applied. Aerated tanks are used to control water consumption and disposal problem. In fact, reuse of water is made possible by aeration, and the offensive odor given off during retting is also reduced.

The completion of retting is judged by a gray-blue color over the whole of the stem, as against yellow-brown. It has been analytically proved that when retting is completed, the bacteria begin to consume the galacturonic acid normally generated by breakdown of the pectinous matter. Thus, the galacturonic acid concentration increases to a maximum and then begins to decrease. In physical terms, as the retting gets completed, the pH of the retting liquor decreases from about 4.6 to 4.9, where it may remain for 6–10 h and then rise again. Rise in pH implies overretting and deterioration in the fiber strength. Despite the efficient removal of fiber, water retting is not very popular as large quantities of waste liquor resulting from the process need to be disposed off (Lewin and Pearce 1998; Hann 2005). This also has ill effects on the environment. Although it was widely practiced in Europe a few decades ago, the process has been largely replaced by dew retting.

5.5.3 Chemical Retting

A number of chemical treatments for retting have been developed. Reagents such as caustic soda, sodium carbonate, soaps, and dilute mineral acids have been employed with some success. Treatment of flax straw, with chemicals in kiers under pressure, has also been tried. The different chemicals tried in separate pressurized conditions are naphtha with water, dilute sulfuric acid, and ethylene (Trotman 1994). Use of chelators such as ethylenediamine tetraacetic acid (EDTA) and sodium tripolyphosphate, both with sodium hydroxide, has been found to be effective in retting flax (Adamsen et al. 2002). An oxalic acid-based system for chemically retting flax has also been claimed to give good retting effects. The presence of the strong detergent such as sodium dodecyl sulfate (SDS) further enhances retting (Henriksson et al. 1998). However, the chemical retting is a costlier process than biological retting and in general chemical retting has been unsatisfactory in terms of fiber yield.

5.5.4 Enzyme Retting

This is a relatively new concept, and a fair amount of research has been carried out in this field. The basic idea has been to replace the anaerobic bacteria with enzyme in the water retting process. This can decrease the time of retting by 4–5 days. Other advantages are production of flax more efficiently and of high and consistent quality. A commercial enzyme mixture known as 'Flaxzyme' has been developed by Novo Nordisk of Denmark. It is claimed that the resultant fiber exhibits fineness, strength, color, and waxiness comparable to those of good-quality water-retted flax (Hann 2005). Enzyme-retting formulations, consisting of Viscozyme L, a pectinase-rich commercial enzyme product and ethylenediamine tetraacetic acid (EDTA), are also claimed to yield fine and clean flax fiber (Akin et al. 2001). A fair amount of research in this field has been carried out in the last decade (Evans et al. 2002; Akin et al. 2003).

The retted flax stalks are then thoroughly dried. Then, they are subjected to mechanical forces namely breaking, and scutching by which complete separation of the woody core and fiber takes place and then hackling is done to comb the flax fiber.

5.6 Properties and Behavior of Flax

Linen is considered to be an elegant, beautiful, durable, and refined luxury fabric. Flax has more wax than that of cotton, which helps to give it a high luster. Luster of flax is quite pronounced and almost silky in appearance. It has somewhat harsh feel, but exhibits great cooling effect. The fiber has exceptionally high resistance to bacteria and fungi including mildew. Like cotton, linen fabrics can also withstand rough laundering treatments, especially under alkaline conditions (Lodish et al. 2004; Pandey and Matthew 1988).

As regards the fiber structure, flax fibers vary in length from 25 to 76 mm. Each fiber is composed of a number of fibrils consisting of several cells joined together. On an average, these are about 25 mm long and are 0.014–0.025 mm in diameter. The fiber has the appearance of a straight tube with thick walls and a narrow but distinct lumen (Batra 1995). The fiber has ordered structure and high crystallinity. Some of the fiber properties are given below (Cook 1968; Trotman 1994; Batra 2007).

5.6.1 Tensile Strength

Flax is considered to be one of the strongest natural fibers. It has an average tensile strength of 6.4 N/tex compared to 3–5 of cotton. The linen yarn is about 20 % stronger when wet than dry.

5.6.2 Elasticity and Elongation

Flax has high elastic modulus, but low extensibility. This is because the flexibility is derived from the nodes or bending points. The elongation at break is approximately 1.8 % (dry) and 2.2 % (wet).

5.6.3 Moisture Regain

Flax has a high moisture regain, and the recognized regain percentage is 13.75 although in general it is often rated as more than 12 %. Because of its high strength in wet condition, linen withstands mechanical treatments well during laundering.

5.6.4 Stability to Heat and Light

The fiber is highly resistant to decomposition up to about 120 °C when the fiber begins to discolor. About 70 % of strength is retained after 20 days of exposure at 100 °C. On exposure to sunlight, flax undergoes a gradual loss of strength. Besides this, lignin present in the fiber may cause yellow or brown coloration on the surface on prolonged exposure to sunlight. The color is because of the oxidation of lignin due to photosensitivity.

5.6.5 Effect of Chemicals

In terms of chemical resistance, flax can withstand dilute, weak acids, but is adversely affected by hot dilute acids or cold concentrated acids. It also has good resistance to alkaline solutions. In general, linen is slightly more resistant to the action of acids and more easily attacked by alkalies than is cotton. As regards effect of organic solvents, flax is not adversely affected by dry cleaning solvents in common use.

5.6.6 Technical Qualities

Flax is a good conductor of heat. This is one reason why linen sheets feel quite cool. With their fresh feel due to good moisture transportation properties, linen sheets provide comfort in summer. Linen clothing is most suitable for hot summer days. Due to high strength, linen is suitable for technical applications such as in sails, mail bags, filters, and water bags. Linen is also considered to be a very hygienic fabric as bacteria grows very slowly on it. A few other mechanical properties and characteristics of the fiber are compiled in Table 5.1.

Table 5.1 Mechanical properties of flax

Mechanical property/characteristics	Value
Density (g/cc)	1.40–1.50
Diameter (μm)	40–620
Modulus of elasticity (N/tex)	18–20
Refractive index	1.515
Birefringence	0.068
Modulus of rupture (mN/tex)	8–9
Degree of crystallinity (X-ray diffraction) (%)	76
Transverse swelling in water (%)	20–24
Degree of polymerization	2200–3000

5.7 Chemical Composition

The composition of flax fiber in retted and unretted form is tabulated below (Peters 1963; Trotman 1994; Batra 2007). In comparison with cotton, the cellulose content is less in this fiber. There is a sizable presence of hemicellulose in flax while lignin is present in small quantity (Table 5.2).

As the chemistry of cellulose, fats and waxes, and pectins is widely known, only lignin and hemicellulose are discussed below.

5.7.1 Lignin

The word lignin owes its genesis to the Latin word *lignum* meaning wood. It is an important component of vascular plants and is second only to polysaccharides in natural abundance. It is a complex amorphous polymer which functions as the structural support material in plants. It imparts rigidity to the cell wall and acts as a permanent binding agent between cells. Besides this, it plays an important role in internal transport of water, nutrients, and metabolites, which are essential to the life

Table 5.2 Chemical composition of flax and cotton

Component	Flax (unretted)	Flax (retted)	Cotton
Cellulose	56–63	64–70	80–85
Hemicellulose	15.4–17	16.7–18.5	2.6–5.7
Lignin	2.5	2.0–2.2	
Pectins	2.5–3.8	1.8–2.0	0.8–1.8
Fats and waxes	1.3	1.5	0.4–1
Moisture	10	10	6–8
Miscellaneous[a]	10–11	3.5 – 4	1–1.5

[a]Pigment, nitrogenous compounds, water-soluble compounds

of plant (Palit et al. 2001; Kirk-Othmer 1992a, b). However, in wet processing of flax, lignin poses problems as its presence in fiber is responsible for the dull grayish look and needs to be removed if a good bleached linen material is to be produced. Lignin is very significantly present in boundary of the middle lamella and primary wall region while it is rather uniformly there across the secondary wall. Thus, lignin serves the dual purpose of binding and stiffening fibers through its distribution between and in the cell walls. Other than rigidity, lignin also offers resistance to biological degradation (Kirk-Othmer 1992a, b).

Chemically speaking, lignin is structurally an ill-defined polymer (as compared to other biopolymers such as cellulose, proteins, and lipids) whose monomeric composition and nature of interunit linkages vary with plant type and even in the same plant from cell to cell. However, generally speaking, lignin is a complex amorphous polymer whose structural units are aromatic alcohols with a phenyl propane backbone, such as *p*-coumaryl alcohol, coniferyl alcohol, and sinapyl alcohol (Ray Maulik 2003; Kirk-Othmer 1992a, b). Figure 5.5 shows the structure of the three lignin precursors.

The main functional groups present in lignin are alcoholic and phenolic hydoxyl groups, methoxy, and dioxymethylene. The presence of hydroxyl groups in the lignin molecule is conformed by its capacity for acetylation and methoxylation. The possibility of phenol hydroxyls is conformed by the fact that on boiling with an alkali solution, lignin is partially dissolved. Lignin is believed to be formed by oxidative polymerization of the phenyl propane units to give large cross-linked molecules containing carbon–carbon and ether linkages (Sadov et al. 1978). Because the interunit carbon–carbon linkages are very strong, degradation of lignin

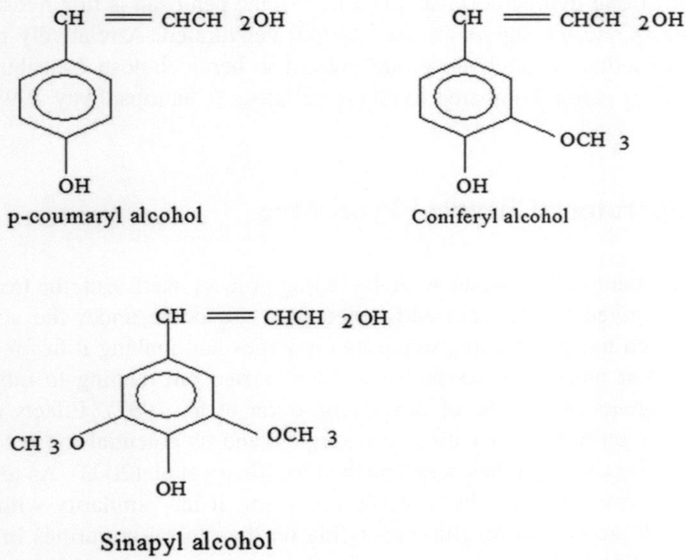

Fig. 5.5 Lignin precursors

is possible by solvolysis of ether linkages. Chlorination is a method to delignify chemical pulp, but evolution of chlorine during the process is considered environmentally unfriendly (Kirk-Othmer 1992a, b; Ray Maulik 2003).

Lignin, a noncrystalline polymer with low degree of polymerization, behaves as a thermoplastic material and undergoes glass transition. It is insoluble in most organic solvents and undergoes chemical changes under some what mild conditions (Peters 1967). It is subject to reactions such as oxidation, reduction, discoloration, and hydrolysis. The phenolic hydroxyl groups of lignin taking part in the formation of reactive intermediates are the obvious reason for above reactions.

5.7.2 *Hemicellulose*

The term hemicellulose was originated by Schulze and is a polymer like cellulose but having short chain length. Hemicelluloses have a lower degree of polymerization (<150) compared to that of cellulose (about 10,000). The lower limit for DP of hemicellulose is however not defined (Kirk-Othmer 1992b). It is found in the cell walls of flax plants in association with lignin. Together with lignin, it appears to provide an amorphous phase enveloping cellulose strands. Hemicellulose can be removed by aqueous sodium hydroxide, before or after removal of lignin. It is in fact soluble in 18 % NaOH in cold. Hemicellulose has a branched structure and is often complex (Peters 1967). Hemicellulose is composed of mainly pentosan (xylan), polyuronide, and a little hexosan.

When pentosan, i.e., a polymer of pentose sugar such as xylan, araban, etc., is heated with strong hydrochloric acid (13.15 %), the pentosan is first hydrolyzed to pentose sugars and the sugars are then further dehydrated. A relatively high concentration of reducing aldehyde group present in hemicellulose contributes to its strong reducing action compared to native cellulose (Chattopadhyay 1998).

5.8 Preparatory Chemical Processing

The wet treatment of flax begins with the retting process itself. But, the treatment at that stage is given to the stem and hence does not come under the actual pretreatment given to linen for improving its properties and making it fit for clothing. However, a fair amount of research has been carried out relating to influence of chemical reagents on retting of flax (Henriksson et al. 1997). Effects of acidic pre-incubation on retting have also been studied, and its potential for development of a commercial technique has been vouched for Zhang et al. (2003). As regards the preparatory process for flax by a textile processor, it has similarity with that for cotton as both are cellulosic. But, depending on the typical impurities in flax, the process is modified. The machines used for cotton processing are also applicable to linen.

5.8.1 Scouring

The initial processes of singe desize are done by the established methods, as per the requirement. In industrial scale, singeing is done in gas singeing machine just as it is done for cotton. Desizing can also be done by conventional enzymatic method. All other alternative desizing techniques applicable for cotton are also suited for linen. As regards scouring of linen, it can be done in any of the forms—rove, sliver, yarn and fabric. Traditionally, lime boil used to be done first, followed by souring with HCl. Then, boiling treatment with soda ash, followed by acid chemick, a further alkali boil and finally hypochlorite bleach at pH 9–10 used to be carried out. Over the years, this complicated sequence has been streamlined. The lime boil has now been replaced by caustic soda and/or soda ash treatment. Use of combination of alkalies is often recommended. An alkali boil with 15 g/l sodium carbonate plus 2 g/l caustic soda is a tried and recommended scouring treatment for linen. Impurity in the form of pectic acid is one of the contributors to the brown coloring matter of linen. Interestingly, this pectic acid does not preexist in flax and is elaborated by the chemical reaction which follows from retting. Boiling with caustic and carbonated alkali eliminates a great amount of pectic acid (Tailfer 1998). Although scouring is invariably done under alkaline conditions, the conditions vary, depending on the end use. In case of linen fabrics for clothing, a relatively soft handle is required and caustic soda treatment is done for removal of impurities in scouring. However, for other applications such as tablecloths where a firm handle is required, removal of all impurities may not be so important, and a soda ash boil will suffice. So there is no fixed recipe for scouring of linen and varies with quality of fiber and end use. Goswami and Mukherjee (1992) studied the effect of alkali on linen, in the preparatory processing stage. They mainly tried with caustic soda and soda ash at varying concentrations, time, and temperature and found that among the options tried, sodium carbonate (10 g/l) along with surfactant (2 g/l) used for 30 min near boil (95 °C) gave the best results without affecting the quality.

5.8.2 Bleaching

Linen can be bleached by any of the methods followed for bleaching cotton. So the well-known bleaching agents such as sodium hypochlorite, hydrogen peroxide, or sodium chlorite can be effectively used on linen. Other oxidizing agents such as peroxy compounds should also be applicable for linen although sufficient information is not available in this regard. But the extent of whiteness achieved in cotton and linen may not be same. The final whiteness of this bast fiber is likely to depend on the actual raw material (type and extent of retting it has undergone) and of course the bleach bath strength. Underretted fiber tends to have a reddish hue, while overretted flax is almost gray in color. Consequently, that will have an impact on the bleach effect.

The prevailing preparatory methods cause lower weight loss (around 15 %) than the older methods (25 %). As flax has cellulose in the range of 60–70 %, there are considerable amount of impurities to remove or decolorize by bleaching. The hemicellulose present in sizable proportion in the fiber is only partially removed by the newer treatment, thus decreasing the vulnerability to subsequent chemical and finishing treatments, particularly in chemical cross-linking under anhydrous acid conditions (Sloan 1974).

Lignin as a component in flax is suspected to hinder in achieving whiteness. The presence of 'sprit' in bleaching can cause problems particularly in peroxide and chlorite treatments. Sprit is basically the remnants of the woody core containing some short fibers and has a high lignin content than the rest of the fibers. It has been claimed that acid hypochlorite is an effective treatment to remove sprit (Sloan 1974). Other than sprit, lignin, wax, and hemicellulose, flax is often contaminated with metallic impurities such as iron either accidentally or through earlier processes. To avoid harmful decomposition of the bleach liquor by iron and other metallic ions, sequestering agents must be included in the bleaching recipe. A thorough demineralization may also be effective prior to bleaching.

It is also believed that sodium chlorite is quite effective in removing sprit. A typicalscouring —bleaching process sequence for linen—is suggested below, where treatment with sodium chlorite as bleaching agent is recommended (Sloan 1997a, b).

- Boil the desized fabric under pressure at a temperature up to 110 °C with sodium carbonate (15 g/l) and caustic soda (2 g/l) along with other scouring assistants for 4 h.
- Neutralize with dilute HCl.
- Treatment with sodium chlorite (4 %) at 60 °C for 1–2 h, maintaining pH at 4–5 with formic acid or cold pad-batch with sodium chlorite (100 g/l) using paraformaldehyde (5 g/l) at a pH of 7.5.
- Wash and further boil in soda ash (20 g/l) for 4 h.
- Peroxide bleach at 80–85 °C for 2–3 h. Bleach bath should have adequate quantity of stabilizer and sequestrant.

Quite a few other techniques have been suggested in the recent times for bleaching of flax. A single bath full bleaching of flax by activated sodium chlorite/hexamethylenetetramine (HTMA) system has been studied and a novel chemical formulation has been suggested. The process is based on activation of sodium chlorite by HTMA in the presence of a nonionic wetting agent (Zahran et al. 2005). El-Rafie et al. (1992) have shown that bleaching of scoured and unscoured linen fabrics with H_2O_2/urea system can be effectively done. The process conditions recommended for this are pH 6, 95 °C temperature, and 2-h treatment time. Considering the ecological aspects of pretreatments, hydrogen peroxide and per acetic acid have been recommended as alternative bleaching agents in place of chlorine compounds. The adsorbable organic halogen (AOX) values of the effluents were found to be very high in case of NaOCl while it was negligible in peroxide bleaching process. An increase in hypochlorite concentration up to 5 g/dm^3 of active chlorine causes the AOX to increase up to 278.3 mgCl/dm^3 (allowable value: 0.01 mgCl/dm^3). On the contrary, hydrogen peroxide bleaching generates no AOX, i.e., <0.01 (Lipp-Symonowicz et al. 2004). This observation further endorses a chlorine-free bleaching of linen.

Peracetic acid (PAA) is also an eco-friendly chemical as it decomposes to acetic acid and oxygen, acetic acid being completely biodegradable. Peracetic acid is well established as an oxidant, notably in the food and paper industries and for the laundering of linen (Krizman et al. 2005). As regards its possibility for application in bleaching of flax, being a strong oxidant, it is likely to remove lignin under relatively mild conditions of temperature and pH. PAA induces epoxidation of colored substances in fabric. It has been used for bleaching of cellulosic as well as synthetic fibers. An advantage with PAA is that it does not usually damage fibers (Muller-Litz 1997; Zhao et al. 2007; Polonca and Tavcer 2008). However, its employment has been limited by its high price and explosive nature.

5.8.3 *Mercerization*

Mercerization is essentially a treatment with high-concentration NaOH to increase luster and dyeability. It can also be carried out for linen although it is not a very common practice. By mercerizing, the swelling takes place at nodes, and as a result, the abrasion resistance significantly improves (Sloan 1997b). So, linen is mercerized not to increase luster, but to strengthen the nodes. Moreover, reediness in cloth due to yarn unevenness—a problem with linen—is covered by mercerization. Just as in cotton, the use of liquid ammonia as an alternative to caustic soda for mercerization is also established for linen. The linen fabric dyed after liquid ammonia mercerization is in fact claimed to have a very clean appearance. In the recent times, liquid ammonia treatment of linen at various stages of processing has been reported by Csiszar et al. (2006). Improvements in wrinkle recovery and abrasion resistance of the linen fabric were found to be the notable changes induced by the treatment.

5.9 Dyeing

Flax being a cellulosic fiber, its dyeing is similar to that of other cellulosic fibers such as cotton. The same class of dyes used to dye cotton can therefore be used for flax. Among the dyes used, reactive colors are the most common industrially, because of good overall fastness, ease of application, and cost. The general structure of reactive colors, for simple understanding, can be represented with the following four components (Patra and Paul 2014):

$$W - D - Q - X$$

where

W = water solubilizing groups (such as SO_3Na or COONa or combination of both), which impart solubility to the dye. Reactive dyes require from one to three (in a few cases four) such groups.
D = chromophoric group usually an azo, metal-complex azo, anthraquinone, triphenodioxazine, formazan, or phthalocyanine residue. This is the color-bearing group.
Q = bridging group which attaches the reactive system X to the chromogen D. This group is usually –NH, $–N(CH_3)$, $–N(C_2H_5)$, –O–, –NHCO, $–OCH_3$, $–SO_3$, etc. Bridging groups determine the stability and also reactivity of the dye. However, reactive dyes belonging to vinylsulfone, epoxide, sulfonyl chloride, etc., do not have any bridging groups.
X = reactive system or group, which reacts chemically with –OH of cellulose or $–NH_2$ of protein fibers.

Reactive dyes are normally applied in two stages on cellulosics, namely exhaustion and fixation. For the exhaustion of dye from solution to substrate, salt is added and the process is carried out at the recommended temperature for the required period depending on the type of reactive dye. Then, the dye is fixed to the fiber using alkali in which a covalent bond is formed between the dye and cellulosic fiber. After the dyeing is over, the material is thoroughly washed and soaped so as to remove the unfixed dye from the surface.

Highly crystalline and highly oriented fiber structure of flax makes the dye diffusion difficult in comparison with that of cotton. Also, lack of complete removal of various noncellulosic associates such as lignin, pectin, and hemicellulose influences the dyeability of flax with reactive dye. Efforts have been made to increase the dyeability by reducing crystallinity and removal of noncellulosics from the bast fibers. It is believed that caustic treatment increases the accessibility in noncrystalline zone of the fiber (Cheek and Russel 1989). Other than this, ultrasonic pretreatment is found to decrease crystallinity of flax, thus improving the reactive dye uptake.

5.10 Finishing

The finishes applicable to the cellulosic fibers, particularly cotton, are also suitable for flax. Linen fabric too has a tendency to wrinkle and can be given anticrease finish like cotton using a suitable cross-linker. Other finishes such as flame retardant and stain repellent can also be given to linen. However, the most common finish of softener treatment is given in the industrial production.

5.11 Use of Enzymes

Biotechnology has made significant inroads in various fields of production. The applications of enzymes range from detergents, textiles, leather and paper to animal feed, starch conversion, dairy, baking, brewing, distillery, and protein hydrolysis. As regards textile applications, the knowledge of specific action of enzymes (amylases) for starch splitting began long back, when malt extract was used to remove size from fabrics. But, the industrial-scale enzymatic desizing was finally launched in the late 1960s (Patra 2008). Although application of enzymes in textile wet processing was confined to desizing, its use in other areas is being aggressively pursued and established in last two decades.

Enzymes are very specific in action, as each type can affect only one chemical bond. Conversely, many chemicals commonly used are very broad in their actions, much less specific. In fact, more control over final effects is available with enzymes. Because of the mild conditions required, working with enzymes is much safer and more energy-efficient.

Enzymes are absolutely eco-friendly with minimal effluent generation as compared to conventional chemicals often used in textile processing.

The microbial process of retting of flax by use of enzymes has already been discussed earlier. Scouring with enzymes, known as bioscouring, has also been tried by researchers varying their concentrations and application conditions (Sampaio et al. 2005; Fakin et al. 2005). Flax in rove form has been bioscoured using different types of enzymes. Scouring with pectinase alone and in combination with hemicellulase and cellulase enzymes has been tried. Combined use of pectinase and lipase has also been attempted and claimed to give good mechanical properties (Abdel-Halim et al. 2007). However, as a process, enzymatic scouring has met with varying degrees of success. Complete removal of pectic substances has also been tried by enzymatic route, which resulted in major strength loss (Desert et al. 1998) while enzymatic hydrolysis of linen and linen blends has been carried out experimentally, with reasonable success (Buschle-Diller et al. 1994; http://www.p2pays.org/ref/08/07098).

Linen treated with xylanase and cellulase enzymes has been found to give results comparable to those by conventional chemical treatment in terms of absorbency and residual lignin content. Xylanase acts on xylan, i.e., hemicellulose, thus removing the adhering ligneous substances. Lignin being chemically and physically associated with hemicellulose, it is likely to be removed to an extent, when xylanase hydrolyzes the hemicellulose (Patra et al. 2010). Multifunctional cultured enzyme containing xylanase and pectinase has also been reported to be effective on linen for wet pretreatment after due optimization of process. Pectinase is likely to act on the natural impurity pectin, thus decreasing the galacturonic acid content in treated linen while xylanase does its usual function (Patra et al. 2013).

Although bioscouring is yet to be popularized in industrial process, use of enzyme as peroxide killer is now an established procedure. Hydrogen peroxide is the most widely used bleaching agent. Before dyeing, the fabric should be completely free from residual peroxide as many dyestuffs are sensitive to oxidation. In such case, the traces of peroxide can cause problems such as loss in color depth and shade change. So to remove peroxide completely, a peroxide killer treatment is given preferably with peroxidase enzyme. The enzyme acts like a catalyst to destroy peroxide residue and is used in very low dosages. The enzyme because of its benign properties does not require a subsequent washing after the treatment and dyeing can be started right away.

Biopolishing is again a clear possibility for finishing of linen. The concept which has gained industrial currency for cotton fabric should also be applicable for linen. This is one area where cellulase enzyme can be used to attain smooth surface and soft feel. However, not much of published literature is available on finishing of linen.

5.12 Conclusion

Linen despite being one of the oldest textile materials has virtually been rediscovered. In the present-day context, it is a niche product. The fiber possesses some outstanding properties such as quick moisture transportation, high heat of sorption, low electrostaticity, and very good resistance to microbial attack. These attributes give linen a high rating for apparel purpose, other than the natural grace and comfort it offers. As regards its chemical processing for wearability and esthetics, the treatment is not cumbersome and has resemblance with that of cotton. However, the initial process of retting for fiber extraction is time-consuming although it is not the job of a textile processor. Other than chemical processing, enzymatic pretreatment is a viable technology option. The biotechnology application makes the wet processing energy-saving and environment-friendly and hence needs further study for wider usage.

References

Abdel-Halim E, Opwis K, Knittel D, Schollmeyer E (2007) Enzymatic pre-treatment of flax fabrics. Melliand Int 13:26–28

Adamsen A, Peter S, Akin DE, Rigsby LL (2002) Chemical retting of flax straw under alkaline conditions. Text Res J 72:789–794

Akin DE, Foulk JA, Dodd RB, McAlister DD (2001) Enzyme-retting of flax and characterization of processed fibre. J Biotechnol 89(2–3):193–203

Akin DE, Morrison WH, Rigsby LL, Evans JD, Foulk JA (2003) Influence of water presoak on enzyme-retting of flax. Ind Crops Prod 17(3):149–159

Batra SK (1995) In: Lewin M, Pearce EM (eds) Handbook of fibre science and technology, vol IV. Marcel Dekker, New York, p 727

Batra SK (2007) In: Lewin M (ed) Handbook of fibre chemistry, 3rd edn, International Fibre Science and Technology Series. CRC Press, Boca Raton

Buschle-Diller G, Zeronian SH, Pan N, Yoon MY (1994) Enzymatic hydrolysis of cotton, linen, ramie and viscose rayon fabrics. Text Res J 64:270–279

Chattopadhyay DP (1998) Introduction, chemistry and preparatory processes of jute. Colourage 45:23–35

Cheek L, Russel L (1989) Mercerization of Ramie: comparison with flax and cotton: Part II: effects on dyeing and behaviour. Text Res J 59(9):541–546

Cook JG (1968) Handbook of textile fibres, vol 1. Marrow Publishing Co, England, p 4

Csiszar E, Dorny B, Somlai P, Boros A (2006) Liquid ammonia treatment of linen fabrics. AATCC Rev 6:61

Desert M, Viallier P, Wattiez D (1998) Continuous control of an enzymatic pretreatment on linen fabric before dyeing. J Soc Dyer Color 114:283

El-Rafie MH, Higazy A, Habeish A (1992) Bleaching of linen fabrics using a hydrogen peroxide/urea system. Am Dyest Rep 81(48–55):67

Evans JD, Akin DE, Foulk JA (2002) Flax-retting by polygalacturonase-containing enzyme mixtures and effects on fibre properties. J Biotechnol 97(3):223–231

Fakin D, Golob V, Kreve T, Marechal AM (2005) Ultrasound in the pretreatment processing of flax fibres. AATCC Rev 5:61–64

Goswami KK, Mukherjee AK (1992) Effects of alkali on linen (*Linum usitatissimum*). Indian J Fibre Text Res 17:136–143

Hann MA (2005) Innovation in linen manufacture. Text Prog 37(3):1–42

Henriksson G, Akin DE, Rigsby LL, Patel N, Eriksson KE (1997) Influence of chelating agents and mechanical pretreatment of enzymatic retting of flax. Text Res J 67:829–836

Henriksson G, Eriksson KL, Kimmel L, Akin DE (1998) Chemical/physical retting of flax using detergent and oxalic acid at high pH. Text Res J 68:942–947

http://www.rootsweb.com. Accessed 18 Feb 2011

http://www.p2pays.org/ref/08/07098. Accessed 25 Jan 2010

http://www.oldandsold.com/articles04/textiles9.shtml. Accessed 6 July 2012

http://www.fabrics.net/linen.asp. Accesses 12 Aug 2010

Kirk-Othmer (1992a) Encyclopedia of chemical technology. vol 15, 4th edn. Wiley, New York, p 268

Kirk-Othmer (1992b) Encyclopedia of chemical technology. vol 13, 4th edn. Wiley, New York, p 54

Krizman P, Kovac F, Tavcer PF (2005) Bleaching of cotton fabric with peracetic acid in the presence of different activators. Color Technol 121:304–309

Lewin M, Pearce EM (1998) Handbook of fiber chemistry, 2nd edn. Marcel Dekker, New York 514

Lipp-Symonowicz B, Tanesaha B, Sapieja A (2004) Ecological aspects of preliminary treatments of flax fibre. Fibres Text East Eur 12:63–66

Lodish A, Berk A, Matsudaira P, Kaiser CA (2004) Molecular cell biology, 5th edn. WH Freeman and Company, New York, p 232

Maulik SR (2003) A mechanistic approach on bleaching of jute fibre. Text Trends 46:31

Muller-Litz W (1997) Peracetic acid bleaching: a chlorine-free and environmentally compatible process. Int Text Bull Dyg/Ptg/Fing 3:37–38

Palit P, Sengupta G, Datta P, Meshram J (2001) Lignin and lignifications with specific reference to its down regulation for the improvement of wood and bast fibre quality. Indian J Plant Physiol 6:217–218

Pandey SN, Matthew MD (1988) Ramie, a versatile industrial fibre. Text Trends 30:49–57

Patra AK (2008) Enzymes-concept, applications and evaluation. Textile Association (India) conference. Bhilwara

Patra AK, Paul R (2014) Reactive dyeing of textiles: practices and developments. In: Fu J (ed) Dyeing: processes, techniques and applications. Nova Science Publishers, New York

Patra AK, Madhu A, Chakraborty JN (2010) Studies on enzymatic pretreatment of linen. Indian J Fibre Text Res 35(4):337–341

Patra AK, Mahish SS, Chakraborty JN (2013) Wet pretreatment of linen by enzyme and alternative bleaching techniques. Indian J Fibre Text Res 38(2):150–155

Peters RH (1963) Textile chemistry, vol I. Elsevier Publishing Co, p 170

Peters RH (1967) Textile chemistry, vol 2. Elsevier Publishing Co, p 91

Polonca P, Tavcer PF (2008) Bioscouring and bleaching of cotton with pectinase enzyme and peracetic acid in one bath. Color Technol 124:36

Sadov F, Korchagin M, Matetsky A (1978) Chemical technology of fibrous materials. MIR Publishers, Moscow, p 54

Sampaio S, Shen J, Bishop D, Miettinen-Oinonen A (2005) Progress on enzymatic preparation of flax and flax/wool blends. AATCC Rev 5:23–28

Sloan FR (1974) Preparation, bleaching, dyeing and finishing of linen. Rev Prog Color 5:12

Sloan FR (1997a) Linen:old as the hills, modern as the hour. J Soc Dyer Color 113(2):46–47

Sloan FR (1997b) Linen:old as the hills, modern as the hour. J Soc Dyer Color 113(3):82–83

Tailfer L (1998) Bleaching of Linen cotton yarn and fabrics. Abhishek Publications, p 255

Trotman ER (1994) Dyeing and chemical technology of textile fibres. 6th edn. Charles Griffin & Co Ltd, p 55

Van den Oever MJA, Bos HL (1999) 2nd international wood and natural fibre composites symposium, in Kassel/Germany

Zahran MK, Rehan MF, El-Rafie MH (2005) Single bath full bleaching of flax fibers using an activated sodium chlorite/hexamethylene tetramine system. J Nat Fibres 2:49–67

Zhang J, Johansson G, Pettersson B, Akin DE, Foulk JA, Khalili S, Henriksson G (2003) Effects of acidic media pre-incubation on flax enzyme retting efficiency. Textile Res J 73:263–267

Zhao X, Wang L, Liu D (2007) Effect of several factors on peracetic acid pretreatment of sugarcane bagasse for enzymatic hydrolysis. J Chem Technol Biotechnol 82:1115–1121

Part II
Biotechnology

Chapter 6
Cotton Regeneration In Vitro

Hamidou F. Sakhanokho and Kanniah Rajasekaran

Abstract Over the years, plant breeders have improved cotton via conventional breeding methods, but these methods are time-consuming. To complement classical breeding and, at times, reduce the time necessary for new cultivar development, breeders have turned to in vitro plant transformation or genetic engineering, relying mostly on two major approaches, *Agrobacterium*-mediated and particle bombardment transformation techniques. Since its adoption in the 1990s, transgenic technology continues to have a tremendous impact on cotton production not only in the United States but also worldwide. Currently, genetically modified cottons, in particular insect and herbicide tolerant cotton, account for over 90 and 80 % of cultivated cotton acreage in the United States and worldwide, respectively. However, efforts in the development of transgenic cotton are hampered by the recalcitrance of most cotton cultivars, particularly the elite cultivars, to regenerate via tissue culture, a step very often necessary for the transformation process. In vitro regeneration of cotton, in particular regeneration via somatic embryogenesis, is highly genotype dependent. In addition, other factors including explant type, composition and type of media (liquid vs. solid) as well as environmental conditions surrounding the cultures affect the in vitro regeneration of cotton. In this chapter, the current status of different regeneration methods and the factors limiting or enhancing these methods are discussed.

Keywords Cotton · Embryogenesis · Meristem culture · Organogenesis · Regeneration · Tissue culture

---

H.F. Sakhanokho
Thad Cochran Southern Horticultural Laboratory, USDA-ARS, 810 Hwy 26 West, Poplarville, MS 39470, USA
e-mail: Hamidou.Sakhanokho@ars.usda.gov

K. Rajasekaran (✉)
USDA-ARS, Southern Regional Research Center, 1100 Robert E. Lee Blvd, New Orleans, LA 70124, USA
e-mail: Kanniah.Rajasekaran@ars.usda.gov

K.G. Ramawat and M.R. Ahuja (eds.), *Fiber Plants*, Sustainable Development and Biodiversity 13, DOI 10.1007/978-3-319-44570-0_6

6.1 Introduction

Cotton is one of the major economically important crops in the world. To the general public, cotton is probably best known for its long, cellulosic, and single-celled fiber, but other cotton products (e.g., cottonseed) are also of great economic importance. Oil, meal, linters, and hulls are obtained from cottonseed. Worldwide, cottonseed is second only to soybean in oilseed production (Kohel 1989; Bell 1989). Refined cottonseed oil is used in the food industry, and in some parts of the world, it is the preferred vegetable oil (Kohel 1989). The cottonseed meal remaining after oil extraction can be used as a protein-rich feed for ruminants whereas the hulls are added to the meal as roughage.

Current world cotton production is made up of *Gossypium hirsutum* or Upland cotton (>90 %), *G. barbadense*, or Extra Long Staple (8 %), with the remaining 2 % consisting of the Old World diploids, *G. arboreum* and G. *herbaceum*, cultivated primarily in African and Asian continents (Juturu et al. 2015). As one of the major world crops, cotton has a long history of improvement through breeding, with sustained long-term yield gains of 7–10 kg lint ha^{-1} $year^{-1}$ (Meredith 1984). Most of this progress has been achieved through conventional breeding, which is often time-consuming. However, the rate of change in US cotton yields has steadily declined since 1985, which appears to be the result of the lack of genetic diversity in cultivated US cotton (Paterson et al. 2004). To help improve yields, current approaches have been directed at improving both agronomic efficiency and product quality using molecular transformation means. For example, the largest gains in farm income worldwide in 2010 were largely due to yield gains as a result of the $5 billion additional income generated by genetically modified (GM) insect resistant cotton (Brookes and Barfoot 2012).

At the beginning, the two major goals of genetic engineering in cotton were to confer insect resistance and tolerance to more environmentally acceptable herbicides (John and Stewart 1992). Insect resistance, herbicide resistance, environmental stress, and hybrid cotton development are topics related to cotton management, and each one has an indirect effect on the quality and quantity of fiber produced. However, none of them directly modifies the quality of fiber, which is a quantitative trait. Recently, however, efforts have been made through genetic engineering to enhance cotton fiber quality (Wilkins et al. 2000; Haigler et al. 2007; Divya et al. 2010; Pasapula et al. 2011; Jiang et al. 2012; Abdurakhmonov et al. 2014; Han et al. 2014; Bai et al. 2014; Chen and Burke 2015; Li et al. 2015). Regardless of the goal, a dependable regeneration system is a prerequisite for an efficient plant transformation scheme. It is noteworthy that cotton transformation, either by *Agrobacterium* or biolistic method, is most often not a limiting factor (Rajasekaran et al. 2001; Rajasekaran 2013a, b), but identification of embryogenic cells and regeneration of plantlets is necessary in order to achieve success.

Two major approaches have been used in cotton genetic engineering, transformation through somatic embryogenesis (Firoozabady et al. 1987; Umbeck et al. 1987; Cousins et al. 1991; Bayley et al. 1992; Rajasekaran 2004; Rajasekaran et al.

1996) and particle bombardment of tissues (Finer and McMullen 1990; McCabe and Martinell 1993; John and Keller 1996) and embryogenic cell cultures (Rajasekaran 2013a). *Agrobacterium*-mediated transformation is an efficient transformation method, but progress in this area is hampered by the recalcitrant nature of most cotton genotypes, particularly the elite cultivars, to regeneration via somatic embryogenesis. Therefore, the search for more elite lines with good regenerability is of the utmost importance.

Successful multiplication of the cotton plants from meristem or shoot proliferation has been reported by several authors (Gould et al. 1991; McCabe and Martinell 1993; Keller et al. 1997; Saeed et al. 1997; Gupta et al. 1997; Agrawal et al. 1997; Hemphill et al. 1998). Both *Agrobacterium*-mediated and particle bombardment transformation methods take advantage of this meristem multiplication system, which is genotype independent. But both methods are less efficient compared to a callus or cell suspension culture-mediated transformation system because of the low frequency of transformation and the production of non-germ line transformed chimeras due to the difficulty in penetrating the physical barrier of the L1 and L2 layers. Layer L1 consists of cells making up the epidermis while L2 constitutes the single cell subepidermal layer. Cells in L2 layers have to be transformed to achieve germ line transformation. However, since the 1990s, successful meristem transformation in cotton has been reported using both the BioRad PDS 1000/He biolistic device (Chlan et al. 1995) and the ACCELL® gene gun technology (McCabe and Martinell 1993; Keller et al. 1997) as well as the *Agrobacterium*-mediated transformation approach (Gould and Magallanes-Cedeno 1998; Zapata et al. 1999; Satyavathi et al. 2002; Katageri et al. 2007; Keshamma et al. 2008; Chen et al. 2014).

6.1.1 Economic Importance

Cotton is one of the major crops of the world as it is commercially grown in about 65 tropical/subtropical and temperate countries, providing income for over 250 million people worldwide and employs close to 7 % of all labor in developing countries (Banuri 1999; WWF 2015; Juturu et al. 2015). The top five cotton-producing countries in 2014 were China, India, the United States, Pakistan, and Brazil. The trading value of cotton in the US was over $8 billion in 2011, and the global world retail value of cotton products was $42 billion in 2010 (Brookes and Barfoot 2012; EPA 2015). Cotton is the most prevalent natural textile fiber, accounting for 96 % of the shared fiber market (John 1995). The apparel industry utilizes 70 % of the total fiber; the remainder is used in the manufacture of furnishings for homes, industry floor coverings, and medical supplies (John 1995). In addition to fiber, cotton has other useful by-products that are of economic importance. Cottonseed is second only to soybean in the amount of oil seed produced worldwide. Most cottonseed contains 30–40 % oil which is rich in linoleic glycerides and the remaining press cake contains approximately 10 % nitrogen (Janick et al. 1981). The global importance of transgenic cotton and its impact on economy, environment, genetic diversity, and safety has been addressed in a recent review in this book (Anderson and Rajasekaran 2016)

6.1.2 Domestication

It is often argued that commercial cotton has at least three centers of origin. Wilkinson (1927), reviewing historical writings by travelers, such as Breyn and Columbus, concluded that at least three centers were known in which the cotton plant was indigenous from very early times and that the people of these centers were acquainted with the properties and uses of the fiber obtained from the cotton plant. These centers of origin were Africa, Asia, and Mesoamerica. Wilkinson (1927) argued that, despite the economic importance of cotton in Egypt, the cultivation of this crop plant in that country was relatively recent. He also indicated that the fiber exclusively used by the ancient Egyptians until the seventeenth century was from flax (*Linum usitatissimum* L.) and that the nature of the garments covering the mummies was exclusively flax. Lee (1984) pointed out that cotton textile technology originated from the ancient knowledge of spinning and weaving of flax.

One country in which the technology of spinning and weaving of flax was well developed was India during Indus Civilization. Consequently, the weaving of cotton first occurred in India where it evolved into a fine craft (Janick et al. 1981; Percival and Kohel 1990). Archeological records of cotton textiles found in excavations at Mohenjo Daro in the valley of the Indus river, in present day Pakistan, date as far back as 2700 B.C. (Percival and Kohel 1990). Percival and Kohel (1990) reported that knowledge and use of cotton fiber spread from India and Arabia to Greece *ca.* 350 B.C. They also stated that cotton culture was spread across North Africa and into Spain by the Moors while the Crusaders (1096–1270 AD) introduced cotton to other parts of Europe. The same authors pointed out that a parallel cotton technology developed in the New World with no known connection to that which occurred in the Old World. In America, cotton was first used by the peoples that inhabited the coastal areas of Peru from 2500 to 1750 B.C. (Percival and Kohel 1990).

6.1.3 Taxonomy and Cytological Features

Cotton is an agricultural and technological term used to describe the cultivated species of the genus *Gossypium*, which is placed in the Malvaceae family. The word "cotton" comes from a corruption of the Arabic word "qutum" or "kutum" (Percival and Kohel 1990). The size of the genus *Gossypium* continues to increase as wild species are discovered and subspecies are elevated to species level. The genus comprises about 50 species (Reinisch et al. 1994), of which only four provide raw material for the textile industry. These four cultivated species include two diploid

($2n = 2x = 26$) species, *G. arboreum* L. and *G. herbaceum* L., and two allotetraploid ($2n = 4x = 52$) species, *G. hirsutum* L. (Upland cotton) and *G. barbadense* L. (Sea Island cotton). Two of these species, *G. herbaceum* and *G. arboreum*, are Old World diploids. *G. arboreum* has been cultivated in India for as long as 5000 years (Janick et al. 1981). *G. herbaceum,* which may have been domesticated in eastern Africa, may be ancestral to the Indian species as well as to the current African domesticates (Janick et al. 1981). The diploid species in the genus are placed in genomic groupings designated A through G and K. There are three theories about the genome parentage of the allotetraploid, which were summarized by Percival and Kohel (1990): (1) the "ancient origin theory" proposed that the A and D species came together on a trans-Pacific land bridge in the late Cretaceous or Tertiary, *ca.* 60 million years ago; (2) the "recent origin theory" argued that *G. herbaceum* was transported to America by people, and that hybridization with more than one D genome species took place less than 12,000 years ago; (3) the "most accepted origin theory" proposed that allotetraploids were of monophyletic (evolved from a single ancestral type) origin and that the event occurred *ca.* one million years ago. To most scientists, however, the cultivated allotetraploid (AADD, $2n = 4x = 52$) species are presumed to have resulted from the combination of an Asian–African A genome diploid species, *G. arboreum* (AA, $2n = 2x = 26$) and a New World D genome diploid, *G. raimondii* (DD, $2n = 2x = 26$).

6.2 Regeneration Systems

As one of the major world crops, cotton has a long history of improvement through breeding, with sustained long-term yield gains of 7–10 kg lint ha^{-1} $year^{-1}$ (Meredith 1984). Most of this progress has been achieved through conventional breeding, which is often time-consuming. To solve this problem, current approaches have been directed at improving both agronomic efficiency and product quality using molecular tools, including plant transformation schemes, which depend on reliable regeneration systems, either through somatic embryogenesis or multiplication of shoots from preformed meristems, often erroneously indicated as de novo organogenesis. Both of these systems have advantages as well as disadvantages.

6.2.1 Callus-Based Regeneration

Somatic embryogenesis resulting in regeneration of whole plants is an important prerequisite in any plant transformation scheme. Somatic embryogenesis is preferred over organogenesis (i.e., adventitious shoot regeneration) because it yields better numbers in terms of regenerated plants per explant and is devoid of chimeric make up of plants due to single cellular origin. Successful stable transformation requires that a single cell give rise to a plant. The ideal transformation scheme is the one done via somatic embryogenesis because from callus cultures (mass of disorganized cells) each transformed cell has the potential to produce a plant. Despite these advantages, plant regeneration via somatic embryogenesis in cotton has a few drawbacks. For example, because of the length of time involved with this method (>6 months on average), the resulting regenerated plants are prone to somaclonal variation resulting in morphologically aberrant regenerants (Trolinder and Goodin 1988; Stelly et al. 1989; Li et al. 1989). In this regard, freshly initiated embryogenic callus and cell suspension cultures of cotton can be cryopreserved long term, and they provide a reliable and uniform source material for experimental use without the need for labor-intensive maintenance procedures. More importantly, the problems associated with long-term cultures due to decrease in embryogenic and regeneration potential and increase in accumulated somatic mutations in callus and suspension cultures over successive subcultures can be avoided (Rajasekaran 1996).

6.2.1.1 Genotype

The first report of regeneration of *G. hirsutum* (accession Coker 310) was by Davidonis and Hamilton (1983), who used polyploid cotyledonary tissue. Somatic embryos developed spontaneously and plants were obtained, but the experimental conditions leading to these results were not well defined. Since then, progress has been made, and somatic embryogenesis and regeneration of plants in cotton have been reported by several authors (Trolinder and Goodin 1987; Trolinder and Xhixian 1989; Firoozabady and DeBoer 1993; Sakhanokho et al. 2001, 2004a, b). However, regeneration via somatic embryogenesis of plants in *Gossypium* species remains highly genotype dependent as demonstrated by Trolinder and Xhixian (1989) who screened 38 genotypes of *G. hirsutum*, *G. barbadense*, and *G. arboreum* using various growth regulator combinations and found that Coker 312 had the highest frequency of somatic embryos, followed by Coker 304, Coker 315, T 25, and Coker 310. Most of the successful regeneration and transformation studies have utilized Coker 312, the standard cotton cultivar for somatic embryogenesis but with poor fiber quality, and related lines. However, screening of 48 cotton accessions including 12 *G. hirsutum*, 12 *G. barbadense*, 12 *G. arboreum*, and 12 *G. herbaceum* using various media types and growth regulator combinations resulted in the induction of somatic embryogenesis and subsequent plant formation in two *G. hirsutum accessions*, Coker 312 and Deltapine 90, an elite cotton cultivar, as well as in *G. barbadense* (accessions GB35, B126) and *G. arboreum* accession A_2-9 (Sakhanokho 2001; Sakhanokho et al. 2001, 2004b). These authors also found that Coker 312 had the highest frequency of somatic embryogenesis, followed by Deltapine 90, GB35B126, and A_2-9. No somatic embryogenesis was achieved with any of the 12 accessions of *G. herbaceum* (Sakhanokho 2001). However, since then reports on somatic embryogenesis in more elite cotton cultivars including the Acala cultivars, highly prized for their superior fiber quality, have been published (Rajasekaran et al. 1996, 2001; Rangan and Rajasekaran 1996; Mishra et al. 2003;

Sakhanokho et al. 2004a; Jin et al. 2006a, b; Khan et al. 2010) (Table 6.1). However, despite these promising results, regeneration from cotton germ plasm via somatic embryogenesis, in general, remains highly genotype dependent and a sought-after goal for the development of more efficient transformation system in *Gossypium* species. This is particularly the case for *G. barbadense*, *G. arboreum*, and *G. herbaceum* for which reports on somatic embryogenesis are extremely scarce. *G. barbadense* is an important cultivated species known for the quality of its extra long staple, but it is notoriously recalcitrant to somatic embryogenesis even though successful plant regeneration via somatic embryogenesis in two *G. barbadense* accessions, GB 35 and B 126, has been reported (Sakhanokho et al. 2001). The diploid species, *G. arboreum* and *G. herbaceum*, account only for about 4 % of the cotton production worldwide because of their poor fiber quality. However, these species, in particular *G. arboreum*, can be valuable to cotton breeders as they exhibit resistance to diseases and pests such as nematodes, which can be exploited for the improvement of the more productive and high yielding tetraploid species (Carter 1981; Sacks and Robinson 2009). Moreover, unlike their tetraploid counterparts, *G. hirsutum* and *G. barbadense*, the diploid species with their less complex genome offer better opportunities to study gene structures and functions through gene knockouts, but a regeneration system is required to achieve these transformation-based goals. Reports on somatic embryogenesis in these two diploid species are scarce to none even though plant regeneration via somatic embryogenesis was achieved in one *G. arboreum* accession, A2-9 (PI 529712) (Sakhanokho et al. 2004b). Figure 6.1 provides some examples of successful regeneration via somatic embryogenesis in several wild species and commercially cultivated upland cotton varieties.

6.2.1.2 Explant Type

The effects of various factors believed to impact regeneration in cotton have been investigated. These factors include source of explants; types of media; amounts, types and combinations of hormones or growth regulators; temperature; and light intensity and dark conditions. There have been various types of explants used to evaluate regeneration in cotton, including hypocotyls (Price et al. 1977; Price and Smith 1979; Trolinder and Goodin 1987; Firoozabady and DeBoer 1993), cotyledons (Smith et al. 1977; Trolinder and Goodin 1988; Finer 1988), leaf tissue (Smith et al. 1977; Finer and Smith 1984; Davidonis and Hamilton 1983), stems (Trolinder and Goodin 1988), and both cotyledon and hypocotyls explants (Rangan and Rajasekaran 1996). Smith et al. (1977) found that hypocotyl tissue was superior to cotyledon or leaf tissue as the explant source for callus proliferation. Further, Sakhanokho et al. (2001) reported excessive root formation when cotton cotyledons were cultured in a medium containing 2 mg/L 1-naphthalene acetic acid (NAA) and 0.1 mg/L kinetin. This may explain the wide use of hypocotyl as a source of explants for somatic embryogenesis and plant regeneration in cotton.

Table 6.1 List of some cotton cultivars and taxa regenerated via somatic embryogenesis

Taxon/cultivar	Explant	Carbon source	Growth regulators	References
G. klotzschianum Anderss	Hypocotyl	30 g/L glucose for callus initiation 20 g/L sucrose for suspension cultures	CIM: 2.0 mg/L IAA + 1.0 mg/L KIN CMM: 10 mg/L 2ip + 1.0 mg/L NAA SEIM: 0.1 mg/L 2,4-D + 10 mM Gln	Price and Smith (1979)
G. hirsutum	Cotyledon	30 g/L glucose	CIM: 2.0 mg/L NAA + 1.0 mg/L KIN SEIM: doubling the concentration of KNO_3 and removal of NH_4NO_3 in MS medium + 0.1 mg/L GA_3	Davidonis and Hamilton (1983)
G. klotzschianum Anderss	Hypocotyl	15 g/L glucose	CIM: 2.0 mg/L IAA + 1.0 mg/L KIN CMM: 10 mg/L 2ip + 1.0 mg/L NAA SEIM: 0.1 mg/L 2,4-D + 10 mM Gln	Finer and Smith (1984)
G. hirsutum L. cv. Acala 1517, Coker 201, 208, 310, 315, Deltapine 311, 61, DES 56, GSA 71, Lankart 57, McNair 235, McDel, Paymaster 145, Quapaw, RC10-3, Stroman 254, Tamcat CAMD-E	Hypocotyl	30 g/L glucose	CIM: 2.0 mg/L IAA + 1.0 mg/L KIN CMM: 10 mg/L 2ip + 1.0 mg/L NAA SEIM and SEMM: 2.0 mg/L NAA + 1.0 mg/L KIN	Shoemaker et al. (1986)
G. hirsutum L. cv. Coker 312, T 25, T 169, Paymaster 303, 784, Stoneville 213 Rqsx (*hirsutum* × *barbadense*)	Hypocotyl	30 g/L glucose	CIM: 0.1 mg/L 2,4-D + 0.1 mg/L KIN CMM: hormone-free MS medium SEIM: hormone-free MS medium with double the concentration of KNO_3 without NH_4NO_3 SEIM + SEMM: 0.1 mg/L GA_3/0.1 mg/L IAA	Trolinder and Goodin (1987)
G. hirsutum L. cv. Coker 310	Cotyledon	CIM: 30 g/L glucose CMM: 30 g/L L sucrose SEIM: 20 g/L sucrose	CIM: 2 mg/L NAA + 1 mg/L KIN SEIM: 0.1 mg/L 2,4-D or 0.5 mg/L picloram SEMM: 5 mg/L 2,4-D	Finer (1988)
G. hirsutum L. cv. Stoneville 215, 453, 506, Acala SJ1, SJ2, SJ5, 44, Lu, Coker 5110, 313, 100s, 304, 315, 310, 312, Deltapine 16, SR5, 15, Paymaster 303, 784, 145 and *G. arboreum* L. var. Jyoti	Hypocotyl	30 g/L glucose	CIM: 0.1 mg/L 2,4-D + 0.1 mg/L KIN CMM: hormone-free MS medium SEIM: 0.1 mg/L GA_3/0.1 mg/L IAA double the concentration of KNO_3	Trolinder and Xhixian (1989)

(continued)

Table 6.1 (continued)

Taxon/cultivar	Explant	Carbon source	Growth regulators	References
G. hirsutum L. cv. Coker 201, 310, 315, 4360, GSA 71, 75, 78, CSC 25, G8160	Cotyledon, hypocotyl, leaf sections	15–50 g/L glucose 15 g/L sucrose	CIM: 5.0 mg/L 2iP + 0.1 mg/L NAA Embryogenic callus induction: 5.0 mg/L NAA + 0.1 to 1.0 mg/L 2ip SEIM: 0.01 mg/L NAA + 0.1 mg/L GA_3	Firoozabady and DeBoer (1993)
G. hirsutum L. cv. Acala: SJ2, SJ4, SJ5, SJC1, GC356, GC510, B1644, B1654-26, B1654-43, B3991, GAM1, Royale (4226), B638, B1810, B2724, B4894, B5002, CSC28; Chembred B2, C4; Coker 312, 315; FC 2017, FC3027, HBX87, HS46, Sicala, Siokra; Stoneville: 506, 825 *G. barbadense*: Pima S-6, MAR SP 37H, Oro Blanco	Cotyledon, hypocotyl, leaf explants	20 g/L sucrose	CIM and SEIM: 1 mg/L KIN + 2–5 mg/L + 2 mg/L IAA or NAA SEMM: MS + 2 mg/L NAA	Rangan and Rajasekaran (1996) Rajasekaran et al. (2001)
G. hirsutum L. cv. MCU 5,7, Khandwa 2, Bikaneri Nerma, F846 *G. barbadense* 1198 and F1s of the listed cultivars crossed with Coker 310	Hypocotyl	30 g/L glucose 15 g/L, 20 g/L sucrose	CIM: 0.1 mg/L 2,4-D + 0.5 mg/L KIN SEIM: double the concentration of KNO_3 without NH_4NO_3 SEMM: hormone-free MS medium + 0.15 % activated carbon.	Kumar et al. (1998)
G. hirsutum L. cv. Coker 201, CRI 12	Cotyledon, hypocotyl	30 g/L sucrose	CIM: 0.1 mg/L zeatin + 0.1 mg/L 2,4-D + 2 g/L activated carbon SEIM: 0.1 mg/L zeatin + 2 g/L activated carbon	Zhang et al. (2000)
G. hirsutum L. cv. Deltapine 90 *G. barbadense* L.: GB 35, B126	Cotyledon, hypocotyl	30 g/L glucose	CIM: 1.0 mg/L KIN + 2.0 mg/L NAA Callus proliferation: 0.1 mg/L KIN + 0.5 mg/L NAA SEIM: hormone-free Ms medium with double the concentration of KNO_3 without NH_4NO_3 SEMM: 0.5 mg/L NAA + 0.05 mg/L KIN	Sakhanokho et al. (2001)
G. hirsutum L. cv. Coker 312, Acala cv. Maxxa, Riata, Ultima	Hypocotyl	30 g/L glucose	CIM:10.7 μM NAA + 0.2 μM 2,4-D SEIM: hormone-free MS medium with double the concentration of KNO_3 without NH_4NO_3	Mishra et al. (2003)

(continued)

Table 6.1 (continued)

Taxon/cultivar	Explant	Carbon source	Growth regulators	References
			SEMM: 5.7 μM IAA	
G. hirsutum L. cv. Coker 310	Hypocotyl, cotyledon	30 g/L maltose	CIM: 0.1 mg/L 2,4-D + 0.5 mg/L KIN SEIM: hormone-free MS medium with double the concentration of KNO_3 without NH_4NO_3 + metabolic stress SEMM: 0.05 mg/L GA_3	Kumria et al. (2003)
G. hirsutum L. cv. Ekang 3, 4, 6, 8, 9, 10, Emian 22, Ejing B1, B11, Coker 201	Hypocotyl	30 g/L glucose	CIM: 2.46 μM IBA + 2.32 μM KIN SEIM: hormone-free MS medium with double the concentration of KNO_3 without NH_4NO_3 SEMM: 7.6 mM Asn/13.6 mM Gln	Wu et al. (2004)
G. hirsutum L. cv. Nazilli M 503, Nazilli 143	Shoot apices, hypocotyl, nodes	30 g/L sucrose	CIM: 1.0 mg/L BAP + 0.5 mg/L KIN + 1.0 g/L PVP (Nazilli M-503) 1.0/2.0 mg/L BAP + 0.5/2.0 mg/L KIN (Nazilli143) SEMM: hormone-free MS medium	Aydin et al. (2004)
G. hirsutum L. cv. Coker 312	Hypocotyl	30 g/L glucose	CIM: 0.1 mg/L 2,4-D + 0.5 mg/L KIN Callus proliferation: 0.1 mg/L 2,4-D + 0.5 mg/L KIN + double the concentration of KNO_3 of MS medium SEIM: hormone-free MS medium with double the concentration of KNO_3 without NH_4NO_3 + metabolic stress SEMM: 0.05 mg/L GA_3	Haq and Zafar (2004)
G. hirsutum L. cv. Coker 312	Hypocotyl, cotyledon	30 g/L glucose	CIM: 2.2 μM 2,4-D + 0.88 μM BAP, SEIM: hormone-free MS medium with double the concentration of KNO_3 without NH_4NO_3 + inositol starvation SEMM: hormone-free MS medium	Kumar and Tuli (2004)

(continued)

Table 6.1 (continued)

Taxon/cultivar	Explant	Carbon source	Growth regulators	References
G. hirsutum: Coker 312, PD 97019, PD 97021, PD 97100, GA 98033	Hypocotyl, cotyledon	30 g/L glucose, 15 g/L sucrose for embryo induction and maturation	CIM: 2.0 mg/L NAA + 1.0 mg/L KIN SEIM: hormone-free Ms medium with double the concentration of KNO_3 without NH_4NO_3	Sakhanokho et al. (2004a)
G. arboreum A2-9 (PI 529712)	Hypocotyl	30 g/L glucose; 20 g/L sucrose for embryo germination	CIM: 1.0 mg/L KIN + 2.0 mg/L NAA CMM: 0.1 mg/L KIN + 0.5 mg/L NAA SEIM: hormone-free Ms medium with double the concentration of KNO_3 without NH_4NO_3 SEMM: 0.5 mg/L NAA + 0.05 mg/L KIN	Sakhanokho et al. (2004b)
G. hirsutum L. cv. SH 131, LH 900, Hybrid H8, Khandwa 2 *G. arboreum* L. cv. BD 1, 6, Sarvottam, Jawahartapti	Hypocotyl, cotyledon	20 g/L sucrose	CIM: 0.452 μM 2,4-D (*G. hirsutum*) and 10.74 μM NAA + 4.64 μM KIN (*G. arboreum*) SEIM: 0.537 μM NAA + 14.9 μM 2iP (*G. hirsutum*) and 0.68 μM 2,4-D + 2.98 μM 2iP (*G. arboreum*)	Khan et al. (2006)
G. hirsutum L. cv. CCR 1521, Zhongzhi 86-6	Hypocotyl	30 g/L sucrose	CIM: 0.1 mg/L 2,4-D + 0.1 mg/L KIN + 1.0 mg/L IBA Callus proliferation: 0.01 mg/L 2,4-D + 0.01 mg/L KIN + 1.0 mg/L IBA/0.05 mg/L IAA + double the concentration of KNO_3 with half the amount NH_4NO_3 of MS medium SEIM + SEMM: ½ strength MS medium without hormones + 0.5 g/L Gln + 0.5 g/L Asn	Wang et al. (2006)
G. hirsutum L. cv. YZ 1, Coker 312, Coker 201	Hypocotyl	25 g/L sucrose	CIM:1.0 mg/L IBA + 0.5 mg/L KIN SEIM: hormone-free MS medium SEMM: 0.5 mg/L IBA + 0.15 mg/L KIN + 1.0 g/L Gln + 0.5 g/L Asn + double the concentration of KNO_3 of MS medium	Jin et al. (2006a& b)
G. hirsutum L. cv. Coker	Hypocotyl	30 g/L sucrose	CIM: 1.0 mg/L BAP + 2.0/0.5 mg/L KIN + 1.0 g/L PVP SEIM: 0.5 μM brassinosteroids SEMM: GA_3	Aydin et al. (2006)

(continued)

Table 6.1 (continued)

Taxon/cultivar	Explant	Carbon source	Growth regulators	References
G. hirsutum L.	Hypocotyl, immature zygotic embryos	30 g/L glucose	CIM + SEIM: 0.1 mg/L 2,4-D + 0.5 mg/L KIN CMM: 0.1 mg/L 2,4-D + 0.1 mg/L KIN SEMM: 0.1 mg/L IAA + 0.1 mg/L GA_3	Hussain et al. (2009)
*G. hirsutum:*Jisheng1	Hypocotyl	30 g/L glucose	CIM: 1.0 mg/L IBA + 0.5 mg/L KIN SEIM + SEMM: Hormone-free medium with 1.0 g/L Gln 0.5 g/L Asp	Sun et al. (2009)
G. hirsutum L. cv. Coker 310, Narasimha	Hypocotyl, cotyledon	05 g/L, 10 g/L sucrose	CIM: 0.1 mg/L 2,4-D + 1.0 mg/L KIN Callus proliferation: 0.01 mg/L 2,4-D + 0.5 mg/L SEIM: hormone-free medium with double the concentration of KNO_3 without NH_4NO_3 SEIM: 0.1 mg/L zeatin	Khan et al. (2010)
G. hirsutum: W10 x TM-1, W10 x CRI12	Leaf petiole	30 g/L glucose	CIM: 0.05 mg/L IAA + 0.05 mg/L KIN + 0.05 mg/L 2,4-D SEIM: 0.06 mg/L KIN + 0.02 mg/L IAA	Zhang et al. (2011)
G. hirsutum: W10 x W10	Hypocotyl	25 g/L glucose	CIM: 0.10 mg/L IAA + 0.10 mg/L KIN + 0.10 mg/L 2,4-D SEIM: 0.08 mg/L KIN + 0.16 mg/L IAA	Xu et al. (2013)
Khandwa-2, Coker 312	Hypocotyl, cotyledon	30 g/L glucose 20 g/L sucrose for somatic embryo maturation	CIM: 0.5 mg/L IBA + 0.5 mg/L KIN SEIM: Hormone-free medium; NH_4NO_3 removed and KNO_3 doubled. SEMM: 1/2 MS medium + 20 g/L sucrose	Kumar et al. (2013)
CCRI24	Leaf petiole	30 g/L glucose	CIM: 0.05 mg/L IAA + 0.05 mg/L KIN + 0.05 mg/L 2,4-D SEIM: 0.06 mg/L KIN + 0.02 mg/L IAA	Yang et al. (2014)

CIM callus induction medium, *CMM* callus maintenance medium, *KIN* kinetin, *SEIM* somatic embryo induction medium, and *SEMM* somatic embryo maturation medium

(a)
(b)
(c)
(d)
(e)
(f)
(g)
(h)
(i)
(j)
(k)
(l)
(m)
(n)
(o)
(p)

◀ **Fig. 6.1** Regeneration in vitro of several species and cultivars of commercial cotton. Commonly used explants include cotyledon (**a**) and hypocotyl (**c**) segments from 7 to 10 days old cotton seedlings. Within a month, these explants produce copious amounts of callus (**b** and **d** from cotyledon and hypocotyl explants, respectively). Callus can then be selectively subcultured to yield friable callus containing small cells with dense cytoplasm (**e**). In another 1–4 months of subculture development of embryogenic callus occurs. Examples of embryogenic callus from *G. barbadense* GB35 (**f**), *G. arboreum* A2-9 (**g**), *G. hirsutum* Acala GC-510 (**h**), *G. hirsutum* Acala B1644 (**i**), *G. hirsutum* DPL-90 (**j**), and *G. hirsutum* GA 98033 (**k**) are provided. Mature embryos of DPL-90 (**l**), *G. arboreum* A2-9 (**m**), or *G. hirsutum* Acala B1644 (**n**) are then transferred to a germination medium. A plantlet of *G. arboreum* A2-9 (**o**) in a *Magenta box* and a potted plant of *G. hirsutum* GA 98033 (**p**) are shown as examples. Details on media composition are provided in Table 6.1

6.2.1.3 Growth Regulators, Carbohydrate Source, and Environmental Conditions

Hormones or growth regulators play an important role in tissue culture of any plant, including cotton. Various auxins including α-naphthaleneacetic acid (NAA), 2,4-dichlorophenoxyacetic acid (2,4-D), indole-3-acetic acid (IAA), indole-3-butyric acid (IBA), and picloram as well as various cytokinins including kinetin (6-furfuryladenine), 2iP (N^6-[2-isopentyl] adenine, zeatin (6-[4-hydroxy-3-methylbut-2-enylamino] purine), and BA (6-benzylaminopurine have been used in different combinations for both callus initiation and maintenance as well as for maturation of embryos and their germination (Price and Smith 1979; Davidonis and Hamilton 1983; Trolinder and Goodin 1987; Firoozabady and Deboer 1993; Sakhanokho et al. 2001; Ganesan and Jayabalan 2005).

Furthermore, carbon source is one of the factors that play major roles in cotton somatic embryogenesis. The early cotton tissue culture protocols for the most part used sucrose as a carbon source; however, because of issues related to the browning of cells during embryogenesis, several authors replaced this carbon source with glucose to reduce the effects of phenolic compounds which cotton callus is notoriously known to produce. Several researchers have used dithiothreitol (DTT) to overcome browning due to phenolic oxidation in cotton explants (Rajasekaran, unpublished) similar to observations by Pedroso and Pais (1993) with *Camellia* cultures. In addition to glucose and sucrose, other carbon sources such as maltose and fructose have been used in cotton somatic embryogenesis studies. For example, Sun et al. (2006) and Kumar et al. (2015) found that maltose was a better source of carbon than glucose in reducing browning of cells.

Additionally, other environmental conditions such as light intensity as well as the composition and types (solid, semi-solid, or liquid) of media influence cotton somatic embryogenesis. For example, Smith et al. (1977) found that callus proliferation in diploid *G. arboreum* was better under high light intensity (8000–9000 lx) than under low light (1000–2000 lx) or under complete dark environment. On the other hand, in a study conducted by Firoozabady and Deboer (1993) in somatic embryogenesis of *G. hirsutum*, high temperature (30 °C) and low light intensity (9 $\mu E\ m^{-2}\ s^{-1}$) were ideal conditions for callus initiation, embryogenic callus induction, and maintenance, whereas lower temperature (25 °C) and high light intensity (90 μE m^{-2} s^{-1}) were the optimal conditions for somatic embryo maturation, germination, and plantlet development. However, they added that light and temperature had less impact than medium composition on different phases of cotton tissue culture. In addition, carbohydrate source and concentration as well as the amount and type of nitrogen can influence cotton somatic embryogenesis. For example, replacing glucose with sucrose followed by exposure of immature somatic embryos to a 3-day desiccation period helped improve somatic embryo initiation and germination in diploid cotton *G. arboreum* (Sakhanokho et al. 2004b). Also, doubling of the concentration of KNO_3 and removing NH_4NO_3 from the culture media enhanced somatic embryogenesis in *G. hirsutum* and *G. barbadense* (Trolinder and Goodin 1987; Sakhanokho et al. 2001).

Furthermore, many plant species including *Prunus persica* L. (Hammerschlag 1982), *Triticum aestivum* L. (Jones and Petolino 1988), and *Gossypium* spp. (Finer 1988; Gawel and Robacker 1990; Sakhanokho et al. 2001) show improved somatic embryogenesis when cultured on liquid media versus semi-solid or solid media. The high medium-to-tissue contact is believed to be one of the major factors behind the greater rate of somatic embryo proliferation in liquid media (Finer 1988; Gawel and Robacker 1990). Further, the continuous shaking of the liquid cultures leads to better aeration due to enhanced oxygen supply to the cultured cells, and probably negates the effects of phenolic browning.

6.2.2 Multiplication of Zygotic Embryo or Shoot Meristems

Because of the recalcitrant nature of cotton to embryogenesis, in vitro meristem culture was investigated in the early 1990s for the development of a non-genotype dependent multiplication system for transformation of elite cotton cultivars (Finer and McMullen 1990; Gould et al. 1991; McCabe and Martinell 1993). Since then, more cotton genotypes have been multiplied through in vitro meristem culture (Chlan et al. 1995; Agrawal et al. 1997; Saeed et al. 1997; Gupta et al. 1997; Hemphill et al. 1998; Zapata et al. 1999; Luo and Gould 1999; Hazra et al. 2000; 2001; Ouma et al. 2004; Nandeshwar et al. 2009; Pathi and Tuteja 2013) (Table 6.2). Apical meristems as well as other explant types such as hypocotyl segments, embryo axes, and cotyledonary nodes have been used to generate cotton plants via organogenesis and multiple shoot production (Agrawal et al. 1997; Gupta et al. 1997; Banerjee et al. 2003; Divya et al. 2008; Yang et al. 2010; Chakravarthy 2013). Multiple shoot production could have the potential to improve the successful recovery rate of cotton germ line transformed plants using the biolistic or *Agrobacterium*-mediated method, as the multiple meristems used as targets could increase the chance of obtaining transformed plants. Shoot proliferation from cotton meristem is very inefficient unlike in other crop plants such as soybean (Rajasekaran

Table 6.2 List of some cotton cultivars and taxa regenerated via organogenesis or shoot tip multiplication

Cultivar/taxon	Explant	Carbon source	Growth regulators (shoot initiation)	Growth regulators (rooting media) and/or method	References
G. hirsutum L. cv. Khandwa *G. hirsutum* L. cv. Pkv081 *G. hirsutum* L. cv. RS 810 *G. hirsutum* L. cv. Pusa 37 *G. hirsutum* L. cv. Pusa 26 *G. hirsutum* L. Stoneville *G. hirsutum* L. cv. F1084 *G. hirsutum* L. cv. CA 1193 *G. arboreum* L. cv. Shyamly *G. arboreum* L. cv. Lolnt	Cotyledonary node with shoot apex	30 g/L glucose	22.2 µM BAP	2.7 µM NAA	Gupta et al. (1997)
G. hirsutum L. cv. Anjali and *G. hirsutum* L. cv. LRK 516	Cotyledonary node with shoot apex devoid of cotyledons	20 g/L sucrose	2.5 mg/L BAP	0.05/0.1 mg/L NAA	Agrawal et al. (1997)
G. hirsutum: GOHAR-87, CIM-70, CIM-109, CIM-240, B-557, SL-41, BH-36, RH-1, COKER-304, MNH-93, A-1-85 A-18/87, AEM-52, N-26, NIAB-78, S-12, FH-87, FH-634, and FH-682. *G. arboreum*: RAVI	Shoot tip	30 g/L glucose	0.46 mM KIN, 0.93 mM KIN, 0.45 mM 2,4-D + 2.32 mM KIN, 0.45 mM 2,4-D + 2.46 mM2iP	1/2MS, MS, MS + IBA (0.98 mM, 2.46 mM, 4.92 mM, 9.84 mM and 24.60 mM) and MS + 2.68 mM NAA + 0.46 mM KIN	Saeed et al. (1997)
G. hirsutum L. cv. Guazuncho II	Embryonic axis	20 g/L glucose	3.0 mg/L BAP	Auxin shock: exposure to 500 mg/L IBA for about 15 s	Morre et al. (1998)
G. hirsutum L. cv. Stoneville 7A and Paymaster HS26	Shoot apices Secondary leaf node Cotyledonary nodes	15 g/L sucrose	0.3 mg/L BAP	1.0 mg/L IBA	Hemphill et al. (1998)

(continued)

Table 6.2 (continued)

Cultivar/taxon	Explant	Carbon source	Growth regulators (shoot initiation)	Growth regulators (rooting media) and/or method	References
G. hirsutum L. cv. MCU-5	Shoot tip	30 g/L glucose	0.1 mg/L BAP + 0.1 mg/L NAA	MS + 15 g/L sucrose	Satyavathi et al. (2002)
G. hirsutum L. cv. NHH 44 *G. hirsutum* L. cv. DCH 32 *G. hirsutum* L. cv. DHY 286 *G. hirsutum* L. cv. H 8 *G. hirsutum* L. cv. LRK 516 *G. hirsutum* L. cv. LRA5166	Embryo axes	20 g/L sucrose	0.4 µM BAP + 0.1 µM NAA	0.5 µM NAA	Banerjee et al. (2003)
G. hirsutum L. cv. NIAB 999	Cotyledonary node with both cotyledons	30 g/L glucose	0.25 mg/L KIN	0.5 mg/L NAA + 0.1 mg/L KIN	Rauf et al. (2005)
G. hirsutum L. cv. Barac(67)B	Cotyledonary node devoid of cotyledons and apical meristems	20 g/L sucrose	2.5 mg/L KIN + 0.1 mg/L BAP + 3.0 mg/L $AgNo_3$ MSB_5 medium	0.1 mg/L NAA	Abdellatef and Khalafalla (2007)
G. hirsutum L. cv. Bharani *G. hirsutum* L. cv. Durga *G. hirsutum* L. cv. JKCH 99	Hypocotyl	30 g/L glucose	2.0 mg/L TDZ + 0.05 mg/L NAA 1.0 mg/L BAP + 2.0 mg/L GA_3	1/2 MS + 30 g/L sucrose + activated charcoal 1.0 mg/L IBA	Divya et al. (2008)
G. hirsutum L. CIM 443 *G. hirsutum* L. CIM 446 *G. hirsutum* L. CIM 473 *G. hirsutum* L. NIAB *G. hirsutum* L. FH 900 *G. hirsutum* L. NIAB 98	Apical meristems	10 g/L sucrose	2.0 mg/L BAP	2.0 mg/L KIN + 1.5 mg/L IAA, grafting	Aslam et al. (2010)

(continued)

Table 6.2 (continued)

Cultivar/taxon	Explant	Carbon source	Growth regulators (shoot initiation)	Growth regulators (rooting media) and/or method	References
G. bickii (wild cotton)	Cotyledonary nodes devoid of cotyledons	30 g/L glucose	4.0 mg/L BAP + 0.1 mg/L TDZ 0.05 mg/LGA_3	½ strength MS medium	Yang et al. (2010)
G. hirsutum L. hybrid H8 *G. hirsutum* L. Khandwa 2 *G. arboreum* L. cv. BD 1 *G. arboreum* L. cv. BD 6; Sarvottam	Cotyledonary node	sucrose	3.0 mg/L BAP 3.0 mg/L BAP 1.5 mg/L BAP 1.5 mg/L BAP 2.0 mg/L BAP	Hormone-free ½ strength MS medium	Obembe et al. (2011)
G. hirsutum L. cv. MCU 11	Cotyledonary node devoid of cotyledons	30 g/L glucose	1.5 mg/L BAP + 0.1 NAA	Hormone-free ½ strength MS Grafting	Mushke et al. (2012)
G. hirsutum L. cv. NA 1325	Embryo axis	30 g/L sucrose	2.0 mg/L BAP + 2.0 mg/L KIN 1.0 mg/L GA_3	1.0 mg/L IBA	Pathi and Tuteja (2013)
G. hirsutum L. cv. NC 601	Cotyledonary node	30 g/L maltose	1.5 mg/L BAP + 1.0 mg/L NAA	½ strength MS 1.5 mg/L IBA	Chakravarthy (2013)

and Pellow 1997) and has not been utilized for production of commercially viable transgenic cotton lines to date.

However, this cotton regeneration through meristem or shoot tip multiplication is not without issues. For example, meristem-derived cotton plantlets sometimes do not readily form roots as induction of roots in cotton shoots is genotype dependent and unreliable (Luo and Gould 1999), thus requiring an extra step in the rooting process, sometimes requiring an auxin treatment or the direct application of a dry rooting agent to shoots and transfer to soil (Gould et al. 1991; Saeed et al. 1997; Hemphill et al. 1998; Divya et al. 2008). Additionally, in vitro grafting has also been used to improve rooting of cotton plantlets regenerated via organogenesis or shoot tip multiplication (Luo and Gould 1999; Mushke et al. 2012; Afolabi-Balogun et al. 2015). Another disadvantage of regeneration through shoot multiplication is that it could lead to the production of chimeric plants resulting in non-germ line expression of transgenes, but the meristem transformation has the advantage of being genotype independent (Rajasekaran 2013b); therefore, any cotton genotype, including the elite cultivars, can potentially be genetically modified using this regeneration scheme.

In summary, cotton has been successfully regenerated from various explants, albeit this success is largely limited to obsolete varieties such as Coker. This heralded the transgenic cotton program in several countries resulting in improved acreage, profit, agronomic, and fiber qualities. The science of regeneration has since been extended to a few elite cultivars and species and this should pave the way for improved cotton varieties with preferred agronomic characteristics without compromising on the most important trait, fiber quality.

References

Abdellatef E, Khalafalla MM (2007) Adventitious shoot and plantlet formation in medium staple cotton cultivar (*Gossypium hirsutum* L. cv. Barac). Int J Agric Biol 9:913–916

Abdurakhmonov IY, Buriev ZT, Saha S, Jenkins JN, Abdukarimov A, Pepper AE (2014) Phytochrome RNAi enhances major fibre quality and agronomic traits of the cotton *Gossypium hirsutum* L. Nat Commun 5:3062

Afolabi-Balogun NB, Inuwa HM, Ume O, Bakare-Odunola MT, Nok AJ, Adebola PA (2015) Optimization of micropropagation protocol for three cotton varieties regenerated from apical shoot. J. Plant Breed Crop Sci 7:38–43

Agrawal DC, Banerjee AK, Kolala RR, Dhage AB, Kulkanmi AV, Nalawade SM, Hazra S, Krishnmurthy KV (1997) In vitro induction of multiple shoots and plant regeneration in cotton (*Gossypium hirsutum* L.). Plant Cell Rep 16:647–652

Anderson DM, Rajasekaran K (2016) The global importance of transgenic cotton. In: Ramawat KG, Ahuja MR (eds) The Global Importance of Transgenic Cotton. Sustainable Development and Biodiversity, vol 13, Fiber Plants, Springer, pp xxx–xxx

Aslam M, Ashfaq M, Saeed T, Ul Allah S, Zafar Y (2010) In vitro response of cotton (*Gossypium hirsutum* L.) from apical meristem cultures. Am-Eurasian J Agric Environ Sci 7:07–11

Aydin Y, Ipekci Z, Talas-Ogras T, Zehir H, Bajrovic K, Gozukirmizi N (2004) High frequency somatic embryogenesis in cotton. Biol Plant 48:491–495

Aydin Y, Talas-Ogras T, Ipekci Z, Gozukirmizi N (2006) Effects of brassinosteroid on cotton regeneration via somatic embryogenesis. Biol Bratisl 61:289–293

Bai WQ, Xiao YH, Zhao J, Song SQ, Hu L, Zeng JY, Li XB, Hou L, Luo M, Li DM, Pei Y (2014) Gibberellin overproduction promotes sucrose synthase expression and secondary cell wall deposition in cotton fibers. PLoS ONE 9:e96537. doi:10.1371/journal.pone.0096537

Banuri T(1999) Pakistan: environmental impact of cotton production and trade. In: Global product chains: northern consumers, southern producers, and sustainability. UNEP

Banerjee AK, Agarwal DC, Nalawade SM, Hazra S (2003) Multiple shoot induction and plant regeneration from embryo axes of six cultivars of *Gossypium hirsutum*. Biol Plant 47:433–436

Bayley C, Trolinder N, Ray C, Morgan M, Quisenberry JE, Ow DW (1992) Engineering 2, 4-D resistance into cotton. Theor Appl Genet 83:645–649

Bell JM (1989) Nutrional characteristics and protein uses of oilseed meals. In: Robbelen G, Downey RK, Ashri A (eds) Oil crops of the world: their breeding and utilization. McGraw-Hill Pub. Co, New York, pp 192–207

Brookes G, Barfoot P (2012) The income and production effects of biotech crops globally 1996–2010. GM Crops Food: Biotech Agric Food Chain 3(4):265–272

Carter W (1981) Resistance and resistant reaction of *Gossypium arboreum* to the reniform nematode, *Rotylenchulus reniformis*. J Nematol 13:368–374

Chakravarthy VSK (2013) Rapid production of multiple shoots from cotyledonary node explants of an elite cotton (*Gossypium hirsutum* L.) variety. Res Plant Biol 3:6–13

Chen J, Burke JJ (2015) Developing fiber specific promoter-reporter transgenic lines to study the effect of abiotic stresses on fiber development in cotton. PLoS ONE 10(6):e0129870. doi:10.1371/journal.pone.0129870

Chen Y, Rivlin A, Lange A, Ye X, Vaghchhipawala Z, Eisinger E, Dersch E, Paris M, Martinell B, Wan Y (2014) High throughout *Agrobacterium tumefaciens*-mediated germline transformation of mechanically isolated meristem explants of cotton (*Gossypium hirsutum* L.). Plant Cell Rep 33:153–164

Chlan CA, Lin J, Cary JW, Cleveland TE (1995) A procedure for biolistic transformation and regeneration of transgenic cotton from meristematic tissue. Plant Mol Biol Rep 13:31–37

Cousins YL, Lyons BR, Llewellyn DJ (1991) Transformation of an Australian cotton cultivar: prospects for cotton improvement through genetic engineering. Aust J Plant Physiol 18:481–494

Davidonis GH, Hamilton RH (1983) Plant regeneration from callus tissue of *Gossypium hirsutum* L. Plant Sci Lett 32:89–93

Divya K, Jami SK, Kirti PB (2010) Constitutive expression of mustard annexin, AnnBj1 enhances abiotic stress tolerance and fiber quality in cotton under stress. Plant Mol Biol 73:293–308

Divya K, Swathi AT, Jami SK, Kirti PB (2008) Efficient regeneration from hypocotyls explants in three cotton cultivars. Biol Plant 52:201–208

Environmental Protection Agency (EPA, United States) (2015) Major crops grown in the United States. http://www.epa.gov/agriculture/ag101/cropmajor.html. Accessed 28 May 2015

Finer JJ (1988) Plant regeneration from somatic embryogenic suspension cultures of cotton (*Gossypium hirsutum* L.). Plant Cell Rep 7:399–402

Finer JJ, McMullen MD (1990) Transformation of cotton (*Gossypium hirsutum* L.) via particle bombardment. Plant Cell Rep 8:586–589

Finer JJ, Smith RH (1984) Initiation of callus and somatic embryos from explants of mature cotton (*Gossypium klotzschianum* Anders). Plant Cell Rep 3:41–43

Firoozabady E, DeBoer DL (1993) Plant regeneration via somatic embryogenesis in many cultivars of cotton (*Gossypium hirsutum* L.). In Vitro Cell Dev Biol-Plant 29:166–173

Firoozabady E, Deboer DL, Merlo DL, Halk EL, Amerson LN, Rashka KE, Murray EE (1987) Transformation of cotton (*Gossypium hirsutum* L.) by *Agrobacterium tumefaciens* and regeneration of transgenic plants. Plant Mol Biol 10:105–116

Ganesan M, Jayabalan N (2005) Carbon source dependent somatic embryogenesis and plant regeneration in cotton, *Gossypium hirsutum* L. cv. SVPR2 through suspension cultures. Ind J Exp Biol 43:921–925

Gawel NJ, Robacker CD (1990) Somatic embryogenesis in two *Gossypium hirsutum* genotypes on semi-solid versus liquid proliferation media. Plant Cell Tiss Org Cult 23:201–204

Gould J, Banister S, Hasegawa O, Fahima M, Smith RH (1991) Regeneration of *Gossypium hirsutum* and *G. barbadense* from shoot apex tissues for transformation. Plant Cell Rep 10:12–16

Gould JH, Magallanes-Cedeno M (1998) Adaption of cotton shoot apex culture to *Agrobacterium*-mediated transformation. Plant Mol Biol Rep 16:1–10

Gupta SK, Srivastava AK, Singh PK, Tuli R (1997) In vitro proliferation of shoots and regeneration of cotton. Plant Cell Tiss Org Cult 51:149–152

Haigler CH, Singh B, Zhang D, Hwang S, Wu C, Cai WX, Hozain M, Kang W, Kiedasisch B, Straus RE, Hequet EF, Wyatt BG, Jividen GM, Holaday AS (2007) Transgenic cotton over-producing spinach sucrose phosphate synthase showed enhanced leaf sucrose synthesis and improved fiber quality under controlled environmental conditions. Plant Mol Biol 63:815–832

Hammerschlag F (1982) Factors affecting establishment and growth of peach shoots in vitro. HortScience 17:85–86

Han J, Tan J, Tu L, Zhang X (2014) A peptide hormone gene, *GhPSK* promotes fibre elongation and contributes to longer and finer cotton fibre. Plant Biotechnol J 12:861–871

Haq IU, Zafar Y (2004) Effect of nitrates on embryo induction efficiency in cotton (*Gossypium hirsutum* L.). Afr J Biotechnol 3:319–323

Hazra S, Agrawal DC, Banerjee AK, Krishnmurthy KV, Nalawade SM (2001) Induction of multiple shoots and plant regeneration from 'accessory buds' of nodal segments from field-grown mature cotton plants (*Gossypium hirsutum* L.). In Vitro Cell Dev Biol-Plant 37:830–834

Hazra S, Kulkarni AV, Nalawade SM, Banerjee AK, Agrawal DC, Krishnamurthy KV (2000) Influence of explants, genotypes and culture vessels on sprouting and proliferation of pre-existing meristems of cotton (*Gossypium hirsutum* L. and *Gossypium arboreum* L.). In Vitro Cell Dev Biol Plant 36:505–510

Hemphill JK, Maier CG, Chapman KD (1998) Rapid in-vitro regeneration of cotton (*Gossypium hirsutum* L.). Plant Cell Rep 17:273–278

Hussain SS, Rao AQ, Hussain T (2009) Cotton somatic embryo morphology affects its conversion to plant. Biol Plant 53:307–311

Janick J, Schery RW, Woods FW, Ruttan VW (1981) Fiber, forest, and ornamental crops. In: Janick J, Schery RW, Woods FW, Ruttan VW (eds) Plant science: an introduction to world crops, 3rd edn. W.H. Freeman and Company, New York, pp 675–734

Jiang Y, Guo W, Zhu H, Ruan YL, Zhang T (2012) Overexpression of *GhSusA1* increases plant biomass and improves cotton fiber yield and quality. Plant Biotechnol J 10:301–312

Jin S, Liang S, Zhang X, Nie Y, Guo X (2006a) An efficient grafting system for transgenic plant recovery in cotton (*Gossypium hirsutum* L.). Plant Cell Tiss Org Cult 85:181–185

Jin S, Zhang X, Nie Y, Guo X, Liang S, Zhu H (2006b) Identification of a novel elite genotype for in vitro culture and genetic transformation of cotton. Biol Plant 50:519–524

John ME (1995) Characterization of a cotton (*Gossypium hirsutum* L.) fiber mRNA (Fb-B6). Plant Physiol 107:1477–1478

John ME, Keller G (1996) Metabolic pathway engineering in cotton: biosynthesis of polyhydroxybutyrate in fiber cells. Proc Natl Acad Sci USA 93:12768–12773

John ME, Stewart JM (1992) Genes for jeans: biotechnological advances in cotton. Trends Biotechnol 10:165–170

Jones AM, Petolino JF (1988) Effects of support medium on embryo and plant production from cultured anthers of soft-red winter wheat. Plant Cell Tiss Org Cult 12:253–261

Juturu VN, Mekala GK, Surabhi GK, Kirti PB (2015) Current status of tissue culture and genetic transformation research in cotton (*Gossypium* spp.). Plant Cell Tiss Org Cult 120:813–839

Katageri IS, Vamadevaiah HM, Udikeri SS, Khadi BM, Kumar PA (2007) Genetic transformation of an elite Indian genotype of cotton (*Gossypium hirsutum* L.) for insect resistance. Curr Sci 93:1843–1847

Keller G, Spatola L, McCabe D, Martinell B, Swain W, John ME (1997) Transgenic cotton resistant to herbicide bialaphos. Trans Res 6:385–392

Keshamma E, Rohini S, Rao KS, Madhusudan B, Kumar MU (2008) Tissue culture-independent in planta transformation strategy: an *Agrobacterium* tumefaciens-mediated gene transfer method to overcome recalcitrance in cotton (*Gossypium hirsutum* L.). J Cotton Sci 12:264–272

Khan T, Singh AK, Pant RC (2006) Regeneration via somatic embryogenesis in different cultivars of cotton (*Gossypium* spp.). In Vitro Cell Dev Biol Plant 42:498–501

Khan T, Reddy VS, Leelavathi S (2010) High-frequency regeneration via somatic embryogenesis of an elite recalcitrant cotton genotype (*Gossypium hirsutum* L.) and efficient *Agrobacterium*-mediated transformation. Plant Cell Tiss Org Cult 101:323–330

Kohel RJ (1989) Cotton. In: Robbelen G, Downey RK, Ashri A (eds) Oil crops of the world: their breeding and utilization. McGraw-Hill Pub. Co, New York, pp 404–415

Kumar S, Sharma P, Pental D (1998) A genetic approach to in vitro regeneration of non-regenerating cotton (*Gossypium hirsutum* L.) cultivars. Plant Cell Rep 18:59–63

Kumar M, Singh H, Shukla AK, Verma PC, Singh PK (2013) Induction and establishment of somatic embryogenesis in elite Indian cotton cultivar (*Gossypium hirsutum* L. cv Khandwa-2). Plant Signal Behav 8:10, e26762, doi:10.4161/psb.26762

Kumar P, Subiramani S, Govindarajan S, Sadasivam V, Manickam V, Mogilicherla K, Thiruppathi SK, Narayanasamy J (2015) Evaluation of different carbon sources for high frequency callus culture with reduced phenolic secretion in cotton (*Gossypium hirsutum* L.) cv. SVPR-2. Biotechnol Rep 7:72–80

Kumar M, Tuli R (2004) Plant regeneration in cotton: a short-term inositol starvation promotes developmental synchrony in somatic embryogenesis. In Vitro Cell Dev Biol Plant 40:294–298

Kumria R, Sunnichan VG, Das DK, Gupta SK, Reddy VS, Bhatnagar RK, Leelavathi S (2003) High frequency somatic embryo production and maturation into normal plants in cotton (*Gossypium hirsutum*) through metabolic stress. Plant Cell Rep 21:635–639

Lee JA (1984) Cotton as world crop. In: Kohel RJ, Lewis CF (eds) Cotton Monograph Series Agronomy No. 2424. American Society of Agronomy, Madison, pp 1–25

Li R, Stelly DM, Trolinder NL (1989) Cytogenetic abnormalities in cotton (*Gossypium hirsutum* L.) cell cultures. Genome 32:1128–1134

Li B, Yang Y, Hu W-R, Li X-D, Cao J-Q, Fan L (2015) Over-expression of *GhUGP1* in upland cotton improves fibre quality and reduces fibre sugar content. Plant Breed 134:197–202

Luo J, Gould JH (1999) *In vitro* shoot-tip grafting improves recovery of cotton plants from culture. Plant Cell Tiss Org Cult 57:211–213

McCabe DE, Martinell BJ (1993) Transformation of elite cotton cultivars via particle bombardment of meristems. Bio/Technol 11:596–598

Meredith WR Jr (1984) Quantitative genetics. In: Kohel RJ, Lewis CF (ed) Cotton. Agronomy No. 24. ASA/CSC/SSSA, Madison, pp 131–150

Mishra R, Wang HY, Yadav NR, Wilkins TA (2003) Development of highly regenerable elite Acala cotton (*Gossypium hirsutum* cv. Maxxa)—a step towards genotype independent regeneration. Plant Cell Tiss Org Cult 73:21–35

Morre JL, Permingeat HR, Romagnoli MV, Heisterborg CM, Vallejos HR (1998) Multiple shoot induction and plant regeneration from embryonic axes of cotton. Plant Cell Tiss Org Cult 54:131–136

Mushke R, Sultana T, Pindi PK (2012) High frequency regeneration and multiple shoot induction in Indian cotton (*Gossypium hirsutum* L.) cultivar. Res J Agric Sci 3:1109–1112

Nandeshwar SB, Moghe S, Chakrabarty PK, Deshattiwar MK, Kranthi K, Anandkumar P, Mayee CD, Khadi BM (2009) *Agrobacterium*-mediated transformation of cry1Ac gene into shoot-tip meristem of diploid cotton *Gossypium arboreum* cv. RG8 and regeneration of transgenic plants. Plant Mol Biol Rep 27:549–557

Obembe OO, Khan T, Popoola J (2011) High frequency multiple shoots induction and plant regeneration in six elite Indian cotton cultivars. Can J Pure Appl Sci 5:1385–1389

Ouma JP, Young MM, Reichert NA (2004) Optimization of in vitro regeneration of multiple shoots from hypocotyl sections of cotton (*Gossypium hirsutum* L.). Afr J Biotechnol 3:169–173

Paterson AH, Boman RK, Brown SM, Chee PW, Gannaway JR, Gingle AR, May OL, Smith CW (2004) Reducing the genetic vulnerability of cotton. Crop Sci 44:1900–1901

Pathi KM, Tuteja N (2013) High-frequency regeneration via multiple shoot induction of an elite recalcitrant cotton (*Gossypium hirsutum* L. cv Narashima) by using embryo apex. Plant Signal Behav. 8:e22763. PMID: 23221745. doi:10.4161/psb.22763

Pasapula V, Shen G, Kuppu S et al (2011) Expression of an *Arabidopsis vacuolar* H^+-pyrophosphatase gene (*AVP1*) in cotton improves drought- and salt tolerance and increases fibre yield in the field conditions. Plant Biotechnol J 9:88–99

Pedroso MC, Pais MS (1993) Direct embryo formation in leaves of *Camellia japonica* L. Plant Cell Rep 12:639–643

Percival EA, Kohel RJ (1990) Distribution, collection, and evaluation of *Gossypium*. Adv Agron 44:225–255

Price HJ, Smith RH (1979) Somatic embryogenesis in suspension cultures of *Gossypium klotzschianum* Anderss. Planta 145:305–307

Price HJ, Smith RH, Grumbles RM (1977) Callus cultures of six species of cotton (*Gossypium* L.) on defined media. Plant Sci 10:115–119

Rajasekaran K (1996) Regeneration of plants from cryopreserved embryogenic cell suspension and callus cultures of cotton (*Gossypium hirsutum* L.). Plant Cell Rep 15:859–864

Rajasekaran K (2004) *Agrobacterium* mediated genetic transformation of cotton. In: Curtis IS (ed) Transgenic crops of the world—essential protocols. Springer, Berlin, pp 243–254

Rajasekaran K (2013a) Biolistic transformation of cotton embryogenic cell suspension cultures. Methods Mol Biol 958:59–70

Rajasekaran K (2013b) Biolistic transformation of cotton zygotic embryo meristem. Methods Mol Biol 958:47–57

Rajasekaran K, Chlan CA, Cleveland TE (2001) Tissue culture and genetic transformation of cotton. In: Jenkins JN, Saha S (eds) Genetic improvement of cotton. Science Publishers Inc, Enfield, pp 269–290

Rajasekaran K, Grula JW, Hudspeth RL, Pofelis S, Anderson DM (1996) Herbicide-resistant Acala and Coker cottons transformed with a native gene encoding mutant forms of acetohydroxyacid synthase. Mol Breed 2:307–319

Rajasekaran K, Pellow JW (1997) Somatic embryogenesis from cultured epicotyls and primary leaves of soybean [*Glycine max* (L.) Merrill]. In Vitro Cell Dev Biol-Plant. 33:88–91

Rangan TS, Rajasekaran K (1996) Regeneration of cotton plant in suspension culture. US patent #5,583,036

Rauf S, Usman M, Fatima B, Khan A (2005) In vitro regeneration and multiple shoot induction in upland cotton (*Gossypium hirsutum* L.). Plant Cell Tiss Org Cult 15:75–81

Reinisch AJ, Dong J, Brubaker CL, Stelly DM, Wendel JF, Paterson AH (1994) A detailed RFLP map of cotton, *Gossypium hirsutum* × *Gossypium barbadense*: chromosome organization and evolution in a disomic polyploid genome. Genetics 138:829–847

Sacks EJ, Robinson AF (2009) Introgression of resistance to reniform nematode (*Rotylenchulus reniformis*) into upland cotton (*Gossypium hirsutum*) from *Gossypium arboreum* and a *G. hirsutum*/*Gossypium aridum* bridging line. Field Crops Res 112:1–6

Saeed NA, Zafar Y, Malik KA (1997) A simple procedure of *Gossypium* meristem shoot tip culture. Plant Cell Tiss Org Cult 51:201–207

Sakhanokho HF (2001) Development of tissue culture and transformation systems in cotton (*Gossypium* spp. L.). Ph.D. Dissertation. Alabama A&M University, Normal, Alabama, p 120

Sakhanokho HF, Zipf A, Rajasekaran K, Saha S, Sharma GC (2001) Induction of highly embryogenic calli and plant regeneration in upland (*G. hirsutum* L.) and Pima (*G. barbadense* L.) cottons. Crop Sci 41:1235–1240

Sakhanokho HF, Ozias-Akins P, May OL, Chee PW (2004a) Induction of somatic embryogenesis and plant regeneration in selected Georgia and Pee Dee cotton lines. Crop Sci 44:2199–2205

Sakhanokho HF, Zipf A, Rajasekaran K, Saha S, Sharma GC, Chee P-W (2004b) Somatic embryo initiation and germination in diploid cotton (*Gossypium arboreum* L.). In Vitro Cell Dev Biol-Plant 40:177–181

Satyavathi VV, Prasad V, Lakshmi GB, Lakshmi S (2002) High efficiency transformation protocol for three Indian cotton varieties via *Agrobacterium tumefaciens*. Plant Sci 162:215–223

Shoemaker RC, Couche LJ, Galbraith DW (1986) Characterization of somatic embryogenesis and plant regeneration in cotton (*Gossypium hirsutum* L.). Plant Cell Rep 3:178–181

Smith RH, Price HJ, Thaxton JB (1977) Defined conditions for the initiation and growth of cotton callus in vitro I. *Gossypium arboreum*. In Vitro 13:329–334

Stelly DM, Altman DW, Kohel RJ, Rangan TS, Commiskey E (1989) Cytogenetic abnormalities of cotton somaclones from callus cultures. Genome 32:762–770

Sun J, Li W, Zhang H, Zhao J, Yin X, Wang L (2009) Somatic embryogenesis and plant regeneration in glandless upland cotton (*Gossypium hirsutum* L.). Front Agric China 3:279–283

Sun YQ, Zhang XL, Huang C, Guo XP, Nie YC (2006) Somatic embryogenesis and plant regeneration from different wild diploid cotton (*Gossypium*) species. Plant Cell Rep 25:289–296

Trolinder NL, Goodin JR (1987) Somatic embryogenesis and plant regeneration in cotton (*Gossypium hirsutum* L.). Plant Cell Rep 6:231–234

Trolinder NL, Goodin JR (1988) I. Somatic embryogenesis in cotton (*Gossypium*). Plant Cell Tiss Org Cult 12:31–42

Trolinder NL, Xhixian C (1989) Genotype specificity of the somatic embryogenesis response in cotton. Plant Cell Rep 8:133–136

Umbeck P, Johnson G, Barton K, Swain W (1987) Genetically transformed cotton (*Gossypium hirsutum* L.) plants. Biotechnology 5:263–266

Wang YX, Wang XF, Zhi-ying MA, Gui-yin Z, Gai-Ying H (2006) Somatic embryogenesis and plant regeneration from two recalcitrant genotypes of *Gossypium hirsutum* L. Agric Sci China 5:323–329

Wilkins TA, Rajasekaran K, Anderson DM (2000) Cotton biotechnology. Crit Rev Plant Sci 19:511–550

Wilkinson F (1927) The story of the cotton plant. The University Society, Inc., New York, pp 9–19

World Wide Fund (WWF) for Nature (2015) Cotton. https://www.worldwildlife.org/industries/cotton. Accessed 30 April, 2015

Wu JH, Zhang XL, Nie YC, Jin SX, Ling SG (2004) Factor affecting somatic embryogenesis and plant regeneration from a range of recalcitrant genotypes of Chinese cottons (*Gossypium hirsutum* L.). In Vitro Cell Dev Biol Plant 40:371–375

Xu ZZ, Zhang CJ, Zhang XY, Liu CL, Wu ZX, Yang ZR, Zhou KH, Yang XJ, Li FG (2013) Transcriptome profiling reveals auxin and cytokinin regulating somatic embryogenesis in different sister lines of cotton cultivar CCRI24. J Integr Plant Biol 55:631–642

Yang Z, Li C, Wang Y, Zhang C, Wu Z, Zhang X, Liu C, Li F (2014) *GhAGL15 s*, preferentially expressed during somatic embryogenesis, promotes embryogenic callus in cotton (*Gossypium hirsutum* L.). Mol Genet Genom 289:873–883

Yang XY, Zhang XL, Fu LL, Min L, Liu GZ (2010) Multiple shoot induction in wild cotton (*Gossypium bickii*) through organogenesis and the analysis of genetic homogeneity of the regenerated plants. Biologia 65:496–503

Zapata C, Park SH, El-Zik KM, Smith RH (1999) Transformation of a Texas cotton cultivar using *Agrobacterium* and the shoot apex. Theor Appl Genet 98:252–256

Zhang BH, Liu F, Yao CB (2000) Plant regeneration via somatic embryogenesis in cotton. Plant Cell Tiss Organ Cult 60:89–94

Zhang CJ, Yu SX, Fan SL, Zhang JF, Li FG (2011) Inheritance of somatic embryogenesis using leaf petioles as explants in upland cotton. Euphytica 181:55–63

Chapter 7
Plant Cell, Tissue, and Organ Culture Approaches to Explore the Functional Cell Differentiation in *Phyllostachys* and *Bambusa* Bamboo Plants

Shinjiro Ogita, Takao Kishimoto, Taiji Nomura and Yasuo Kato

Abstract We present two case studies developed in our laboratory that involve the establishment of efficient plant cell tissue and organ culture (PCTOC) models to explore functional differentiation of cells responsible for cell wall and fiber formation in *Phyllostachys* and *Bambusa* bamboo plants and their application. First, we standardized a novel xylogenic suspension culture model in order to understand the process of xylogenic cell differentiation during lignification in living *Phyllostachys nigra* (Pn). The Pn cells rapidly formed secondary cell wall components that were highly lignified, making up approximately 25 % of the dry weight of the cells under the xylogenic differentiation condition (1/2 MS medium supplemented with 10 μM BA). Two types of xylogenic differentiation were observed-fiber-like elements (FLEs) with cell wall thickening and tracheary elements (TEs). We systematically maintained in vitro node culture stocks of two prominent *Bambusa* bamboo species: *B. multiplex* (Bm), which shows a normal 'hollow' culm, and *B. glaucescens* f. houraikomachi (Bg), which has a thick-walled 'solid' culm. As node portions have apical and intercalary meristems, they could directly be used as explant sources for the establishment of callus and organ cultures without sterilization. When node portions were cultured on an optimized proliferation medium (MSp680 medium supplemented with 10 μM picloram), active callus induction and organ differentiation were seen. Although calli usually proliferate as irregular tissue masses and vary widely in texture, it is still possible to generate different cell lines such as whitish and greenish callus, and bunches of adventitious roots, under the same medium conditions by carefully separating these

S. Ogita (✉) · T. Kishimoto · T. Nomura · Y. Kato
Biotechnology Research Center and Department of Biotechnology,
Toyama Prefectural University, Imizu, Toyama 939-0398, Japan
e-mail: ogita@pu-hiroshima.ac.jp

S. Ogita
Faculty of Life and Environmental Sciences, Department of Life Sciences,
Prefectural University of Hiroshima, Shoubara 727-0023, Japan

K.G. Ramawat and M.R. Ahuja (eds.), *Fiber Plants*, Sustainable Development and Biodiversity 13, DOI 10.1007/978-3-319-44570-0_7

cultures from the explant. The growth patterns of these two PCTOC model cultures during maintenance and functional cell differentiation cultures are discussed in this chapter.

Keywords Bamboo · *Bambusa* · Callogenesis · Plant cell tissue and organ culture · *Phyllostachys* · Xylogenesis

7.1 Introduction

As commonly known, a plant mainly consists of three types of biopolymers, namely cellulose (ca. 45–55 %), hemicellulose (25–35 %), and lignin (20–30 %). These estimations are important to explain the quality of plant biomass. Our societies continuously focus on the application of plant biomass for multi-purposes such as building materials, paper production, fine chemicals, fiber materials, and energy sources. In order to increase quality and quantity of plant biomass, it is essential to understand the complexity and flexibility of these polymers within plant cells. There are many kinds of plant cells, i.e., meristematic cells (procambial cells/vascular cambial cells), parenchyma cells, phloem cells, and xylem cells (tracheary elements (TEs) and xylem fiber cells), with unique shapes, sizes, and functions. Therefore, it is very important to evaluate the characteristics of each cell during a specific biological event by using the appropriate methodologies. However, which of the methods is more appropriate? This is a fundamental but sophisticated question. Nowadays, we can choose many methodologies to investigate plant cells. Anatomical observation is one essential method to understand cell characteristics. Physiological analysis is an another vital approach to investigate detailed phenomena, like enzymatic activities of a target biosynthetic pathway during cell proliferation and differentiation processes. Molecular analysis is also important to elucidate the fundamental principles of a given phenomenon in the target plant organs or cells. Furthermore, new and substantial biological information derived from omics methodologies, e.g., DNA sequencing, metabolome, transcriptome, and proteome analysis, is now compiled into a mega-data bank, the National Center for Biotechnology Information (NCBI), which can be accessed through the World Wide Web [http://www.ncbi.nlm.nih.gov/]. The next important question is how can we systematically regulate plant cell proliferation and differentiation processes in a target plant? One efficient methodology is plant cell, tissue, and organ culture (PCTOC), which is thought to be the most flexible foundation for morphological, physiological, and molecular biological applications on target plants (Ogita 2015). Once an appropriate cell culture system of a target plant is established, we can manipulate cell proliferation and differentiation in a more logical way. In this chapter, we focus on general information regarding plant cell wall formation from a PCTOC approach and then describe two case studies of bamboo plants grown in our laboratory to demonstrate the advantage of using PCTOC for understanding TEs and fiber cell formation. One case study comprises the use of a novel xylogenic suspension culture model in order to understand the process of fiber formation during the lignification process of living *Phyllostachys nigra* (Pn) bamboo cells (Ogita et al. 2012a). The suspension-cultured cells used in this study are currently available as Pn (rpc00047), from the RIKEN Bioresource Center [http://www.brc.riken.jp/lab/epd/Eng]. The other case study concerns the use of systematic node and callus culture approaches for growing two prominent bamboo species: *Bambusa multiplex* Raeushel (Bm) that has a normal 'hollow' culm and *B. glaucescens* f. houraikomachi (Bg) that has a thick-walled 'solid' culm. These technologies are useful for better understanding bamboo plants fiber formation as a renewable source of biomass production.

7.2 PCTOC Approaches for Understanding Plant Cell Wall Formation, General Information

It is a well-known fact that we can observe primary and secondary cell wall formation with transdifferentiation of TEs and fiber cells via PCTOC approaches. Protoplast, callus, and cell suspension cultures have been performed in numerous species of both herbaceous and woody plants, and thus, when using the prominent keywords 'cell wall formation' and 'callus' in a search engine (e.g., Google Scholar), more than 50,000 related documents appear. For example, TEs formation via callus and cell suspension cultures is available for *Arabidopsis thaliana* (Oda et al. 2005), *Cryptomeria japonica* (Mehra and Anand 1979), *Cupressus sempervirens* (Havel et al. 1997), *Daucus carota* (Aloni 1980), *Glycine max* (Aloni 1980), *Pinus radiata* (Möller et al. 2006), *Populus tremula* × *P. tremuloides* (Ohlsson et al. 2006). How can we select the 'necessary' information from these searching results? One important suggestion is to pick up an efficient PCTOC method with high reproducibility of the target biological event, which is quite useful for understanding plant cell wall formation. Fukuda and Komamine (1980) reported the successful achievement of in vitro xylem cell formation using a xylogenic cell culture of *Zinnia elegans*, which is a unique system to transdifferentiate isolated mesophyll cells into tracheary elements (TEs). Since the 1980s, numerous PCTOC studies have been carried out to characterize the fiber formation process in this leading xylogenic cell culture model (i.e., the *Zinnia* model). Another important method for investigating the functional differentiation process of cell walls is the new *in planta* approach, using model plants such as *Arabidopsis* and *Brachypodium*. The *in planta* study of xylem cell formation of a target plant is normally intricate because the process is organ/tissue specific and also a long life cycle-dependent phenomenon. However, recent (from 2000 onwards) advances on model plants' genome sequencing and assembly, mutation and transformation, in conjunction with their small stature, rapid life cycle, and convenient cultivation, have allowed gathering a hoard of biological information concerning xylem cell formation. Here, we provide a brief introduction to these two important models.

7.2.1 Identification of Regulatory Factors of Xylem Cells Formation via the Zinnia Model

The entire process, from a mesophyll cell to TE transdifferentiation, is frequently completed within 96 h in the *Zinnia* model. Using this highly reproducible model, knowledge on cytological characteristics, effects of plant hormones and small molecules interactions, as well as the identification and characterization of key enzymes involved on cell wall-related biosynthetic pathways and gene expression patterns during TEs differentiation have been expanded. The following three major processes are proposed in the *Zinnia* model (Fukuda 1997): Stage I, during which mesophyll cells dedifferentiate to become pluricompetent cells; Stage II, during which dedifferentiated cells restrict their differentiation competence to TE precursors via procambium-like cells; and Stage III, during which TE precursors complete secondary wall formation, execute programmed cell death, and maturate into TEs. A number of exogenous and endogenous regulatory factors, e.g., wounding, auxin-cytokinin interaction, brassinolid, xylogen (an arabinogalactan protein that induces xylem differentiation), and transcription factors (TFs), are known to influence TEs differentiation (see Fukuda 1997; Motose et al. 2001, 2004; Demura et al. 2002; Yoshida et al. 2009; Novo-Uzal et al. 2013) and have been recently applied to in vitro TE systems of angiosperms and gymnosperms model plants (Devillard and Walter 2014).

7.2.2 Molecular Information for Xylem Cell Formation via Model Plants

The publication of *A. thaliana* genome sequence (The Arabidopsis Genome Initiative 2000) provided a well-standardized model plant system with various advantages, new ideas, and breakthroughs. Nowadays, the regulatory network for secondary cell wall (SCW) biosynthesis is well explained via the *Arabidopsis* model (e.g., Taylor-Teeples et al. 2014). According to the latest review article by Nakano et al. (2015), the SCW biosynthesis in higher plants is mainly regulated via the NAC-MYB-based transcriptional network. The plant-specific NAC transcription factors, VASCULAR-RELATED-NAC-DOMAIN 1-7 (VND1-7), were identified as master regulators of xylem vessel cell differentiation (Kubo et al. 2005). NAC SECONDARY WALL THICKENING PROMOTING FACTOR1 and FACTOR2 (NST1 and NST2) were also identified as master regulators of anther cells SCW biosynthesis (Mitsuda et al. 2005). NST1 and NST3 (also called SECONDARY WALL-ASSOCIATED NAC-DOMAIN PROTEIN1 [SND1]) function as important switches in fiber cells differentiation (Zhong et al. 2006; Mitsuda et al. 2007). Other VND-related proteins, such as SOMBRERO (SMB), BEARSKIN1 (BRN1), and BEARSKIN2 (BRN2), were mentioned to be involved in SCW formation, since they induce ectopic SCW deposition when overexpressed (Willemsen et al. 2008; Bennett et al. 2010). Based on *Arabidopsis* model studies, VNS (VND, NST/SND, and SMB related) catabolized proteins function as first-layer master switches of SCW formation. The transcriptional regulatory network of MYB proteins and the second-layer master switches of SCW formation are also well documented by Nakano et al. (2015).

Although VNS proteins are well-conserved among vascular plants, both angiosperms and gymnosperms, continuous efforts have been made to establish new model plant systems. The whole genome sequence of *Brachypodium distachyon*, a member of the Pooideae subfamily, was released in 2010 (The International Brachypodium Initiative 2010), and this new model is developing similarly to the *Arabidopsis* model. For example, Mochida et al. (2013) constructed a full-length cDNA library from a large-scale collection of 21 different tissues from *B. distachyon* Bd21 (seed, shoot, leaves [vegetative stage, post-flowering, and stressed conditions], root, crown, spikelet [flowering to DAPs 1-30], and callus) and the obtained 78,163 high quality expressed sequence tags (ESTs) from both ends of ca. 40,000 clones; 16,079 of those clones were contigs. This information was then integrated with wheat (*Triticum aestivum*) and barley (*Hordeum vulgare*) full-length cDNAs, increasing the genomic resources of Pooideae crops. Detailed information is available in the RIKEN *Brachypodium* full-length cDNA database (RBFLDB) at [http://brachy.bmep.riken.jp/ver.1/index.pl].

7.3 Classification and Characterization of Bamboo Plants

Bamboo plants are classified into the family Poaceae (subfamily Bambusoideae), which comprises one tribe of woody bamboos, Bambuseae, including 1447 species. This tribe is further divided into the following 10 subtribes: Arundinariinae, Thamnocalaminae, Racemobambosinae, Shibataeinae, Bambusinae, Melocanninae, Hickeliinae, Guaduinae, Chusqueinae, and Arthrostylidiinae. The herbaceous bamboos in the family Poaceae are distributed within five tribes: Olyreae, 76 species; Parianeae, 40 species; Buergerslochloeae, one species; Puelieae, five species; and Guaduelleae, six species. Bamboo plants are naturally distributed within tropical and subtropical area such as eastern and southern Asia, and southern and central America. The distribution of woody bamboos extends far into the cool temperate zones of both hemispheres, whereas the herbaceous bamboos are confined to the tropics (Ohrnberger 1999). Since woody and herbaceous bamboos altogether comprise 1575 species, several options are available to select the appropriate bamboo species for a given use, such as house construction, common objects, agricultural and fisheries tools, food, crafting materials, and as a renewable bio-resource. In the following section, we focus on two important woody bamboo genera, *Phyllostachys* (76 species), which belongs to Shibataeinae and propagates through a monopodial type rhizome, and *Bambusa* (139 species), which belongs to Bambusinae and propagates through a sympodial type rhizome.

7.3.1 Anatomical Characteristics of Bamboo Culm and Rhizome

As commonly known, the culm is one important organ of the bamboo plant and significant component of its unused biomass. The quality of the rhizome is also quite important for the sustainable production of bamboo biomass. Here, we describe anatomical and histochemical features of *Phyllostachys nigra* (Pn) bamboo's culm and rhizome. Two staining solutions, Lugol's iodine and phloroglucinol-HCl reagent, were used for the detection of starch and lignin, respectively (Ogita et al. 2012a; Carciofi et al. 2012). As shown in Fig. 7.1, seasonal- and tissue-specific accumulation patterns of starch and lignin were observed in the culm and in the rhizome of Pn bamboo. In the culm of ca. 3-year-old plants, for example, whereas the phloroglucinol-HCl reaction produced stains in the nodes (red) and internodes (stripes of red and pale pink), the few stains produced by Lugol's reaction were only detected in the nodes (black). In addition, a large amount of starch grains (black stains) were densely distributed in the rhizome, which can actively form bamboo shoots. On the other hand, in immature (1 year old) and mature (more than 4 year old) rhizomes, none or little starch grains were detected, but a gradual increase in lignin deposition (red stains) could be seen in older rhizome. The complexity and diversity of cells in the Pn bamboo culm were visualized under a fluorescent microscope in cross sections (Fig. 7.2). Small vascular bundle units surrounded by abundant and thick-walled fiber cells (F), which had a pale-bluish auto-fluorescent signal, could be seen in the outer region of the culm (Fig. 7.2a). On the other hand, large vascular bundle units with developed xylem (Mx and Px) and phloem cells (Ph) were distributed in ground tissues consisting of parenchyma cells (Pa), as shown in Fig. 7.2b. These observations strongly suggest the necessity of

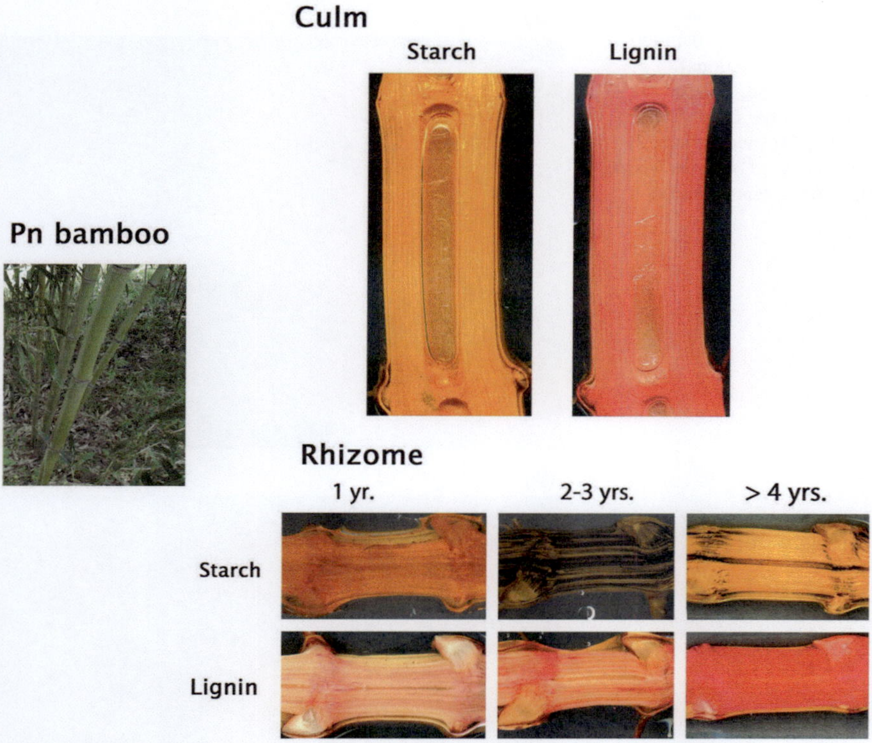

Fig. 7.1 Anatomical observations of Pn bamboo culm and rhizome. Two staining solutions, Lugol's iodine, and phloroglucinol-HCl, were used for detection of starch and lignin, respectively

establishing of a scheme with detailed information to elucidate the biological aspects of bamboo living cells. Therefore, we standardized a novel xylogenic suspension culture model to unveil the sequential biological processes occurring during cell wall formation and lignification in living bamboo cells.

7.3.2 Xylogenic Cell Suspension Culture in Phyllostachys Bamboo

Previously, we established an efficient callus and suspension culture system for Pn bamboo (Ogita 2005; Ogita et al. 2011) and found that these cultured cells proliferated with highly synchronous morphological features (Fig. 7.3a–d), under the proliferation (PR) condition (modified Murashige and Skoog (MS) liquid medium

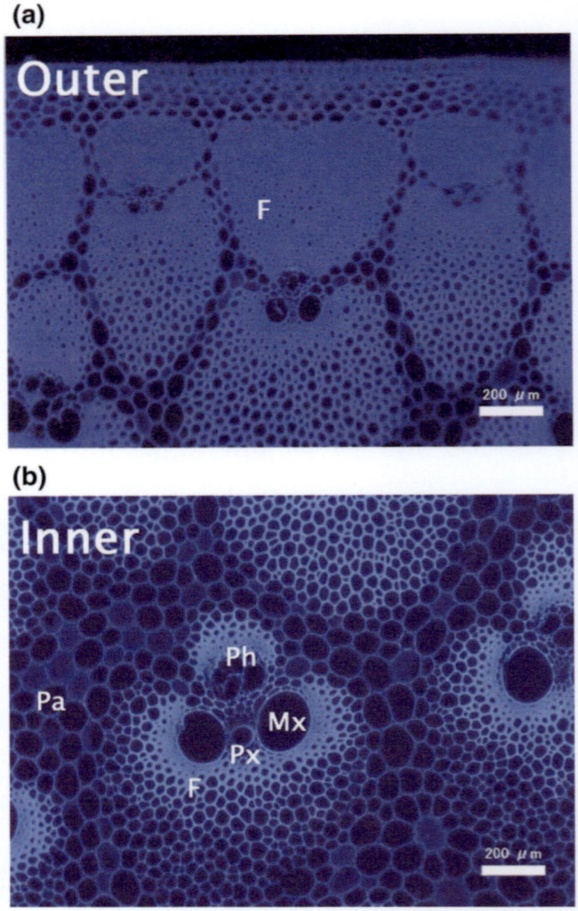

Fig. 7.2 Microscopic observations of Pn bamboo culms (cross sections). **a** Outer region, **b** inner region. *F* fiber cells, *Ph* phloem, *Mx* meta-xylem, *Px* proto-xylem, *Pa* parenchyma cells. *Scale bars* indicate 200 μm

(Murashige and Skoog 1962) supplemented with 680 mg l^{-1} KH_2PO_4 and 10 μM picloram). This culture system was applied to investigate several metabolic features involved in the growth and cell wall development of bamboo cells (Ogita et al. 2008b, 2012a, b; Nomura et al. 2013). Under lignification (LG) conditions (1/2 MS liquid medium supplemented with 10 μM BA), Pn cells rapidly formed secondary cell wall components that were highly lignified making up approximately 25 % of the cells dry weight within two weeks. Two types of xylogenic differentiation were observed (Fig. 7.3e–h): fiber-like elements (FLEs) with cell wall thickening and tracheary elements (TEs). Changes in the levels of xylogenesis-related gene transcripts were assessed by reverse transcription (RT)-PCR analysis. The transcription of key genes associated with early stages of lignin biosynthesis, such as

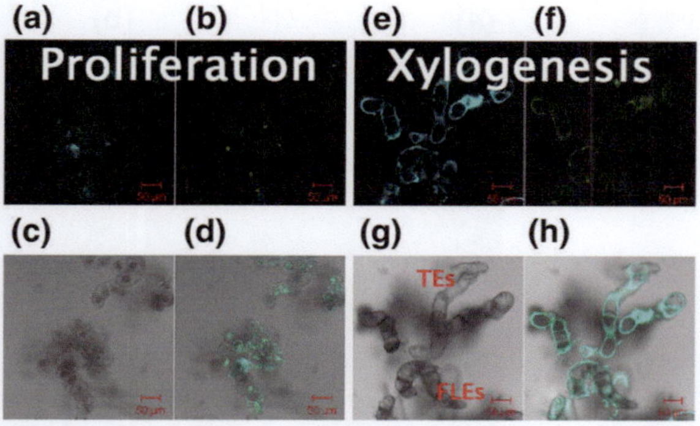

Fig. 7.3 Microscopic observations of Pn cells during proliferation (**a**–**d**) and xylogenesis (**e**–**h**). **a** and **e** are autofluorescent images of the cell wall. **b** and **c** are localization images of the nuclei stained with Sytox Green. **c** and **g** are phenotypic images of cells. **d** and **h** are merged images of all channels. *Scale bars* indicate 50 μm

phenylalanine ammonia-lyase (PAL), *cinnamate 4-hydroxylase* (C4H), *caffeoyl-CoA 3-O-methyltransferase* (CCoAOMT), and *cinnamoyl-CoA reductase* (CCR), was enhanced under LG conditions. Heteronuclear single quantum coherence (HSQC) NMR spectroscopy was used to compare the interunit linkage of lignins between mature bamboo culms and xylogenic suspension cells. HSQC NMR identified the most common interunit linkages, including β-aryl ether (β-O-4), phenylcoumaran (β-5), and resinol (β-β) structures, in the bamboo cultured cell lignin (BCCL). In addition to these common features of lignin, *several* differences in lignin substructures were also found between the BCCL and the bamboo milled wood lignin (BMWL). In addition to these common features of lignin, several differences in lignin substructures were also found between the BCCL and the BMWL. The *p*-hydroxyphenyl, guaiacyl, and syringyl (H/G/S) units could be identified in the BCCL. These results suggest the efficacy of our xylogenic cell culture model as a powerful tool for exploring the dynamics of the lignification process in *Phyllostachys* bamboos. Further investigations with metabolomics and next-generation sequencing technologies are now in progress.

7.3.3 Node Culture and Selective Induction of Callogenesis in Bambusa Bamboo

The other example considers the systematic node and callus culture approaches of two prominent bamboo species: both *Bambusa multiplex* (Bm) that show a normal 'hollow' culm (Fig. 7.4a, c) and *B. glaucescens* f. houraikomachi (Bg) that has a thick-walled 'solid' culm (Fig. 7.4b, d). *B. glaucescens* is a synonym of *B.*

Fig. 7.4 Phenotypes of bamboo plants and bamboo culms of Bm (**a** and **c**) and Bg (**b** and **d**)

multiplex, and Bg is considered as a closely related genotype. The complexity and diversity of vascular bundles and fiber cells were different among the three bamboo species culms, as shown in the cross sections of Pn (Fig. 7.5a, b), Bm (Fig. 7.5c, d), and Bg (Fig. 7.5e–g). Small vascular bundle units surrounded by abundant and thick-walled fiber cells could be seen in the outer region of the three bamboo species. Interestingly, this feature was also recognized in the inner regions of Bg culms. This fact suggests that Bg has potentially useful characteristics for renewable biomass production. In order to understand the sequential biological processes of 'solid' culm development in Bg, we systematically maintained in vitro node culture stocks of Bm and Bg, according to the method previously reported (Ogita et al. 2008a). Hence, a number of clonal small plantlets, with a high ability to form multiple shoots in a short period (ca. every 1–2 months), were available for investigations. Node portions of these culture stocks that had apical and intercalary meristems were excellent explants for the initiation of callus culture in Bm and Bg. Active callus induction and organ differentiation from node explants could be seen when node portions were cultured on PR medium (Ogita et al. 2011) as shown in Fig. 7.6. Although calli usually proliferate in irregular tissue masses with a wide variation in texture, it is still possible to generate different cell lines under the same medium conditions by carefully separating the different cultures originating in the explant. The following variation in morphology can be seen in the callus or organ

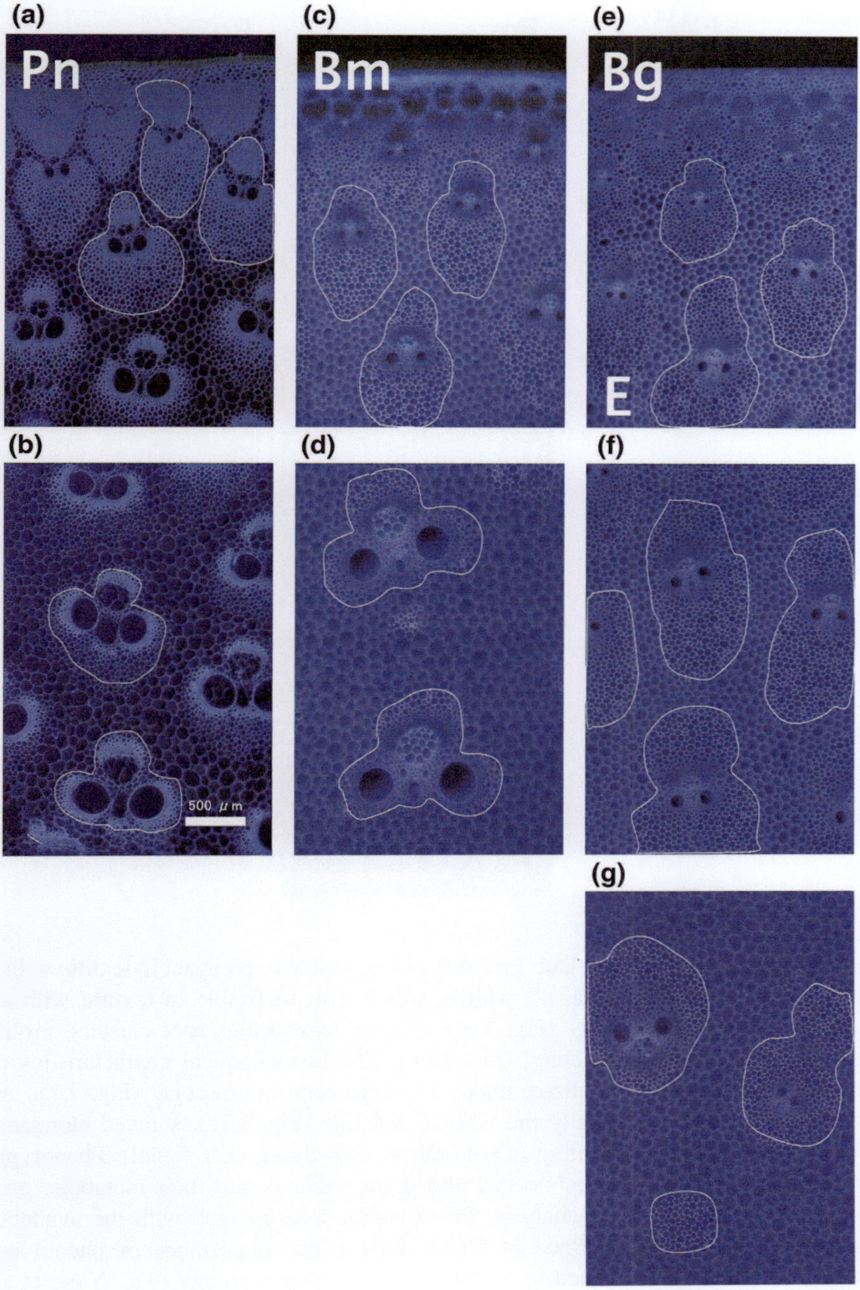

Fig. 7.5 Microscopic observations of vascular bundles with fiber cells in Pn (**a** and **b**), Bm (**c** and **d**), and Bg (**e**–**g**) culms. *Scale bar* indicates 500 μm

Fig. 7.6 Phenotypes of bamboo calli and organs derived from node explants of Bm (**a–c**) and Bg (**d** and **e**). *Scale bars* indicate 1 cm

originating from a node explant: greenish callus, friable to compact in texture with a moderate growth (Fig. 7.6a, d); whitish callus, soft to friable in texture with an active proliferation capacity (Fig. 7.6b, e); and adventitious root clusters, proliferating by elongation/branching (Fig. 7.6c). The histochemical characteristics of these cultures were visualized under a fluorescent microscope (Fig. 7.7). As expected, Bg cells, especially the whitish cell line (Fig. 7.7e), showed elongated FLEs with cell wall thickening (a pale-bluish auto-fluorescent signal). Phenotypic information from these bamboo cell and organ cultures and their metabolic profiling, obtained via HPLC analysis, for example, were merged with the available omics-based data as described in Ogita (2015). The importance of phenotypic approach is currently claimed in the field of plant biotechnology (e.g., Yang et al. 2013).

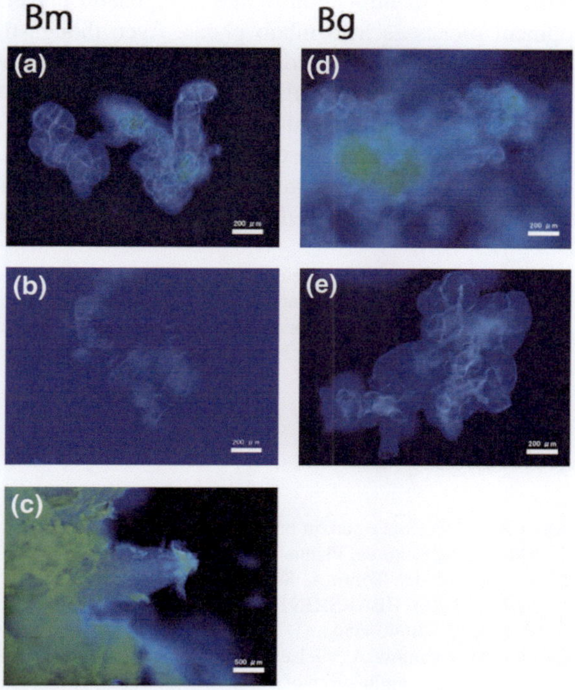

Fig. 7.7 Microscopic observations of bamboo calli and organs derived from node explants of Bm (**a**–**c**) and Bg (**d** and **e**). *Scale bars* indicate 200 μm in **a**, **b**, **d**, and **e**, and 500 μm in **c**

7.4 Conclusion and Perspectives

In order to increase knowledge on bamboo biological development and its detailed mechanisms, molecular approaches to these subjects have recently focused in comparison with model plants. For example, after comparing the anatomical structures of rhizome bud, rhizome shoot (early form of the rhizome), and bamboo shoot (early form of the bamboo culm), Wang et al. (2010) examined the difference in gene expression between the rhizome bud and the leaf in *Phyllostachys praecox* using a cross-species microarray with 7500 rice unigenes. cDNAs isolated from leaf and internodes were used to investigate the fiber development process in the bamboo (*Bambusa balcooa*), through PCR-based suppressive subtractive hybridization (PCR-SSH) (Rai et al. 2011). Omics methodologies such as DNA sequencing and transcription profiling were also applied for further characterization of numerous biological events in bamboo plants. For example, high-throughput sequencing of bamboo chloroplast genomes is carried out in six woody bamboos: *Acidosasa purpurea*, *Bambusa emeiensis*, *Ferrocalamus rimosivaginus*, *Indocalamus longiauritus*, *Phyllostachys edulis* (also called as *P. heterocycla* or P. *pubescens*), and *Phyllostachys nigra* (Zhang et al. 2011). In 2013, the draft genome of *P. heterocycla* was released as the first of a member of the Bambusoideae subfamily (Peng et al. 2013). These omics-based platforms will become important tools for the identification of the key genetic players involved in targeted development processes in bamboo plants. Even though these platforms contain useful information, we should consider translating such knowledge into applied studies using bamboo plants, mainly in the field of functional engineering of cell wall formation. To overcome this issue, the author believes that new PCTOC applications using well-standardized phenotypic information, as mentioned in the previous sections, are quite important for the functional engineering of bamboo cell wall. Establishment of such methodologies is underway in our laboratory.

Acknowledgments This research was partially supported by a Grant-in-Aid for scientific research B (25292106 to TK) and C (22580387 and 20254237 to SO; 25450134 to YK) from the Japan Society for the Promotion of Science. This study was also partially supported by a plant research grant to SO from the New Technology Development Foundation (2013–2014).

References

Aloni R (1980) Role of auxin and sucrose in the differentiation of sieve and tracheary elements in plant tissue cultures. Planta 150:255–263

Bennett T, van den Toorn A, Sanchez-Perez GF, Campiho A, Willemsen V, Snel B et al (2010) SOMBRERO, BEARSKIN1, and BEARSKIN2 regulate root cap maturation in *Arabidopsis*. Plant Cell 22:640–654

Carciofi M, Blennow A, Nielsen MM, Holm PB, Hebelstrup KH (2012) Barley callus: a model system for bioengineering of starch in cereals. Plant Methods 8:36. doi:10.1186/1746-4811-8-36

Demura T, Tashiro G, Horiguchi G, Kishimoto N, Kubo M, Matsuoka N, Minami A, Nagata-Hiwatashi M, Nakamura K, Okamura Y, Sassa M, Suzuki S, Yazaki J, Kikuchi S, Fukuda H (2002) Visualization by comprehensive microarray analysis of gene expression programs during transdifferentiation of mesophyll cells into xylem cells. PNAS 99: 15794–15799

Devillard C, Walter C (2014) Formation of plant tracheary elements in vitro—a review. N Z J For Sci 44:22. doi:10.1186/s40490-014-0022-7

Fukuda H (1997) Tracheary element differentiation. Plant Cell 9:1147–1156

Fukuda H, Komamine A (1980) Establishment of an experimental system for the study of tracheary element differentiation from single cells isolated from the mesophyll of *Zinnia elegans*. Plant Physiol 65:57–60

Havel L, Scarano MT, Durzan DJ (1997) Xylogenesis in *Cupressus* callus involves apoptosis. Adv in Hort Sci 11:37–40

Kubo M, Udagawa M, Nishikubo N, Horiguchi G, Yamaguchi M, Ito J, Mimura T, Fukuda H, Demura T (2005) Transcription switches for protoxylem and metaxylem vessel formation. Genes Dev 19:1855–1860

Mehra PN, Anand M (1979) Cytology of callus of *Cryptomeria japonica*. Physiol Plant 45: 127–131

Mitsuda N, Iwase A, Yamamoto H, Yoshida M, Seki M, Shinozaki K, Ohme-Takagi M (2007) NAC transcription factors, NST1 and NST3, are key regulators of the formation of secondary walls in woody tissues of *Arabidopsis*. Plant Cell 19:270–280

Mitsuda N, Seki M, Shinozaki K, Ohme-Takagi M (2005) The NAC transcription factors NST1 and NST2 of *Arabidopsis* regulate secondary wall thickenings and are required for anther dehiscence. Plant Cell 17:2993–3006

Mochida K, Uehara-Yamaguchi Y, Takahashi F, Yoshida T, Sakurai T, Shinozaki K (2013) Large-scale collection and analysis of full-length cDNAs from Brachypodium distachyon and integration with pooideae sequence resources. PLoS ONE 8:e75265. doi:10.1371/journal.pone.0075265

Motose H, Fukuda H, Sugiyama M (2001) Involvement of local intercellular communication in the differentiation of *Zinnia* mesophyll cells into tracheary elements. Planta 213:121–131

Motose H, Sugiyama M, Fukuda H (2004) A proteoglycan mediates inductive interaction during plant vascular development. Nature 429:873–878

Möller R, Ball RD, Henderson AR, Modzel G, Find J (2006) Effect of light and activated charcoal on tracheary element differentiation in callus cultures of *Pinus radiata* D. Don. Plant Cell Tis Org Cult 85:161–171

Murashige T, Skoog F (1962) A revised medium for rapid growth and bioassays with tobacco tissue culture. Physiol Plant 15:473–497

Nakano Y, Yamaguchi M, Endo H, Rejab NA, Ohtani M (2015) NAC-MYB-based transcriptional regulation of secondary cell wall biosynthesis in land plants. Front Plant Sci 6:288. doi:10.3389/fpls.2015.00288

Nomura T, Shiozawa M, Ogita S, Kato Y (2013) Occurrence of hydroxycinnamoylputrescines in xylogenic bamboo suspension cells. Plant Biotechnol 30:447–453

Novo-Uzal E, Fernández-Pérez F, Herrero J, Gutiérrez J et al (2013) From *Zinnia* to *Arabidopsis*: approaching the involvement of peroxidases in lignification. J Exp Bot 64:3499–3518

Oda Y, Mimura T, Hasezawa S (2005) Regulation of secondary cell wall development by cortical microtubules during tracheary element differentiation in *Arabidopsis* cell suspensions. Plant Physiol 137:1027–1036

Ohlsson AB, Djerbi S, Winzell A, Bessueille L, Ståldal V, Li X, Blomqvist K, Bulone V, Teeri TT, Berglund T (2006) Cell suspension cultures of *Populus tremula* × *P. tremuloides* exhibit a high level of cellulose synthase gene expression that coincides with increased in vitro cellulose synthase activity. Protoplasma 228:221–229

Ohrnberger D (1999) The bamboos of the world: annotated nomenclature and literature of the species and the higher and lower taxa. Elsevier, Amsterdam

Ogita S (2005) Callus and cell suspension culture of bamboo plant, *Phyllostachys nigra*. Plant Biotechnol 22:119–125

Ogita S (2015) Plant cell, tissue and organ culture: the most flexible foundations for plant metabolic engineering applications. Nat Prod Commun 10:815–820

Ogita S, Kashiwagi H, Kato Y (2008a) In vitro node culture of seedlings in bamboo plant, *Phyllostachys meyeri* McClure. Plant Biotech 25:381–385

Ogita S, Ohki S, Kato Y (2008b) Uptake of carbohydrates by suspension cultured cells of Bamboo plants In: Teixeira da Silva JA (ed) Floriculture, ornamental and plant biotechnology 5. Global Science Books, UK, pp 240–244

Ogita S, Kikuchi N, Nomura T, Kato Y (2011) A practical protocol for particle bombardment-mediated transformation of *Phyllostachys* bamboo suspension cells. Plant Biotechnol 28:43–50

Ogita S, Nomura T, Kishimoto T, Kato Y (2012a) A novel xylogenic suspension culture model for exploring lignification in *Phyllostachys* bamboo. Plant Methods 8:40. doi:10.1186/1746-4811-8-40

Ogita S, Ohki S, Nomura T, Kato Y (2012b) A β-glucosidase activity potentially involved in cell division and wall development of *Phyllostachys* bamboo suspension cells. Am J Plant Sci 3:1066–1072

Peng Z, Lu Y, Li L, Zhao Q, Feng Q, Gao Z, Lu H et al (2013) The draft genome of the fast-growing non-timber forest species moso bamboo (*Phyllostachys heterocycla*). Nat Genet 45:456–461

Rai V, Ghosh JS, Pal A, Dey N (2011) Identification of genes involved in bamboo fiber development. Gene 478:19–27

Taylor-Teeples M, Lin L, de Lucas M, Turco G et al (2014) An Arabidopsis gene regulatory network for secondary cell wall synthesis. Nature 517:571–575

The Arabidopsis Genome Initiative (2000) Analysis of the genome sequence of the flowering plant *Arabidopsis thaliana*. Nature 408:796–815

The International Brachypodium Initiative (2010) Genome sequencing and analysis of the model grass *Brachypodium distachyon*. Nature 463:763–768

Wang K, Peng H, Lin E, Jin Q, Hua X, Yao S, Bian H, Han N, Pan J, Wang J, Deng M, Zhu M (2010) Identification of genes related to the development of bamboo rhizome bud. J Exp Bot 61:551–561

Willemsen V, Bauch M, Bennett T, Campiho A, Wolkenfelt H, Xu J et al (2008) The NAC domain transcription factors FEZ and SOMBRERO control the orientation of cell division plane in *Arabidopsis* root stem cells. Dev Cell 15:913–922

Yang W, Duan L, Chen G, Xiong L, Liu Q (2013) Plant phenomics and high-throughput phenotyping: accelerating rice functional genomics using multidisciplinary technologies. Curr Opin Plant Biol 16:180–187

Yoshida S, Iwamoto K, Demura T, Fukuda H (2009) Comprehensive analysis of the regulatory roles of auxin in early transdifferentiation into xylem cells. Plant Mol Biol 70:457–469

Zhang Y-J, Ma P-F, Li D-Z (2011) High-throughput sequencing of six bamboo chloroplast genomes: phylogenetic implications for temperate woody bamboos (Poaceae: Bambusoideae). PLoS ONE 6:e20596. doi:10.1371/journal.pone.0020596

Zhong R, Demura T, Ye ZH (2006) SND1, a NAC domain transcription factor, is a key regulator of secondary wall synthesis in fibers of *Arabidopsis*. Plant Cell 18:3158–3170

Chapter 8
Cotton Fiber Biotechnology: Potential Controls and Transgenic Improvement of Elongation and Cell Wall Thickening

Michael R. Stiff, J. Rich Tuttle, Benjamin P. Graham and Candace H. Haigler

Abstract Cotton is grown on five continents as an economically important crop. Its long, fine, seed fibers are one of the most highly used natural fibers, providing a high-quality spinnable fiber to the textile industry. The cotton fiber undergoes a complex, staged developmental program, resulting in a single cell that is 1.8–5 cm long with a thick wall composed of about 95 % cellulose. Biotechnological improvements have either directly or indirectly enhanced the fiber properties that are important for spinning, including length, bundle strength, and maturity. These experiments have generally targeted carbohydrate metabolism, cell wall structure, and hormone signaling. In this chapter, we present a brief review of cotton fiber development with a focus on processes affecting elongation and cell wall thickening. We discuss rigorous criteria for evaluating studies on transgenic cotton fiber and mention the challenges of performing such research in the public sector. We highlight selected genetic engineering experiments that have resulted in improved cotton fiber quality and discuss future prospects for use of biotechnology to improve cotton fiber and its competitiveness with synthetic fibers.

Keywords Cotton fiber development · Fiber improvement · Fiber quality · Primary cell wall · Secondary cell wall

M.R. Stiff · J.R. Tuttle · C.H. Haigler (✉)
Department of Crop Science, North Carolina State University,
Raleigh, NC 27695, USA
e-mail: candace_haigler@ncsu.edu

B.P. Graham · C.H. Haigler
Department of Plant and Microbial Biology, North Carolina State University,
Raleigh, NC 27695, USA

K.G. Ramawat and M.R. Ahuja (eds.), *Fiber Plants*, Sustainable Development and Biodiversity 13, DOI 10.1007/978-3-319-44570-0_8

8.1 Introduction

Cotton (*Gossypium* spp.) seed fibers are among the most widely used natural fibers, providing renewable, high-quality fiber to the textile industry. The worldwide cotton fiber crop accounts for approximately one-third of all fibers used in clothing (Shui and Plastina 2013). The long, spinnable "lint" fibers reach lengths of up to 5 cm depending on species and cultivar, while much shorter "fuzz" (or linter) fibers often remain on the seed coat after ginning. Cotton fiber differentiation includes several overlapping developmental stages: initiation, elongation, transitional cell wall remodeling, secondary cell wall synthesis, and maturation. Fiber morphogenesis begins approximately when the flower blooms, referred to as 0 days post-anthesis (DPA), and continues for at least 50 days until the fruit (or boll) opens (Stiff and Haigler 2012; Haigler et al. 2012; Kim 2015).

Biological processes that influence cell wall properties during development largely determine final fiber quality (Haigler et al. 2012). Key fiber quality parameters include length, strength, and fineness, and these factors affect the selling price and final use of a bale. Long fiber with higher uniformity is preferred for modern spinning and processing technologies that lead to high-quality textiles. The lengthening of the fiber results from balancing new primary wall synthesis, wall extensibility, and high turgor in the growing cell. High strength helps to preserve fiber length during processing and enhances the production of the strong yarns need for weaving smooth textile fabrics. Both the "winding" cell wall layer that is deposited during transitional cell wall remodeling and the cellulose-rich secondary wall make positive contributions to fiber strength. Fiber fineness (weight per unit length) is determined by the extent of secondary wall thickening and fiber perimeter. Smaller perimeter allows more fibers to be included in a silky strong yarn for high-quality fabrics. Fiber quality as measured at the point of sale must fall within an acceptable range in order for the producer to receive the best price per pound. Occasionally, price premiums are available for premium fiber, which tends to be progressively defined at a higher standard to reflect breeding successes. Therefore, the continuing improvement of traditional fiber qualities and future innovations of novel fiber properties are essential to keep cotton fiber competitive in the marketplace with synthetic fiber (Wakelyn et al. 2007; Haigler et al. 2012; Constable et al. 2015).

8.1.1 Cotton Breeding for Fiber Improvement

A recent review by Constable and coworkers provides an excellent overview of the cotton industry, cotton improvement strategies, and fiber quality (Constable et al. 2015). The *Gossypium* genus, inclusive of more than 50 wild and domesticated species, is well characterized (Wendel et al. 2009). More detailed information related to the overview below can be found in these sources. Cotton fiber was used to make thread for stringing beads more than 8000 years ago in present-day Pakistan. Cotton species from this region were domesticated as the basis for the *G. arboreum* or *G. herbaceum* (diploid A genome) fiber crops grown to a minor extent in Asia (Moulherat et al. 2002). The wild forms of the diploid D-genome *Gossypium* species grow in Central America, including *G. raimondii* that provided the first cotton genome sequence (Paterson et al. 2012). The D-genome cotton fibers are rudimentary, non-spinnable, and therefore not useful to humans. The allotetraploid species, including the current commercial species *G. hirsutum* (upland cotton) and *G. barbadense* (Pima, Sea Island, or Egyptian cotton), originated and were domesticated in Central and South America long before the arrival of European explorers. Humans used cotton to make products more than 5000 years ago in modern-day Peru, near the point of origin of *G. barbadense* (Dillehay et al. 2007). The AD allotetraploid genome is composed of the A and D sub-genomes. The allotetraploid species have longer and stronger fibers and higher yield than their diploid progenitors, and they were the primary focus of domestication and breeding in modern times.

After 1900, cotton producers began to apply theories of genetics and Mendelian inheritance to select desirable traits that could increase productivity through cotton breeding. The qualities most sought after by cotton producers were cotton fiber length and strength. In the middle twentieth century, cotton fiber fineness became a target for enhancement through breeding (May 2002). Selection of traits for temperature tolerance, early boll development, high germination rate, and pest resistance made it possible to produce regional cultivars and improved overall yield. In the 1950s, growers began crossing wild relatives with *G. hirsutum* to improve fiber quality by incorporating previously uncultivated genetic material with existing commercial cultivars (Meyer 1974). Collection of germplasm from around the world, introgression programs, traditional breeding within a species, and marker-assisted selection continue in an effort to improve commercial cotton.

As summarized for recent years (Cotton Incorporated 2014), these efforts have enhanced cotton fiber length from an average of about 1.07–1.14 in. (27–29 mm) between 1985 and 2013. Average fiber bundle strength for upland cotton grown in the USA increased from 29.25 to 30.25 g/tex between 2003 and 2013. In the same time period, micronaire increased from approximately 4.3–4.5 on average, which is not necessarily a desirable change for cotton fiber (micronaire is a unitless industrial measurement that is affected by both fiber perimeter and maturity and of these two only an increase in maturity could potentially be desirable). These values are similar to those reported on average for four continents, along with indications that fiber quality improvement through traditional methods has recently slowed down (Constable et al. 2015).

Table 8.1 Biological processes affecting elongation and secondary wall thickening during cotton fiber development

Biological processes	Affected fiber qualities	References
Elongation stage		
Hormones (growth factors)		
Gibberellins	Promotes length	Beasley and Ting (1973), Aleman et al. (2008), Xiao et al. (2010)
Brassinosteroids	Promotes length	Sun et al. (2005), Luo et al. (2007)
Ethylene	Promotes length	Shi et al. (2006), Pang et al. (2010)
Abscisic acid	Inhibits length	Beasley and Ting (1973), Dhindsa et al. (1976), Kosmidou-Dimitropoulou (1986), Nayyar et al. (1989), Dasani and Thaker (2006)
Phytosulfokine	Promotes length	Han et al. (2014)
Second messengers		
Basal H_2O_2 production	Promotes length	Potikha et al. (1999), Yang et al. (2008), Li et al. (2007)
Basal O_2^- production	Promotes length	Mei et al. (2009)
Basal Ca^{2+}	Promotes length	Huang et al. (2008), Taliercio and Haigler (2011)
Excess Ca^{2+}	Inhibits length	Taliercio and Haigler (2011)
Secondary metabolites		
VLCFAs	Promotes length	Qin et al. (2007), Pang et al. (2010)
Flavonoids	Inhibits length	Tan et al. (2013)
Cytoskeleton		
Microtubules	Influences cellulose microfibril orientation and fiber shape; promotes length	Seagull (1990, 1995)
Actin microfilaments	Influences microtubule orientation and fiber shape; promotes length	Seagull (1990), Wang et al. (2005), Li et al. (2005)
Primary walls		
Pectic precursors	Promotes length	Pang et al. (2010)
Pectate lyase activity	Promotes length	Wang et al. (2010a)

(continued)

Table 8.1 (continued)

Biological processes	Affected fiber qualities	References
Cell wall thickening		
Hormones		
Auxins	Alters length and timing of secondary wall synthesis	Singh et al. (2009b)
Second messengers		
Excess H_2O_2	Decreases length, early secondary wall synthesis	Potikha et al. (1999), Yang et al. (2008)
Cytoskeleton		
GhPFN2, profilin 2	Decreases length, early secondary wall synthesis	Wang et al. (2010b)
Temperature		
Cooler temperature	Alters length and strength	Hinchliffe et al. (2010, 2011)

There is room for further improvement in cotton fiber quality using targeted genetic engineering strategies. For example, fiber strength and length have been enhanced through breeding, while cotton fiber perimeter has remained largely unchanged. Fiber perimeter is the primary contributing factor to fiber fineness, and lower perimeter fibers can be used to produce stronger, silkier, and softer textiles (Naylor et al. 2014). *G. barbadense* has longer and finer fiber at the cost of high yield as compared to *G. hirsutum*. Today, *G. hirsutum* accounts for 90 % of fiber production worldwide (Constable et al. 2015). Ultimately, the cotton industry will benefit from a cultivar that combines high fiber quality as currently found in *G. barbadense* with high yield in diverse growing environments as currently found in *G. hirsutum.*

In this chapter, we discuss the application of biotechnology to manipulate fiber elongation and cell wall thickening for improvement of fiber quality. To provide biological context for these studies, we first review the cellular processes known to directly affect the elongation and secondary cell wall thickening of cotton fibers (summarized in Table 8.1). Emphasis is placed on data emerging from the modern domesticated allotetraploid species *G. hirsutum*, although comparative studies and other species are mentioned. Unless otherwise noted, all the data descriptions are for *G. hirsutum*. In the following section, we describe stringent criteria for evaluating reports of current applications of biotechnology for fiber improvement. Selected experiments that targeted carbohydrate metabolism, cell wall synthesis, and hormone homeostasis are critically analyzed (summarized in Table 8.2). We conclude with future prospects for biotechnological improvement of cotton fiber quality.

Table 8.2 Transgenic modifications that improved cotton fiber quality

Gene name	Biological function	NCBI accession	Method	Length	Bundle strength	Micronaire	References
Carbohydrate and cell wall							
SusA1	Sucrose cleavage	HQ702185	OE	+8 %	+6 to 11 %	ND	Jiang et al. (2012)
SPS	Sucrose phosphate synthase	L04803	OE	+14 %	NS	+12 %	Haigler et al. (2007)
XTH1	Xyloglucan endotransglycosylase/hydrolase	HM749062	OE	+10 to 14 %	NS	NS	Lee et al. (2010)
UGP1	UDP-glucose pyrophosphorylase	FJ415164	OE	+15 %	+12 %	NS	Li et al. (2015)
PRP5	Secreted proline-rich protein	EF095706	RNAi	+6 %	ND	ND	Xu et al. (2013)
Cytoskeleton							
WLIM1a	LIM-domain protein, actin bundler, transcription factor	JX648310	OE	+10 %	+9 %	−20 %	Han et al. (2013)
Signaling							
BRI1	Brassinosteroid receptor	BG441661	AS	NS	NS	−13 %	Sun et al. (2015)
Ga20ox1	Gibberellic acid oxidase	FJ623272	OE	+14 %	ND	ND	Xiao et al. (2010)
PHYA1	Phytochrome red/far-red photoreceptor	HM143735	RNAi	+5 %	NS	−4 to 8 %	Abdurakhmonov et al. (2014)

All genes were from *G. hirsutum*, except SPS was from spinach

ND not determined, *NS* no significant difference, *OE* overexpressed, *RNAi* RNA interference, *AS* antisense suppression

8.2 Cellular Processes Affecting Elongation and Secondary Wall Signaling

Here, we discuss selected aspects of cotton fiber morphogenesis. More extensive information can be found in other reviews (Stiff and Haigler 2012; Haigler et al. 2012; Kim 2015).

8.2.1 Cotton Fiber Elongation

Fiber length is biologically determined during the period of polar cell growth known as elongation. The onset of elongation is recognized by the tapering of a large proportion of fibers around 2 DPA (Farr 1931; Stewart 1975). Fibers begin to wind around one another to form bundles joined together by the cotton fiber middle lamella (CFML; Singh et al. 2009a; Avci et al. 2013). The greater level of organization afforded by this bundling likely aids the relatively uniform elongation of the >10,000 fibers on the ovule surface (Seagull 1998; Zhang et al. 2011). Elongation ends around 20 DPA as the CFML is enzymatically degraded, the primary wall begins to remodel and secondary wall synthesis commences.

Turgor changes dynamically as the fiber elongates (Ruan et al. 2001). Indirect evidence supports the contribution of cotton vacuolar invertase (*GhVIN1*) to high osmotic pressure within the vacuole during early elongation. *GhVIN1* expression and activity levels correlated positively with higher hexose levels during early fiber elongation (<10 DPA). The fiber of *G. barbadense* showed higher GhVIN1 activity from 5 to 20 DPA in correlation with its longer length as compared to *G. hirsutum* (Wang et al. 2010c). Rapid fiber elongation from 10 to 16 DPA is facilitated by a transient turgor increase, which is correlated with expression of sucrose and potassium transporters and closure of plasmodesmata at the fiber base (Ruan et al. 2001, 2004; Ruan 2007).

8.2.1.1 Phytohormone Signaling

The effect of phytohormone signaling on fiber elongation has been studied for decades, especially through the use of cotton ovules cultured in vitro (Beasley and Ting 1973; Liao et al. 2010). Gibberellins (GAs), brassinosteroids (BRs), ethylene (ET), and abscisic acid (ABA) have been shown to influence fiber elongation when added to in vitro cotton ovule culture. In addition, endogenous hormone levels correlate with differences in length, and many genes involved in hormone biosynthesis or perception have been characterized in cotton fiber as discussed further below.

GA is an integral component of the medium in in vitro cultures of *G. hirsutum* fibers (Beasley and Ting 1973). More recently, GA_3 addition to 2 DPA cultured ovules followed by analysis at 12 DPA showed enhanced ovule growth, fiber elongation, and the expression of genes related to fiber elongation such as expansion and xyloglucan endotransglucosylase/hydrolase (*EXP*, *XTH1*, and *XTH2*, Aleman et al. 2008). The authors identified cotton homologs for rice GA receptor genes (*OsGID1*, *GhGID1a*, and *GhGID1b*) and DELLA-type genes (*GhSLR1a* and *GhSLR1b*). The latter repress GA-mediated effects, permitting fine control of GA responses. The proposed molecular functions for *GhGID1a* and *GhSLR1b* were supported by mutant rescue and overexpression (leading to a dwarf phenotype), respectively, in Arabidopsis. GA function during fiber elongation correlates with a peak in in vivo levels at 10 DPA (Xiao et al. 2010).

Evidence for BR activity during fiber elongation arose from the effects of BR on fiber growth in ovule culture. The addition of brassinolide, the most biologically active BR, to ovule culture medium promoted fiber length by 12.7 % after 14 days in culture and promoted elongation-associated gene expression (*EXP*, *XTH*, *AGP*, and *GhTUB1*). Inhibition of BR biosynthesis using brassinazole 2001 reduced fiber length (Sun et al. 2005). Similarly, another BR biosynthesis inhibitor, finasteride, reduced fiber length (Luo et al. 2007).

ET participates in fruit ripening, flower development, and stress responses (Bleecker and Kende 2000; Merchante et al. 2013). In cotton fiber, ET addition to cultured ovules promoted fiber length while inhibition of ET biosynthesis by L-(2-aminoethoxyvinyl)-glycine (AVG) resulted in shorter fibers in a concentration-dependent manner. Analyses with microarrays and polymerase chain reaction showed enrichment of ET biosynthesis gene expression (*ACO1*, *ACO2*, and *ACO3*) from 5 to 15 DPA, further supporting the participation of ET in fiber elongation (Shi et al. 2006). ET addition also induced expression of pectin biosynthesis genes and proteins, while the addition of pectin precursors could reverse the short fibers induced by ET biosynthesis inhibition (Pang et al. 2010). Fiber growth effects mediated by ET are regulated by the upstream expression of the *GhMYB109*, a transcription factor that is structurally similar to regulators of Arabidopsis trichome development, *AtGL1* and *AtWER*. The expression of*GhMYB109* was fiber-specific, and suppressing its expression with antisense technology resulted in shorter fibers with decreased expression of genes related to ET biosynthesis (*GhACO1* and *GhACO2*) and the cytoskeleton (*GhTUB1* and *GhACT1*) (Pu et al. 2008).

ET effects in fiber provide an example of possible coordinated signaling among other hormones and second messengers. ET treatment reversed the effect of BR biosynthesis inhibition, leading to longer fibers than untreated controls. Treatment with either phytohormone promoted expression of genes involved in the biosynthesis of the other. There was no additive effect on fiber length when ET and BR were used together, supporting their participation within an integrated signaling pathway (Shi et al. 2006). ET treatment of cultured ovules increased hydrogen peroxide (H_2O_2) in elongating fibers, which also exhibited higher ascorbate peroxidase activity and fiber length in the presence of exogenous H_2O_2 (Li et al. 2007). Likewise, cultured ovules treated with H_2O_2 exhibited higher ET production, which inspired a model for ET stimulation of H_2O_2 production and fiber elongation while H_2O_2 reinforces additional ET synthesis (Qin et al. 2008). In another example of integrated ET signaling, fiber ET levels were shown to be sensitive to extracellular adenosine triphosphate (ATP). Elongating fibers release ATP into growth medium that must be hydrolyzed by exogenous apyrases (GhAPY1 and 2) to permit fiber growth (Clark et al. 2009). Low concentrations of poorly hydrolysable nucleoside triphosphate (ATPγS) stimulated fiber elongation, while high concentrations inhibited it. The participation of ET in exogenous ATP perception/signaling was supported by the following: (1) the loss of ATPγS growth effects when ET biosynthesis was blocked, (2) increased ET production after ATPγS treatment, and (3) lowered ATPγS concentration required to promote fiber length when the ET precursor, 1-aminocyclopropane-1-carboxylic acid, was added to the medium (Clark et al. 2009). Finally, fiber ET levels were increased by very-long-chain fatty acids (VLCFA), which are required for and promote fiber elongation (Qin et al. 2007).

Abscisic acid (ABA) inhibits fiber elongation in ovule culture (Beasley and Ting 1973; Dhindsa et al. 1976; Kosmidou-Dimitropoulou 1986) as it does in other plant cells (Lee et al. 1994). To determine the stage at which fibers were most sensitive to ABA, *G. hirsutum* ovules were grown in reciprocal transfer cultures: 0 DPA ovules were grown in medium with or without ABA then transferred to the reciprocal medium at each day up to 14 DPA. Within this interval, *G. hirsutum* fibers were sensitive to the addition of ABA between 0–4 DPA (Dhindsa et al. 1976), corresponding to initiation and the earliest stages of elongation. Elongation of the domesticated diploid *G. arboreum* fiber in ovule culture was improved by the addition of the ABA biosynthesis inhibitor fluridone (Nayyar et al. 1989). Decreased fiber length was correlated with increasing ABA concentration on ovules cultured from a short, medium, and long fiber genotype (Dasani and Thaker 2006). The in vivo ABA concentrations of field-grown fiber were lower during elongation than in later stages of development for all three genotypes.

Phytosulfokine-α (PSK-α) is a growth-stimulating pentapeptide growth factor (Sauter 2015). Its expression peaks at 10–15 DPA in rapidly elongating cotton fiber, and it promoted a 27 % increase in fiber length by 10 DPA when added along with GA_3 to the ovule culture. The addition of IAA had a synergistic effect, leading to maximum fiber length in culture. When PSK-α was constitutively overexpressed in cotton, three transgenic lines had 6–7 % longer fiber with 5–6 % lower micronaire as compared to wild type. However, no reduction in fiber length was observed when an RNAi strategy was used to reduce the level of PSK-α (Han et al. 2014).

8.2.1.2 Ca^{2+} and Reactive Oxygen Species as Second Messengers

Second messengers are intermediary molecules in signal transduction pathways that serve to amplify the signal and generate opportunities for fine control of signal-induced responses and cross talk between diverse signaling pathways. One such second messenger, calcium or Ca^{2+}, has broad-ranging roles in plant development and stress signaling (Kudla et al. 2010). Ca^{2+} and the activity of calmodulin were required for fiber elongation in ovule culture (Huang et al. 2008). High concentrations of Ca^{2+} (1 mM) in culture medium reduced the expresin of expansin and fiber length (as compared to 0.1 mM Ca^{2+}), with sensitivity to Ca^{2+} addition observed at 0–3 DPA. Experiments with Ca^{2+} signaling antagonists were consistent with the effect occurring through Ca^{2+} signaling pathways rather than pectin cross-linking (Taliercio and Haigler 2011).

The homeostasis of reactive oxygen species (ROS) also influences fiber elongation. ROS act as signaling molecules or cause oxidative damage at high concentrations (Bhattacharjee 2012; Sharma et al. 2012). Transcriptomic and proteomic analyses point to improved capacity for ROS management as a key factor in the domestication of long-fibered cottons (Hovav et al. 2008; Chaudhary et al. 2009; Hu et al. 2013; Yoo and Wendel 2014). Similarly, longer fiber is correlated with enhanced capacity for ROS management in a commercial cultivar of *G. barbadense* as compared to *G. hirsutum* (Tuttle et al. 2015).

ROS in lower concentrations have been experimentally shown to positively influence fiber growth, similarly to the stimulation of elongation in maize roots (Liszkay et al. 2004), Arabidopsis root hairs (Monshausen et al. 2007), and tobacco pollen tubes (Potocký et al. 2007). Hydrogen peroxide (H_2O_2) is the most stable ROS and acts as a transmissible signaling molecule. H_2O_2 accumulates at low levels during early cotton fiber elongation until it peaks at 20 DPA (Potikha et al. 1999; Yang et al. 2008). Addition of H_2O_2 to cultured ovules induced expression of an ascorbate peroxidase, *GhAPX1*, within 6 h and longer fibers after 6-day growth compared to untreated ovules (Li et al. 2007). Ascorbate peroxidases use ascorbate to reduce H_2O_2 to water, thereby helping to prevent or moderate the damaging effects of high ROS concentrations. Other peroxidases generate ROS in the form of superoxide ($O_2^{\cdot-}$), and their activity is necessary for fiber growth. Cultured ovules treated with inhibitors of NADPH oxidase or peroxidase activity produced less $O_2^{\cdot-}$ and shorter fibers in a concentration-dependent manner (Mei et al. 2009). Expression of *GhPOX1* correlated with peak peroxidase activity during late fiber elongation. This Class III plant peroxidase is homologous to AtPOX13, which is required for Arabidopsis lateral root initiation and elongation. GhPOX1 may promote fiber elongation through the production of a basal level of ROS (Mei et al. 2009).

8.2.1.3 Secondary Metabolites

Secondary metabolites are not required for the basic life processes of plants, but they help to distinguish cell and tissue types and aid in the control of differentiation, adaption, and/or defense. In this class, both very-long-chain fatty acids (VLCFAs) and flavonoids directly affect fiber elongation. VLCFAs, with aliphatic chains greater than 22 carbons long, are constituents of sphingolipids and waxes. Endogenous VLCFA and transcripts for their biosynthesis (*KCS2*, *KCS6*, *KCS12*, and *KCS13*) accumulated during fiber in vitro elongation (Qin et al. 2007). Addition of VLCFA (C24:0 and C26:0) to cultured ovules stimulated the accumulation of ET and sphingolipids and increased fiber length. Inhibiting VLCFA biosynthesis using

2-chloro-*N*-[ethoxymethyl]-*N*-[2-ethyl-6-methyl-phenyl]-acetamide led to a concentration-dependent decrease in cultured fiber length. Adding exogenous VLCFA or ET mitigated this negative effect, and the data supported ET acting downstream of VLCFA to promote fiber elongation (Qin et al. 2007).

Comparisons of flavonoid biosynthesis gene expression and accumulating flavonoids showed that the higher quality fiber of *G. barbadense* accumulated fewer flavonoids and in a different pattern compared to *G. hirsutum* fiber during early elongation (<10 DPA, Tan et al. 2013). From a panel of flavonoids tested by Tan and coworkers, naringenin and dihydrokampferol significantly reduced fiber growth in ovule culture. A high concentration of dihydrokaempferol was also detected in the shorter fiber of *G. hirsutum* as compared to *G. barbadense* (Tuttle et al. 2015). Flavanone 3-hydoxylase (F3H) is responsible for the conversion of naringenin to dihydrokaempferol in plants. Silencing *F3H* by RNAi resulted in greater accumulation of naringenin and eriodictyol with a concurrent decrease in dihydrokaempferol among other flavonoids in the downstream pathway. Both field- and greenhouse-grown F3H silenced T3 plants produced significantly shorter fiber with lower micronaire. Short fiber and lower micronaire persisted in backcrossed brown-fibered progeny. The short fiber phenotype could not be rescued by treatment with downstream flavonoids in ovule culture, demonstrating a specific negative effect of naringenin accumulation. In non-targeted metabolomic analysis of 10–28 DPA *G. hirsutum* and *G. barbadense* cultivars, naringenin was not detected but the higher quality *G. barbadense* fiber contained a higher concentration of naringenin 7-O-glucoside (Tuttle et al. 2015). Analogously to animal cells (Han et al. 2012), this flavanone molecule may contribute to the superior ROS management capacity of *G. barbadense* fiber, which correlates with its extended elongation period and greater final fiber length as compared to *G. hirsutum* fiber (Tuttle et al. 2015).

8.2.1.4 Primary Wall Synthesis and Structure

New primary wall synthesis is required to support plant cell elongation. The primary wall of elongating cotton fiber is typical of other expanding cells, being composed of about 16 % para-crystalline (ß-1, 4-glucan) cellulose fibrils embedded in a matrix of protein and other polysaccharides, especially pectin and xyloglucan (Meinert and Delmer 1977; Avci et al. 2013; Lee et al. 2015). Cellulose microfibrils are synthesized transversely to the fiber axis, providing strong resistance to turgor pressure to constrain fiber perimeter as the fiber lengthens. Elements of the cytoskeleton, microtubules, and actin microfilaments, contribute to cell wall synthesis by regulating aspects of cellulose synthesis and the delivery of cell wall matrix polymers to the exoplasmic space where cellulose microfibrils and the nascent cell wall are forming. Microtubules control the orientation of cellulose microfibrils, with relevant mechanisms during primary wall synthesis in Arabidopsis having been recently revealed (Lei et al. 2014). Both microtubules and cellulose microfibrils are oriented transversely to the fiber axis during cotton fiber elongation, corresponding to the highly anisotropic expansion of the cell (Seagull 1993). Inhibition of microtubule assembly in elongating cultured fibers led to less ordered microfibril orientation (Seagull 1990), while microtubule stabilization hindered their reorientation to the helical pattern characteristic of the transition stage. Actin microfilaments in cotton fiber are oriented axially or parallel to the transverse cortical microtubules during fiber elongation (Seagull 1993). Disruption of actin microfilaments with cytochalasin D induced early reorientation of microtubules from transverse to helical orientation (Seagull 1990) and resulted in shorter fibers with ruffled rather than smooth cell walls (Wang et al. 2005). Possibly the changed cell wall morphology is relevant to the role of actin in modulating the insertion and lifetime of cellulose synthases in the plasma membrane, as demonstrated for Arabidopsis hypocotyl cells (Sampathkumar et al. 2013). Suppression of *GhACT1* expression by RNAi reduced fiber length at 0–3 DPA in transgenic cotton, although fiber lengths from later days during elongation were not reported (Li et al. 2005).

Pectins are galacturonic acid-rich polysaccharides that include homogalacturonan and rhamnogalacturonan I, each of which can have a variety of substituent side chains (Mohnen 2008). Xyloglucan is a ß-1, 4 glucan with side chains made of α-1, 6 xylose and sometimes galactose and fucose (O'Neill and York 2003). The largely non-covalent interactions among cellulose and matrix polysaccharides that contribute to strength and developmental flexibility of the primary wall remain under intense investigation (e.g., Abasolo et al. 2009; Boyer 2009; Park and Cosgrove 2012). Early elongating fiber has an outer pectin sheath surrounding an inner wall layer enriched in cellulose and xyloglucan (Vaughn and Turley 1999; Bowling et al. 2011). Matrix polysaccharide distribution within fiber walls changes with development as shown by immunolabeling of cell wall polysaccharide epitopes. Early elongating fibers (2 DPA) lost a characteristic epitope of (1–6)-*ß*-D-galactan carrying arabinose and gained an epitope representative of (1–4)-*ß*-D-galactan, modifications associated with a developmental change in rhamnogalacturonan I (Bowling et al. 2011).

Cultured ovules treated with precursors of pectin synthesis (UDP-L-rhamnose, UDP-D-galacturonic acid, or UDP-D-glucuronic acid) produced longer fibers than untreated ovules (Pang et al. 2010). Pectin biosynthesis was connected to a regulatory process involving VLCFA and ET, which increased the expression of pectin biosynthesis genes and promoted fiber elongation (Qin et al. 2007). The short fiber effect of inhibiting ET biosynthesis in ovule culture was reversed by the addition of UDP-rhamnose or UDP-galacturonic acid. Similar stimulatory effects on root hair length from Arabidopsis mutants defective in pectin biosynthesis, VLCFA biosynthesis, or ET signaling further supported the interaction of these pathways to promote cell elongation (Pang et al. 2010).

Cell wall extensibility may also be modulated by the enzymatic modification of polysaccharides already present in the cell wall (Wolf and Greiner 2012). Xyloglucan is a common matrix polymer within dicot primary cell walls. Cellular expansion may be aided by the activity of xyloglucan endotransglycosylase/ hydrolase (XET/XTH) enzymes that join newly secreted and older xyloglucan polymers together or cleave xyloglucan to increase cell wall plasticity (Cosgrove 2005), and positive effects on cotton fiber elongation are discussed in Sect. 8.3. Homogalacturonan is often secreted and integrated into the cell wall with a relatively high degree of methyl-esterification. Pectin methylesterase (PME) may subsequently remove a number of methyl groups and facilitate wall rigidification through more extensive Ca^{2+} cross-linking between homogalacturonan molecules. Demethylated pectin is also susceptible to degradation by pectate lyase (PEL) to increase wall extensibility. A role for pectate lyase in stimulating fiber elongation was demonstrated through use of antisense technology to down-regulate *GhPEL* using the fiber-specific E6 promoter (Wang et al. 2010a). Transformants exhibited decreased expression of *GhPEL* and lower pectate lyase activity during elongation (10 and 15 DPA) as well as shorter fibers compared to wild type. These results correlated with increased amounts of de-methylated pectin in *GhPEL*-suppressed fiber, highlighting the importance of pectin modification for sustained fiber growth.

8.2.2 Secondary Wall Thickening

The thickening of the cotton fiber cell wall includes synthesis of the relatively thin winding layer followed by the deposition of the thick secondary cell wall, which contains mainly cellulose. Major structural changes occur in the thickening cell wall, and these can be detected with analytical methods sensitive to cell wall polymer composition and/or cellulose content and structure (Hsieh 1999; Singh et al. 2009a; Abidi et al. 2010; Avci et al. 2013; Lee et al. 2015). Fiber wall thickening is essential for the development of fiber strength and maturity, which describes the degree of filling of the fiber interior space with cell wall material as the secondary wall is deposited. A strong fiber is fundamentally important for spinning yarn, and stronger fibers also retain greater length after processing. The degree of fiber maturity must fall within a middle range that optimizes strength and dyeing potential while still allowing the fibers to collapse into the "kidney bean" shape required for yarn spinning (Haigler et al. 2005).

Transitional cell wall remodeling is a distinct developmental stage during which the cotton fiber transcriptome and metabolome are different and relatively stable as compared to the primary and secondary wall stages (Tuttle et al. 2015). Multifaceted changes in cell wall structure occur during this stage, which lasts several days before massive secondary wall thickening begins. For example, transitional cell wall remodeling persists between 18 and 21 DPA under moderate temperature greenhouse conditions (Tuttle et al. 2015). The cellular events occurring during transitional cell wall remodeling have been reviewed previously (Stiff and Haigler 2012; Haigler et al. 2012) and include the following: (a) reorientation of microtubules and microfibrils from transverse to a shallow helical arrangement relative to the fiber axis, (b) degradation of selected primary wall polymers, (c) synthesis of the winding layer with composition similar to the primary wall, except for an increased concentration of cellulose, and (d) various shifts in the fiber metabolic state. The "ply" structure conferred by the web-like winding layer, with its differently oriented cellulose microfibrils as compared to the primary wall, substantially increases fiber strength even though this intermediary cell wall layer is relatively thin (Hsieh 1999). Pectin remodeling by pectin methylesterase (PME) may also contribute to differences in length and strength between cultivars (Liu et al. 2013). When compared to *G. hirsutum*, a *G. barbadense* cultivar had longer, stronger fiber and (1) expressed higher levels of *PME4*, (2) had higher PME activity, and (3) contained more demethylesterified pectin during transitional cell wall remodeling.

The main part of secondary wall thickening is controlled by a genetic program adapted from an ancient one required for the synthesis of xylem and other sclerenchyma cells in vascular plants (Haigler et al. 2009; Betancur et al. 2010; Hinchliffe et al. 2010; Zhong et al. 2010; Tuttle et al. 2015). However, the ancient secondary wall genetic program has been changed in a major way in cotton fiber due to the repression of hemicellulose and lignin synthesis, with the repression of lignin arising through transcriptional control (Tuttle et al. 2015). The molecular regulation of SCW thickening in cotton fiber occurs differently than in Arabidopsis leaf trichomes, which have thick cell walls with a composition dissimilar to cotton fiber (Marks et al. 2008, Betancur et al. 2010). The shift to synthesis of the cellulose-rich secondary wall in cotton fiber occurs along with extensive down-regulation of overall transcript levels at 28 DPA as compared to 21 DPA, which represents the end of transitional cell wall remodeling. Conversely, numerous metabolites become more concentrated when high-rate cellulose synthesis dominates at 28 DPA (Tuttle et al. 2015). Secondary wall thickening via intensive cellulose synthesis lasts about 20 days, e.g., from 22 to 45 DPA under moderate temperature greenhouse conditions (Lee et al. 2015). The major part of cotton fiber SCW thickening occurs while both microtubules and cellulose microfibrils adopt an acute angle relative to the fiber axis (Seagull 1993), and this angle affects the resistance of cotton fiber to breaking upon stretching. Although about 95 % of the cotton cell wall is cellulose on a weight basis, other components such as callose are also associated with the secondary wall (Salnikov et al. 2003; Avci et al. 2013).

Fewer attempts have been made to manipulate cotton fiber thickening experimentally as compared to elongation, possibly because documenting the relevant phenotypes is more difficult. The use of a synthetic auxin (naphthalene acetic acid, NAA) delayed the onset of high-rate cellulose synthesis by at least 6 days in ovule culture, as compared to natural auxin (indoleacetic acid, IAA). This effect was shown through imaging the microfibril angle and the relative levels of secondary wall cellulose birefringence and profiling the expression of selected secondary wall biosynthetic genes over the time-course of fiber development (Singh et al. 2009b). Several types of experiments show the importance of ROS in signaling the onset of secondary wall thickening, specifically H_2O_2 that increases in concentration during transitional cell wall remodeling (Potikha et al. 1999; Yang et al. 2008). The appearance of a birefringent secondary wall was delayed when an inhibitor of ROS generation or antioxidants were added to the medium in ovule cultures. Conversely, the addition of H_2O_2 or stimulating H_2O_2 accumulation induced an earlier transition to secondary wall thickening. A small GTPase from cotton, *GhRAC13*, was up-regulated at the transition stage in cotton fiber and induced ROS accumulation when ectopically expressed in soybean and Arabidopsis. GhRAC13 likely participates in generation of the ROS signal through activation of NADPH oxidase (Potikha et al. 1999).

Different cotton cultivars of *G. hirsutum* grown in parallel may show different timing of transitional cell wall remodeling, e.g., 17–18 DPA or 21–24 DPA based on FTIR analysis (Abidi et al. 2010). Alternatively, cultivars representing *G. hirsutum* and *G. barbadense* grown in parallel may show nearly the same timing (Tuttle et al. 2015; Lee et al. 2015), indicating that the timing of transitional cell wall remodeling is under genetic control. This timing affects fiber strength, as shown in a multiseason field study of two near-isogenic lines that have different bundle strengths (Hinchliffe et al. 2010, 2011). The line with higher bundle strength entered the transition stage earlier in most seasons, being less affected by cooler temperatures that delay the temporal progression of fiber development. Overexpressing the actin-bundling protein profilin 2 (*GhPFN2*) under the control of the cauliflower mosaic virus 35S promoter (called hereafter the 35S promoter), in *G. hirsutum* resulted in 14–15 % shorter fibers in T3–T5 generation transgenic plants as compared to wild type. The transgenic fibers showed microtubule reorientation and cell wall birefringence 2 days earlier (Wang et al. 2010b). How the various factors influencing the timing of the transition stage are coordinated and how this timing mechanistically affects fiber bundle strength remain to be determined.

8.3 Applications of Biotechnology for Cotton Fiber Improvement

The methods for cotton transformation have been well documented for nearly three decades. However, their use for the improvement of cotton fiber properties has been somewhat limited, in part due to the difficulty and length of the cotton transformation process (Trolinder 2009; Zhang 2013). In this section, we will highlight selected transgenic experiments in which cotton fiber length and/or secondary cell wall deposition was improved directly or indirectly. The highlighted studies are broadly grouped in two areas: (a) carbohydrate metabolism and cell wall structure and (b) hormones and signaling.

The studies mentioned typically include at least several of the factors supporting sound conclusions: (a) statistical analysis of the results by appropriate methods, (b) analysis of fiber traits in the T1 or further generations in multiple lines, (c) use of a null-segregant control to guard against phenotypes arising from somaclonal variation during a lengthy cotton transformation process, and (d) indication of any pleiotropic effects in cases where a constitutive promoter was used. The use of null-segregant control lines is preferred, but acceptable alternatives (or additions) include the correlation of gene expression level with phenotype severity and demonstration of opposite phenotypes for overexpressing versus silenced lines. Preferably, changes in fiber phenotypes that are consistent in multiple lines in repeated growth chamber or greenhouse experiments should also be assessed in multiple growing cycles in the field. In rigorous experiments, changes in fiber properties should be assessed on uniformly harvested bolls of multiple plants planted in a randomized design under equivalent management conditions. The fiber quality parameters should be fully reported for all measured parameters using one or both of the automated systems standard in the industry: the advanced fiber information system (AFIS) or the high-volume instrument (HVI). The use of AFIS is preferred because of its ability to reveal more attributes of individual fibers as compared to the bundle of fibers tested in HVI. For subtle phenotypes and precise single fiber measurements, other analytical methods should be used in addition (Constable et al. 2015; Kim 2015). It is difficult to conduct ideal experiments on transgenic cotton in the public sector due to numerous factors including the following: (a) substantial time, personnel, growth chamber, and greenhouse requirements for producing multiple fertile transgenic lines, (b) the relatively long growing cycle and large size of cotton plants, (c) the need for controlled fields and regulatory procedures to conduct field tests of transgenic cotton, and (d) challenges in sustaining grant funding for long-term experiments. Nonetheless, biotechnological strategies are leading to improvements in cotton fiber quality, as highlighted below.

8.3.1 Carbohydrate Metabolism and Cell Wall Synthesis

Cotton fiber quality arises through progressive stages of cell wall synthesis as reviewed previously (Haigler et al. 2012). Fiber quality directly depends on cell wall synthesis: Primary wall synthesis supports elongation, synthesis of the winding layer abruptly increases fiber strength, and secondary wall cellulose synthesis increases fiber strength and the richness of dyeing. Transgenic cotton has been produced using strategies to alter cell wall polysaccharides and structural proteins in elongating fiber and during secondary wall thickening.

Modification of cell wall matrix polymers is critical for cotton fiber elongation. In the cotton fiber, the expression of genes encoding several *XTH* family members peaks during elongation (Michailidis et al. 2009). The XTH proteins modify xyloglucan through changes in interchain connections or degradation. The longer-fibered domesticated cotton species have higher expression of XTH genes than their short-fibered wild ancestors (Rapp et al. 2010). When *GhXTH1* was overexpressed under the control of the constitutive 35S promoter, the XTH enzyme activity increased as compared to wild-type plants (Lee et al. 2010). The segregating T1 plants had fibers that were approximately 16 % longer than those of null segregants, and length increases of approximately 5 mm (14 %) were still evident in the T3 generation. Segregation analysis indicated that the *XTH* transgene exerted a dominant effect on fiber length and even heterozygous lines had fibers that were approximately 10 % longer than null segregants.

Modification of structural proteins within the expanding primary cell wall represents another strategy for improving fiber length. Lightly glycosylated hydroxyproline-rich glycoproteins (PRP) are thought to become insoluble within the cell wall through the formation of protein and carbohydrate cross-linkages, which increases cell wall rigidity (Xu et al. 2013). Expression of *GhPRP5* in Arabidopsis under control of the 35S promoter resulted in reduced cell size and general dwarfing relative to wild-type plants. RNA interference with *GhPRP5* expression in transgenic cotton stimulated the expression of expansion-related genes such as alpha-expansin, *XET/XTH*, and alpha-tubulin and resulted in 6 % longer lint fibers and longer fuzz fibers compared to wild-type plants (Xu et al. 2013).

The nearly irreversible deposition of the cellulose-rich secondary wall in cotton fiber is a strong sink for sugar produced in photosynthesis. Secondary wall cellulose synthesis in cotton fiber has been associated with the activity of sucrose synthase (Amor et al. 1995), including a novel isoform that is partly distributed to the apoplast (Salnikov et al. 2003; Brill et al. 2011). This enzyme typically operates to cleave sucrose, resulting in the release of UDP-glucose (the substrate for cellulose and callose synthesis) and fructose. The released fructose represents half of the available carbon, and it can be recycled to sucrose phosphate via the activity of sucrose phosphate synthase (SPS) using fructose phosphate and UDP-glucose as substrates (Haigler et al. 2001; Haigler 2007). These pathways have been manipulated in transgenic cotton, reinforcing the importance of adequate carbohydrate supply for development of high-quality cotton fiber.

Transgenic cotton has been generated with up- and down-regulated cotton sucrose synthase (*GhSusA1*) (Jiang et al. 2012). When gene expression was constitutively repressed using antisense technology and the 35S promoter, reductions in fiber length, boll size, seed weight, and dry seedling weight were observed as compared to both wild-type and control transformants. Fiber length was also reduced when the fiber-specific E6 promoter was used (Jiang et al. 2012), and the negative effect on length was likely due to diminished hexoses and turgor in the central vacuole and/or impaired primary cell wall synthesis (Ruan et al. 2003). Conversely, constitutive overexpression of 35S::*GhSusA1* caused increased plant dry weight, SUS activity (in both leaves and fibers), as well as increased fructose, glucose and sucrose in leaves compared to wild type. The overexpressor with the highest *GhSusA1* transcript level had fibers that were approximately 8 % longer, 6–11 % stronger, and they deposited secondary cell wall material approximately 1.7 times faster than wild-type controls (Jiang et al. 2012). Similarly, increases in fiber length at 20 DPA were also observed when a sucrose synthase gene from potato was constitutively overexpressed in cotton, although data from mature fibers were not provided and other fiber qualities were not evaluated (Xu et al. 2012).

To address the problem of low cotton fiber maturity (secondary wall thickening) under cool night temperatures by potentially increasing the sucrose available for cellulose synthesis, SPS from cool-tolerant spinach was constitutively overexpressed under control of the 35S promoter in transgenic cotton (Haigler et al. 2007). The transgenic plants had increased SPS activity, sucrose to starch ratio in leaves, and leaf sucrose in two of three overexpressing lines. Fiber quality was enhanced in one line with the highest SPS expression, as compared to wild type and segregating nulls, when plants were grown in a growth chamber with cool nights. Specifically, the fiber length and maturity ratio were increased by 14 and 6 % and the short fiber content was 47 % lower. However, micronaire was also increased by 12 %, presumably due to the increased secondary wall thickness.

Low growth temperatures in the field and the consequent low rates of cellulose biosynthesis (Roberts et al. 1992) can also result in sticky fiber due to the accumulation of unused sugars. To address this problem, a uridine diphosphate glucose pyrophosphorylase (*UGP*) was overexpressed under the control of the 35S promoter (Li et al. 2015). UGP synthesizes UDP-glucose from glucose-1-phosphate and uridine triphosphate. UDP-glucose is the immediate substrate for cellulose and callose synthesis, and it is also required to cycle fructose released by sucrose synthase activity back to sucrose, as previously discussed. The UGP overexpressing cotton plants were larger with more bolls when grown in the field. The fibers of three transgenic lines were 15 % longer, 12 % stronger, and had higher cellulose content compared to wild-type plants, while the content of soluble and reduced sugars was reduced. Whether secondary wall thickening and fiber maturity were increased in the transgenic fibers was not evaluated in this study (Li et al. 2015).

Several processes involved in cell wall synthesis depend on the cytoskeleton, inclusive of microtubules and actin. The expression of an actin-bundling LIM-domain protein was up- and down-regulated in transgenic cotton under control of the 35S promoter (Han et al. 2013). (The LIM domain is a double-zinc-finger motif, and the name derives from the original discovery in proteins called LIN-11, Isl1, and MEC-3.) The cotton WLIM1 protein colocalized with phalloidin labeled F-actin structures in vitro and increased the amount of actin precipitating in cosedimentation experiments. Fibers from *GhWLIM1* overexpressors were 10 % longer and had 20 % lower micronaire and 9 % higher bundle strength as compared to wild-type controls. Preliminary results in the supplemental materials associate the micronaire and bundle strength improvements with lower fiber perimeter. These phenotypes were not reversed in antisense lines with lower transcript levels, possibly due to gene redundancy. Interestingly, WLIM1 was also shown in tobacco to act as a transcription factor for genes with promoters containing phenylalanine ammonia lyase (PAL)-box elements, which are associated with phenylpropanoid and lignin biosynthesis. The nuclear localization of WLIM1 was stimulated by H_2O_2, which also signals the beginning of cotton fiber secondary wall thickening (Potikha et al. 1999). This dual function of WLIM1 explains the elevated expression of lignin biosynthetic genes (*4CL1*, *CCR1*, and *CAD6*) in the overexpressing cotton lines, as well as the higher but still low level of lignin-like phenolics in the transgenic cotton fiber (Han et al. 2013).

8.3.2 Hormones and Signaling

Fiber quality has been improved through manipulation of gibberellins, brassinosteroids, and a photoreceptor. These factors represent high-level regulatory processes, so they impact downstream processes already discussed. For example, the expression of endogenous *GhXTH*1 was elevated by treatment with NAA, brassinolide, and GA (Lee et al. 2010), and GhPRP5 interacted with an auxin response protein, as shown by bimolecular fluorescence complementation assay (Xu et al. 2013).

Gibberellins stimulate elongation of plant cells, and their level was highest in the cotton fiber during a period of rapid elongation (Xiao et al. 2010). GA20-oxidase is needed for the conversion of GA12 or GA53 precursors to bioactive gibberellin such as GA_4. When *GhGA20ox1* was constitutively expressed using a 35S promoter, the overexpressing cotton plants had more fiber initials, longer internodes, smaller flowers, smaller bolls, and fibers that were approximately 14 % longer as compared to null segregants (Xiao et al. 2010). The 10 DPA fibers of overexpressing lines had higher levels of GA_4, but less GA_1. A subsequent experiment in which the GA_4 level increased due to overexpression of *GhGA20ox1* under the control of the fiber-specific *SCFP* promoter did not result in increased fiber length, although there were increases in micronaire (8–11 %) and fiber weight per 1000 fibers (4–7 %) as compared to wild-type plants (Bai et al. 2014). Potentially, the positive effects on fiber length (Xiao et al. 2010) arose indirectly through whole-plant changes or due to greater GA_4 content when the 35S promoter was used as compared to the *SCFP* promoter. The GA_4 concentration data cannot be precisely compared in the two studies due to experimental differences.

Following up the observation that increased expression of the BR receptor gene *AtBRI1*-enhanced Arabidopsis cell elongation (Wang et al. 2001), the effect of altering the expression level of the cotton homolog was tested (Sun et al. 2015). Transgenic cotton seedlings with *GhBRI1* constitutively down-regulated by antisense expression driven by the 35S promoter had more robust root growth in the presence of exogenous brassinolide compared to wild-type plants. The fiber of these antisense plants at T3 generation or beyond also showed 13 % reduction in micronaire. Conversely, the constitutive overexpression of *BRI1* resulted in a 13 % increase in micronaire. The imprecise indicator provided by micronaire measurement was clarified through demonstrating 20 % thicker or 38 % thinner secondary wall in the *GhBRI1* overexpression or repression lines, respectively. The expression level of *GhCesA1*, which in normally up-regulated for cotton fiber cell wall thickening (e.g., see Singh et al. 2009b), was increased or decreased in the overexpression or repression lines, respectively. These data support the role of BR molecules in signaling the onset of secondary wall thickening in cotton fiber.

Phytochrome is a red/far-red reversible chromoprotein that is an important signaling molecule in a number of developmental processes in plants including seed germination, hypocotyl elongation, stem elongation, leaf expansion, and flowering (Briggs and Olney 2001). QTL mapping of a cross *G. hirsutum* and *G. barbadense* revealed an association between the PHYA1 locus and fiber length. Cotton plants that received increased R/FR had longer fibers (Kasperbauer 2000). The effect of down-regulating the cotton phytochrome A1 transcript (*PHYA1*) was investigated in transgenic cotton using an RNAi construct driven by the 35S promoter (Abdurakhmonov et al. 2014). The resulting plants had enhanced vegetative growth, compared to wild-type plants. Compared to null segregants, fiber length was increased by 5 %, and micronaire was decreased by 4–8 %. Seed weight, lint index, and lint percent all differed between null-segregant and untransformed control plants indicating that the transformation process alone may influence these properties. Although this experiment successfully reduced the levels of *PHYA1* in cotton, levels of the other phytochrome genes *PHYA2*, *PHYC*, and *PHYE* were elevated in RNAi lines compared to a wild-type control. This makes it difficult to ascribe the improved fiber and plant growth qualities to down-regulation of *PHYA1* alone.

8.4 Summary and Future Prospects for Biotechnological Improvement of Cotton Fiber

Experimental results showing direct effects on cotton fiber quality (Table 8.1), theoretical considerations, or gene expression analyses provide a basis for improvement of fiber quality in transgenic cotton (Table 8.2). Single gene manipulations have resulted in up to 15 % improvement in fiber length and 12 % improvement in fiber bundle strength, with both of these maximum results observed for overexpression of *UGP1* as compared to wild-type plants. In this and all similar cases of the use of only the wild-type control, it will be interesting to see the results repeated when compared to segregating nulls or multiple other lines that have arisen through the same transformation process. The fiber bundle strength also only partly predicts the usefulness of fiber. Positive and negative changes (−20 to +12 %) in micronaire have been reported (Table 8.2). Generally, lower micronaire is preferred within the range of 3.5–4.9, but an increase in micronaire in a particular transgenic plant family may be acceptable if it falls within this range and arises through increased secondary wall thickness and not an increase in fiber perimeter (Constable et al. 2015). However, the micronaire measurement alone is not sufficient to predict the usefulness of transgenic fiber because it confounds the effects of fiber perimeter and secondary wall maturity. Future work should aim to characterize these two parameters directly and in multiple lines with rigorous control of boll harvesting and analytical replication. In general, reduced fiber perimeter while fiber maturity (secondary wall thickness) remains in the optimal range will increase the competitiveness of cotton with synthetic fiber. The overexpression of WLIM1a has so far moved length, bundle strength, and micronaire collectively in promising directions as compared to wild type (Han et al. 2013). These transgenic lines provide an exception to commonly observed parallel increases in micronaire (not preferred) and strength (Cotton Incorporated 2014), possibly due to the increase in lignin-like phenolics that could act as cross-linkers in the fiber cell wall. Hopefully, additional results will be forthcoming on these interesting lines to fully characterize multiple length parameters, single fiber strength, fiber perimeter, secondary wall thickness, and yarn strength as compared to transgenic control lines.

Typically the fiber properties of transgenic cotton have been measured in comparison with the parental line, or outdated transformable cultivars with wild-type fiber properties that may be inferior to elite modern lines. The absolute values for fiber quality parameters in transgenic cotton highlighted here (Table 8.2), especially length, sometimes fell near or slightly above the price premium thresholds for *G. hirsutum* fiber in Australia and the USA (32 mm fiber length, bundle strength of 26 g/tex, and 3.8–4.2 micronaire) (Constable et al. 2015). Future experiments will be required to determine whether fiber quality improvements will also be observed upon introducing the trait into elite modern cultivars. Any genetically engineered trait that increased the uniformity of length, strength, maturity, and fineness, especially under diverse or stressful growing conditions, will be particularly useful and increase the competitiveness of cotton with synthetic fibers. To our knowledge, no published report on genetically engineered cotton fiber has characterized the performance of the transgenic fibers in yarns or textiles. Ultimately, we want to optimize all of the traditional cotton fiber qualities in terms of their usefulness in diverse products while gaining the ability to generate novel fibers with unique characteristics (Constable et al. 2015).

The "omics" era of scientific research (genomics, transcriptomics, proteomics, metabolomics) provides new clues for directed strategies to improve cotton fiber quality through biotechnology. Complete genome sequences of diploid and cultivated allotetraploid cotton (Paterson et al. 2012; Li et al. 2014; Zhang et al. 2015) provide essential resources in this new era. The ability to compare genotypes with large-scale, relatively high-throughput methods provides a powerful discovery framework for genes and other processes that control cotton fiber morphogenesis and quality. Given the difficulty in regenerating cotton transformants and the logistical challenges of analyzing a transgenic family with appropriate methods in comparison with transgenic controls, future genetic engineering targets should be selected with care. Targeted genome editing technologies such as Crispr/Cas9, meganucleases, and TALENs have the potential to simplify this process by allowing multiple traits to be stacked within a single locus making it easier to modify several aspects of fiber development simultaneously and simplifying multitrait introgression into modern cotton cultivars (D'Halluin et al. 2013). Currently, no cotton with fiber quality improved through biotechnology has entered the marketplace (Constable et al. 2015), and we can look forward to such improvements reaching the consumer in the future.

Acknowledgments For support of research in this area, the authors thank the National Science Foundation (Grant # 1025947) and Cotton Incorporated, Cary, NC, USA.

References

Abasolo W, Michaela E, Yamauchi K, Obel N, Reinecke A, Neumettzler L, Dunlop JW, Mouille G, Pauly M, Höfte H et al (2009) Pectin may hinder the unfolding of xyloglucan chains during cell deformation: Implications of the mechanical performance of Arabidopsis hypocotyls with pectin alterations. Mol Plant 2:990–999

Abdurakhmonov IY, Buriev ZT, Saha S, Jenkins JN, Abdukarimov A, Pepper AE (2014) Phytochrome RNAi enhances major fibre quality and agronomic traits of the cotton *Gossypium hirsutum* L. Nat Commun 5:3062

Abidi N, Cabrales L, Hequet E (2010) Fourier transform infrared spectroscopic approach to the study of the secondary cell wall development in cotton fiber. Cellulose 17:309–320

Aleman L, Kitamura J, Abdel-Mageed H, Lee J, Sun Y, Nakajima M, Ueguchi-Tanaka M, Matsuoka M, Allen RA (2008) Functional analysis of cotton orthologs of GA signal transduction factors GID1 and SLR1. Plant Mol Biol 68:1–16

Amor Y, Haigler CH, Johnson S, Wainscott M, Delmer DP (1995) A membrane-associated form of sucrose synthase and its potential role in synthesis of cellulose and callose in plants. Proc Natl Acad Sci USA 92:9353–9357

Avci U, Pattahil S, Singh B, Brown VL, Hahn MG, Haigler CH (2013) Cotton fiber cell walls of *Gossypium hirsutum* and *Gossypium barbadense* have differences related to loosely bound xyloglucan. PLoS ONE 8:e56315

Bai W, Xiao Y, Zhao J, Song S, Hu L, Zeng J, Li X, Hou L, Luo M, Li D (2014) Gibberellin overproduction promotes sucrose synthase expression and secondary cell wall deposition in cotton fibers. PLoS ONE 9:e96537

Beasley CA, Ting IP (1973) The effects of plant growth substances on in vitro fiber development from fertilized cotton ovules. Am J Bot 60:130–139

Betancur L, Singh B, Rapp RA, Wendel JF, Marks MD, Roberts AW, Haigler CH (2010) Phylogenetically distinct cellulose synthase genes support secondary wall thickening in Arabidopsis shoot trichomes and cotton fiber. J Integr Plant Biol 52:205–220

Bhattacharjee S (2012) The language of reactive oxygen species signaling in plants. J Bot. Article ID 985298

Bleecker AB, Kende H (2000) Ethylene: a gaseous signal molecule in plants. Annu Rev Cell Dev Biol 16:1–18

Bowling AJ, Vaughn KC, Turley RB (2011) Polysaccharides and glycoprotein distribution in the epidermis of cotton ovules during early fiber initiation and growth. Protoplasma 248:579–590

Boyer JS (2009) Cell wall biosynthesis and the molecular mechanism of plant enlargement. Funct Plant Biol 36:383–394

Briggs WR, Olney MA (2001) Photoreceptors in plant photomorphogenesis to date. Five phytochromes, two cryptochromes, one phototropin, and one superchrome. Plant Physiol 125:85–88

Brill E, van Thournout M, White RG, Llewellyn D, Campbell PM, Engelen S, Ruan Y-L, Arioli T, Furbank RT (2011) A novel isoform of sucrose synthase is targeted to the cell wall during secondary cell wall synthesis in cotton fiber. Plant Physiol 157:40–54

Chaudhary B, Hovav R, Flagel L, Mittler R, Wendel JF (2009) Parallel expression evolution of oxidative stress-related genes in fiber from wild and domesticated diploid and polyploid cotton (Gossypium). BMC Genom 10:378

Clark G, Torres J, Finlayson S, Guan X, Handley C, Lee J, Kays JE, Chen ZJ, Roux SJ (2009) Apyrase (nucleoside triphosphate-diphosphohydrolase) and extracellular nucleotides regulate cotton fiber elongation in cultured ovules. Plant Physiol 152:1073–1083

Constable G, Llewellyn D, Walford SA, Clement JD (2015) Cotton breeding for fiber quality improvement. In: Cruz VMV, Dierig DA (eds) Industrial crops, handbook of plant breeding, vol 9. Springer, New York, pp 191–232

Cosgrove DJ (2005) Growth of the plant cell wall. Nat Rev Mol Cell Bio 6:850–861

Cotton Incorporated (2014) 2013 Cotton crop quality summary. http://www.cottoninc.com/fiber/quality/Crop-Quality-Reports/2013-Cotton-Crop-Quality-Summary/. Accessed 29 June 2015

D'Halluin K, Vanderstraeten C, Hulle J, Rosolowska J, Den Brande I, Pennewaert A, D'Hont K, Bossut M, Jantz D, Ruiter R, Broadhvest J (2013) Targeted molecular trait stacking in cotton through targeted double-strand break induction. Plant Biotechnol J 11(8):933–941

Dasani SH, Thaker VS (2006) Role of abscisic acid in cotton fiber development. Russ J Plant Physiol 53:62–67

Dhindsa RS, Beasley CA, Ting IP (1976) Effects of abscisic acid on in vitro growth of cotton fiber. Planta (Berl) 130:197–201

Dillehay TD, Rossen J, Andres T, Williams DE (2007) Preceramic adoption of peanut, squash, and cotton in Northern Peru. Science 316:1890–1893

Farr WK (1931) Cotton fibers. I. Origin and early stages of elongation. Contrib Boyce Thompson Inst 3:441–458

Haigler CH, Ivanova-Datcheva M, Hogan PS, Salnikov VV, Hwang S, Martin K, Delmer DP (2001) Carbon partitioning to cellulose synthesis. Plant Mol Biol 47:29–51

Haigler CH, Zhang D, Wilkerson CG (2005) Biotechnological improvement of cotton fiber maturity. Physiol Plant 124:285–294

Haigler CH, Singh B, Zhang D, Hwang S, Wu C, Cai WX, Hozain M, Kang W, Kiedaisch B, Strauss RE, Hequet EF, Wyatt BG, Jividen GM, Holaday AS (2007) Transgenic cotton over-producing spinach sucrose phosphate synthase showed enhanced leaf sucrose synthesis and improved fiber quality under controlled environmental conditions. Plant Mol Biol 63 (6):815–832

Haigler CH (2007) Substrate supply for cellulose synthesis and its stress sensitivity in the cotton fiber. In: Brown RM Jr, Saxena I (eds) Cellulose: molecular and structural biology. Springer, New York, pp 145–166

Haigler CH, Singh B, Wang G, Zhang D (2009) Genomics of cotton fiber secondary wall deposition and cellulose biogenesis. In: Paterson AH (ed) Genetics and genomics of cotton. Plant genetics/genomics. Springer, New York, pp 385–417

Haigler CH, Betancur L, Stiff MR, Tuttle JR (2012) Cotton fiber: a powerful single-cell model for cell wall and cellulose research. Front Plant Sci 3:1–7

Han LB, Li YB, Wang HY, Wu XM, Li CL, Luo M, Wu SJ, Kong ZS, Pei Y, Jiao GL, Xia GX (2013) The dual functions of WLIM1a in cell elongation and secondary wall formation in developing cotton fibers. Plant Cell 25:4421–4438

Han J, Tan J, Tu L, Zhang X (2014) A peptide hormone gene, GhPSK promotes fibre elongation and contributes to longer and finer cotton fibre. Plant Biotechnol J 12:861–871

Han XZ, Gao S, Cheng YN, Sun YZ, Liu W, Tang LL, Ren DM (2012) Protective effect of naringenin-7-O-glucoside against oxidative stress induced by doxorubicin in H9c2 cardiomyocytes. Biosci Trends 6:19–25

Hinchliffe DJ, Meredith WR, Yeater KM, Kim HJ, Woodward AW, Chen ZJ, Triplett BA (2010) Near-isogenic cotton germplasm lines that differ in fiber-bundle strength have temporal differences in fiber gene expression patterns as revealed by comparative high-throughput profiling. Theor Appl Genet 120:1347–1366

Hinchliffe DJ, Meredith WR, Delhorn CD, Thibodeaux DP, Fang DD (2011) Elevated growing degree days influence transition stage timing during cotton fiber development resulting in increased fiber-bundle strength. Crop Sci 51:1683–1692

Hovav R, Udall JA, Chaudhary B et al (2008) The evolution of spinnable cotton fiber entailed prolonged development and a novel metabolism. PLoS Genet 4:e25

Hsieh Y-L (1999) Structural development of cotton fibers and linkages to fiber quality. In: Basra AS (ed) Cotton fibers: developmental biology, quality improvement, and textile processing. Haworth Press, New York, pp 137–165

Hu G, Koh J, Yoo M-J, Grupp K, Chen S, Wendel JF (2013) Proteomic profiling of developing cotton fibers from wild and domesticated *Gossypium barbadense*. New Phytol 2013:1–13

Huang QS, Wang HY, Gao P, Wang G-Y, Xia G-X (2008) Cloning and characterization of a calcium dependent protein kinase gene associated with cotton fiber development. Plant Cell Rep 27:1869–1875

Jiang Y, Guo W, Zhu H, Ruan Y, Zhang T (2012) Overexpression of GhSusA1 increases plant biomass and improves cotton fiber yield and quality. Plant Biotechnol J 10:301–312

Kasperbauer MJ (2000) Cotton fibre length is affected by far-red light impinging on developing bolls. Crop Sci 40:1673–1678

Kim HJ (2015) Fiber biology. In: Fang DD, Percy RG (eds) Cotton, 2nd edn. ASA, CSSA, and SSSA, Madison, pp 1–31

Kosmidou-Dimitropoulou K (1986) Hormonal influences in fiber development. In: Mauney JR, Stewart JM (eds) Cotton physiology. The Cotton Foundation, Memphis, pp 361–374

Kudla J, Batistič O, Hashimoto K (2010) Calcium signals: the lead currency of plant information processing. Plant Cell 22:541–563

Lee CM, Kafle K, Belias DW, Park YB, Glick RE, Haigler CH, Kim SH (2015) Comprehensive analysis of cellulose content, crystallinity, and lateral packing in *Gossypium hirsutum* and *Gossypium barbadense* cotton fibers using Sum Frequency Generation, Infrared and Raman Spectroscopy, and X-ray diffraction. Cellulose 22:971–989

Lee J, Burns TH, Light G, Sun Y, Fokar M, Kasukabe Y, Fulisawa K, Maekawa Y, Allen RD (2010) Xyloglucan endotransglycosylase/hydrolase genes in cotton and their role in fiber elongation. Planta 232:1191–1205

Lee TM, Lur HS, Shieh YJ, Chu C (1994) Levels of abscisic acid in anoxia- or ethylene-treated rice (*Oryza sativa* L.) seedlings. Plant Sci 95:125–131

Lei L, Li S, Bashline L, Gu Y (2014) Dissecting the molecular mechanism underlying the intimate relationship between cellulose microfibrils and cortical microtubules. Front Plant Sci 5:1–8

Li F, Fan G, Wang K et al (2014) Genome sequence of the cultivated cotton *G. arboreum*. Nat Genet 46:567–572

Li B, Yang Y, Hu W, Li X, Cao J, Fan L (2015) Over-expression of GhUGP1 in Upland cotton improves fibre quality and reduces fibre sugar content. Plant Breed 134:197–202

Li H, Qin Y, Pang Y, Song W-Q, Mei W-Q, Zhu Y-X (2007) A cotton ascorbate peroxidase is involved in hydrogen peroxide homeostasis during fibre cell development. New Phytol 175:462–471

Li XB, Fan XP, Wang XL, Cai L, Yang WC (2005) The cotton ACTIN1 gene is functionally expressed in fibers and participates in fiber elongation. Plant Cell 17:859–875

Liao W, Zhang J, Xu N, Peng M (2010) The role of phytohormones in cotton fiber development. Russ J Plant Physiol 57:462–468

Liszkay A, van der Zalm E, Schopfer P (2004) Production of reactive oxygen intermediates ($O_2^{\cdot-}$, H_2O_2, and $^{\cdot}OH$) by maize roots and their role in wall loosening and elongation. Plant Physiol 136:3114–3124

Liu Q, Talbot M, Llewellyn D (2013) Pectin methylesterase and pectin remodelling differ in the fibre walls of two *Gossypium* species with very different fibre properties. PLoS ONE 8:e65131

Luo M, Xiao Y, Li X, Lu X, Deng W, Li D, Hou L, Hu M, Li Y, Pei Y (2007) GhDET2, a steroid 5α-reductase, plays an important role in cotton fiber cell initiation and elongation. Plant J 51:419–430

Marks MD, Betancur L, Gilding E, Chen F, Bauer S, Wenger J, Dixon RA, Haigler CH (2008) A new method for isolating large quantities of Arabidopsis trichomes for transcriptome, cell wall and other types of analyses. Plant J 56:483–492

Merchante C, Alonso JA, Stepanova AN (2013) Ethylene signaling: simple ligand, complex regulation. Curr Opin Plant Biol 16:554–560

May OL (2002) Quality improvement of Upland cotton (*Gossypium hirsutum* L.). J Crop Prod 5:371–394

Mei W, Qin Y, Song W, Li J, Zhu Y (2009) Cotton GhPOX1 encoding plant class III peroxidase may be responsible for the high level of reactive oxygen species production that is related to cotton fiber elongation. J Genet Genomics 36:141–150

Meinert MC, Delmer DP (1977) Changes in biochemical composition of the cell wall of the cotton fiber during development. Plant Physiol 59:1088–1097

Meyer VG (1974) Interspecific cotton breeding. Econom Bot 28:56–60

Michailidis G, Argiriou A, Darzentas N, Tsaftaris A (2009) Analysis of xyloglucan endotransglycosylase/hydrolase (XTH) genes from allotetraploid (*Gossypium hirsutum*) cotton and its diploid progenitors expressed during fiber elongation. J Plant Physiol 166:403–416

Mohnen D (2008) Pectin structure and biosynthesis. Curr Opin Plant Biol 11:266–277

Monshausen GB, Bibikova TN, Messerli MA, Shi C, Gilroy S (2007) Oscillations in extracellular pH and reactive oxygen species modulate tip growth of Arabidopsis root hairs. Proc Natl Acad Sci USA 104:20996–21001

Moulherat C, Tengberg M, Haquet JF, Mille B (2002) First evidence of cotton at Neolithic Mehrgarh, Pakistan: analysis of mineralized fibres from a copper bead. J Archaeol Sci 29:1393–1401

Naylor GR, Stanton JH, Speijers J (2014) Skin comfort of base layer wool garments. Part 2: fiber diameter effects on fabric and garment prickle. Textile Res J84:1506–1514

Nayyar H, Kaur K, Basra AS, Malik CP (1989) Hormonal regulation of cotton fibre elongation in *Gossypium arboreum* L. In vitro and in vivo. Biochem Physiol Pflazen 185:415–421

O'Neill MA, York WS (2003) The composition and structure of primary cell walls. In: Rose JKC (ed) The plant cell wall. Annual Plant Review, CRC Press, Boca Raton, pp 1–54

Pang CY, Wang H, Pang Y, Xu C, Jiao Y, Qin Y-M, Western TL, Yu S-X, Zhu Y-X (2010) Comparative proteomics indicates that biosynthesis of pectic precursors is important for cotton fiber and Arabidopsis root hair elongation. Mol Cell Proteomics 9:2019–2033

Park S, Cosgrove D (2012) A revised architecture of primary cell walls based on biochemical changes induced by substrate-specific endoglucanases. Plant Physiol 158:1933–1943

Paterson AH, Wendel JF, Gundlach H et al (2012) Repeated polyploidization of *Gossypium* genomes and the evolution of spinnable cotton fibres. Nature 492:423–427

Potikha T, Collins C, Johnson D, Delmer DP, Levine A (1999) The involvement of hydrogen peroxide in the differentiation of secondary walls in cotton fibers. Plant Physiol 119:849–858

Potocký M, Jones M, Bezvoda R, Smirnoff N, Žárský V (2007) Reactive oxygen species produced by NADPH oxidase are involved in pollen tube growth. New Phytol 174:742–751

Pu L, Li Q, Fan X, Yang W, Xue Y (2008) The R2R3 MYB transcription factor GhMYB109 is required for cotton fiber development. Genetics 180:811–820

Qin YM, Hu CY, Pang Y, Kastanoitis AJ, Hiltunen JK, Zhu Y-X (2007) Saturated very-long-chain fatty acids promote cotton fiber and Arabidopsis cell elongation by activating ethylene biosynthesis. Plant Cell 19:3692–3704

Qin YM, Hu C, Zhu Y (2008) The ascorbate peroxidase regulated by H_2O_2 and ethylene is involved in cotton fiber cell elongation by modulating ROS homeostasis. Plant Signal Behav 3:194–196

Rapp R, Haigler C, Flagel L, Hovav RH, Udall JA, Wendel JF (2010) Gene expression in developing fibres of Upland cotton (*Gossypium hirsutum* L.) was massively altered by domestication. BMC Biol 8:139

Roberts EM, Nunna RR, Huang JY, Trolinder NL, Haigler CH (1992) Effects of cycling temperatures on fiber metabolism in cultured cotton ovules. Plant Physiol 100:979–986

Ruan Y-L, Llewellyn DJ, Furbank RT (2001) The control of single-celled cotton fiber elongation by developmentally reversible gating of plasmodesmata and coordinated expression of sucrose and K^+ transporters and expansin. Plant Cell 13:47–60

Ruan Y-L, Llewellyn DJ, Furbank RT (2003) Suppression of sucrose synthase gene expression represses cotton fiber cell initiation, elongation, and seed development. Plant Cell 15:952–964

Ruan Y-L, Xu S-M, White R, Furbank RT (2004) Genotypic and developmental evidence for the role of plasmodesmatal regulation in cotton fiber elongation mediated by callose turnover. Plant Physiol 136:4104–4113

Ruan Y (2007) Rapid cell expansion and cellulose synthesis regulated by plasmodesmata and sugar: insights from the single-celled cotton fibre. Funct Plant Biol 34:1–10

Salnikov V, Grimson MJ, Seagull RW, Haigler CH (2003) Localization of sucrose synthase and callose in freeze substituted, secondary wall stage, cotton fibers. Protoplasma 221:175–184

Sampathkumar A, Gutierrez R, McFarlane HE, Bringmann M, Lindeboom J, Emons A-M, Samuels L, Ketelaar T, Ehrhardt DW, Persson S (2013) Patterning and lifetime of plasma membrane-localized cellulose synthase is dependent on actin organization in Arabidopsis interphase cells. Plant Physiol 162:675–688

Sauter M (2015) Phytosulfokine peptide signaling. J Expt Bot. Advanced access. doi:10.1093/jxb/erv071

Seagull RW (1990) The effects of microtubule and microfilament disrupting agents on cytoskeletal arrays and wall deposition in developing cotton fibers. Protoplasma 159:44–59

Seagull RW (1993) Cytoskeletal involvement in cotton fiber growth and development. Micron 24:643–660

Seagull RW (1995) Cotton fiber growth and development: evidence for tip synthesis and intercalary growth in young fibers. Plant Physiol (Life Sci Adv) 14:27–38

Seagull RW (1998) Cytoskeletal stability affects cotton fiber initiation. Int J Plant Sci 159:590–598

Sharma P, Jha AB, Dubey RS, Pessarakli M (2012) Reactive oxygen species, oxidative damage, and antioxidative defense mechanism in plants under stressful conditions. J Bot. Article ID 217037

Shi Y, Zhu S, Mao X, Feng J-X, Qin Y-M, Zhang L, Cheng J, Wei L-P, Wang Z-Y, Zhu Y-X (2006) Transcriptome profiling, molecular biological, and physiological studies reveal a major role for ethylene in cotton fiber cell elongation. Plant Cell 18:651–664

Shui S, Plastina A (2013) FAO/ICAC World Apparel Fiber Consumption Survey. International Cotton Advisory Committee, Washington DC, https://www.icac.org/cotton_info/publications/statistics/world-apparel-survey/FAO-ICAC-Survey-2013-Update-and-2011-Text.pdf. ISBN 9780979390395

Singh B, Avci U, Eichler Inwood SE, Grimson MJ, Landgraf J, Mohnen D, Sørensen I, Wilkerson CG, Willats WGT, Haigler CH (2009a) A specialized outer layer of the primary cell wall joins elongating cotton fibers into tissue-like bundles. Plant Physiol 150:684–699

Singh B, Cheek HD, Haigler CH (2009b) A synthetic auxin (NAA) suppresses secondary wall cellulose synthesis and enhances elongation in cultured cotton fiber. Plant Cell Rep 28:1023–1032

Stewart J McD (1975) Fiber initiation on the cotton ovule (*Gossypium hirsutum*). Am J Bot 62:723–730

Stiff MR, Haigler CH (2012) Recent advances in cotton fiber development. In: Oosterhuis D, Cothren T (eds) Cotton flowering and fruiting. Cotton Physiology Book Series, National Cotton Council, Memphis, pp 163–192

Sun Y, Veerabomma S, Abdel-Mageed H, Fokar M, Asami T, Yoshida S, Allen RA (2005) Brassinosteroid regulates fiber development on cultured cotton ovules. Plant Cell Physiol 46:1384–1391

Sun Y, Veerabomma S, Fokar M, Abidi N, Hequet E, Payton P, Allen RD (2015) Brassinosteroid signaling affects secondary cell wall deposition in cotton fibers. Ind Crops Prod 65:334–342

Taliercio E, Haigler CH (2011) The effect of calcium on early fiber elongation in cotton ovule culture. J Cotton Sci 15:154–161

Tan J, Tu L, Deng F, Hu H, Nie Y, Zhang X (2013) A genetic and metabolic analysis revealed that cotton fiber cell development was retarded by flavonoid naringenin. Plant Physiol 162:86–95

Trolinder N (2009) Genetic engineering of cotton. In: Paterson AH (ed) Genetics and genomics of cotton. Plant genetics/genomics. Springer, New York, pp 187–207

Tuttle JR, Nah G, Duke MV, Alexander DC, Guan X, Song Q, Chen ZJ, Scheffler BE, Haigler CH (2015) Metabolomic and transcriptomic insights into how cotton fiber transitions to secondary wall synthesis, represses lignification, and prolongs elongation. BMC Genom 16:477

Vaughn KC, Turley RB (1999) The primary walls of cotton fibers contain an ensheathing pectin layer. Protoplasma 209:226–237

Wakelyn PJ, Bertoniere NR, French AD, Thibodeaux DP, Triplett BA, Rousselle MA, Goynes Jr WR, Edwards JV, Hunter L, McAlister DD et al (2007) Cotton fiber chemistry and technology. International Fiber Science Technical Series, vol 162. CRC Press, Boca Raton

Wang HY, Yu Y, Chen ZL, Xia GX (2005) Functional characterization of *Gossypium hirsutum* profilin 1 gene (GhPFN1) in tobacco suspension cells. Planta 222:594–603

Wang H, Guo Y, Lv F, Zhu H, Wu S, Jiang Y, Li F, Zhou B, Guo W, Zhang T (2010a) The essential role of *GhPEL* gene, encoding a pectate lyase, in cell wall loosening by depolymerization of the de-esterified pectin during fiber elongation in cotton. Plant Mol Biol 72:397–406

Wang J, Wang H, Zhao P, Han L-B, Jiao G-L, Zheng Y-Y, Huang S-J, Xia G-X (2010b) Overexpression of a profilin (GhPFN2) promotes the progression of developmental phases in cotton fibers. Plant Cell Physiol 51:1276–1290

Wang L, Li X, Lian H, Ni D-A, He Y, Chen X-Y, Ruan Y-L (2010c) Evidence that high activity of vacuolar invertase is required for cotton fiber and arabidopsis root elongation through osmotic dependent and independent pathways, respectively. Plant Physiol 154:744–756

Wang ZY, Seto H, Fujioka S, Yoshida S, Chory J (2001) BRI1 is a critical component of a plasma-membrane receptor for plant steroids. Nature 410:380–383

Wendel JF, Brubaker C, Alvarez I, Cronn R, Stewart J McD (2009) In: Paterson AH (ed) Genetics and genomics of cotton. Plant genetics/genomics. Springer, New York, pp 3–22

Wolf S, Greiner S (2012) Growth control by cell wall pectins. Protoplasma 249(2):169–175

Xiao Y, Li D, Yin M, Li X, Zhang M, Wang Y, Dong J, Zhao J, Luo M, Luo X (2010) Gibberellin 20-oxidase promotes initiation and elongation of cotton fibers by regulating gibberellin synthesis. J Plant Physiol 167:829–837

Xu S, Brill E, Llewellyn DJ, Furbank RT, Ruan Y (2012) Overexpression of a potato sucrose synthase gene in cotton accelerates leaf expansion, reduces seed abortion, and enhances fiber production. Mol Plant 5(2):430–441

Xu W, Zhang D, Wu Y, Qin L, Huang G, Li J, Li L, Li X (2013) Cotton PRP5 gene encoding a proline-rich protein is involved in fiber development. Plant Mol Biol 82:353–365

Yang Y, Bian S, Yao Y, Liu J (2008) Comparative proteomic analysis provides new insights into the fiber elongating process in cotton. J Proteome Res 7:4623–4637

Yoo M, Wendel JF (2014) Comparative evolutionary and developmental dynamics of the cotton (*Gossypium hirsutum*) fiber transcriptome. PLoS Genet 10:e1004073

Zhang M, Zheng X, Song S, Zeng Q, Hou L, Li D, Zhao J, Wei Y, Li X, Luo M et al (2011) Spatiotemporal manipulation of auxin biosynthesis in cotton ovule epidermal cells enhances fiber yield and quality. Nat Biotechnol 29:453–459

Zhang B (2013) From biotransformation to agricultural application. In: Zhang B (ed) Transgenic cotton: methods and protocols, methods in molecular biology, vol 958. Springer, New York, pp 3–15

Zhang T, Hu Y, Jiang W et al (2015) Sequencing of allotetraploid cotton (*Gossypium hirsutum* L. acc. TM-1) provides a resource for fiber improvement. Nat Biotechnol 33:531–537

Zhong R, Lee C, Ye Z-H (2010) Evolutionary conservation of the transcriptional network regulating secondary wall biosynthesis. Trends Plant Sci 15:625–632

Chapter 9
Jute Genomics: Emerging Resources and Tools for Molecular Breeding

Debabrata Sarkar, Pratik Satya, Nur Alam Mandal, Debajeet Das, Pran Gobinda Karmakar and Nagendra Kumar Singh

Abstract Perhaps no other natural fibre crop is as versatile as jute (*Corchorus capsularis* L. and *C. olitorius* L.; Malvaceae s. l.) in the industrial and engineering uses of textiles. Yet, genomics research was rather delayed in this typical ligno-cellulosic bast fibre crop, which has been characterized by slow responses to classical genetic selection due to several limitations, the most important being narrow genetic base, strong sexual incompatibility and dearth of compatible genetic and/or genomic resources. Recent depreciation of the genome sizes of *Corchorus* spp. concomitant with a drastic fall in the cost of next-generation sequencing has now made it possible to explore and characterize jute genomes at affordable prices. This has resulted in the development and validation of expressed sequence tag-derived simple sequence repeat (EST-SSR) markers from transcriptomic unigenes, genome-wide discovery of single nucleotide polymorphisms (SNPs) using reduced-representation restriction-site-associated DNA (RAD) sequencing de novo, construction of a dense RAD-SNP-based genetic linkage map, detection of QTL and candidate gene analysis for bast fibre yield and its components, and generation and characterization of reference bast transcriptomes using next-generation RNA-seq. A structured association mapping panel has been developed for *C. olitorius*, and a large number of RAD-SNP markers are being used for genome-wide association mapping of complex bast fibre quality traits. Current results have further addressed some of the basic issues of jute genome biology, ranging from chromosomal evolution to comparative genomics to cellulose and lignin biosynthesis in bast fibres. Here, we have discussed the present status of jute genomics, with a historical perspective on DNA markers development and utilization. The potentials of genomic selection to accelerate the rate of genetic gain in selection for bast fibre quality traits

---

D. Sarkar (✉) · P. Satya · N.A. Mandal · D. Das · P.G. Karmakar
Biotechnology Unit, Division of Crop Improvement, ICAR-Central Research Institute for Jute and Allied Fibres (CRIJAF), Nilganj, Barrackpore, Kolkata, West Bengal 700 120, India
e-mail: debabrata_s@yahoo.com; d.sarkar@excite.com; debabrata.sarkar@icar.gov.in

N.K. Singh
ICAR-National Research Centre on Plant Biotechnology (NRCPB), Pusa, New Delhi 110 012, India

K.G. Ramawat and M.R. Ahuja (eds.), *Fiber Plants*, Sustainable Development and Biodiversity 13, DOI 10.1007/978-3-319-44570-0_9

and mutagenesis-based reverse genetic approach for developing low-lignin jute are also discussed.

Keywords Bast fibre · Genomics · Jute · Molecular breeding · Next-generation sequencing · RAD-seq · RNA-seq Single nucleotide polymorphism · Transcriptomics

9.1 Introduction

Jute (*Corchorus* spp.) is an annual, herbaceous bast fibre crop that belongs to the subfamily Grewioideae of Malvaceae s. l. (The Angiosperm Phylogeny Group 2009), grows in tropical lowland areas with 60–90 % relative humidity and 24–37 °C temperature (Rowell and Stout 2007) and consumes 15 t of CO_2 and releases 11 t of O_2 ha^{-1} (Palit and Meshram 2008). It is represented by two cultivated species, viz. *C. capsularis* L. (white jute, $2n = 2x = 14$) and *C. olitorius* L. (dark jute, $2n = 2x = 14$) and ~50–60 wild species (Edmonds 1990; Kar et al. 2009; Benor 2011; Benor et al. 2011; Sinha et al. 2011). A jute plant has a vegetative growth period of about 3–4 months and is distinguished by its morphological plasticity (Palit et al. 1996; Benor et al. 2011). Long day fosters its vegetative growth that supports bast (secondary phloem) fibre production, whereas short day induces its reproductive growth (flowering) that requires a critical day-length of 12.5 hours (Palit and Meshram 2008). At harvest, white jute attains a height of about 5–12 feet, while dark jute about 5–15 feet or more (Kundu 1956). Both the cultivated jute species are mostly self-fertilized, however, as high as 15 % natural outcrossing occurs in *C. olitorius* (Mir et al. 2008b; Sinha et al. 2011). Seeds of white jute are coppery in colour and weigh ~500–600 seeds g^{-1}, whereas those of dark jute are greyish in colour with ~1000 seeds g^{-1} (Benor 2011); they have no dormancy, however (Palit and Meshram 2008). A single jute fibre (i.e. ultimate fibre cell) which is 0.75–5.0 mm in length with an average of 2.3 mm and 15–25 μm in width (Rowell and Stout 2007) is extracted from the bark of the plant by stepping and retting in water, usually 120 days after sowing, followed by stripping, washing, squeezing and drying and bailing (Palit et al. 2006b). However, industrial jute fibres represent fibre bundles consisting of thousands of such ultimate fibre cells (Palit et al. 2006a; Meshram and Palit 2013a). Anatomically, jute fibre is a composite of bast fibre bundles, characterized by short sclerenchymatous fibre cells with thick wall and narrow lumen (Kundu and Rao 1975; Kundu et al. 2012). Thick wall of these sclerenchyma cells results from the deposition of the secondary cell wall, which is added to the inner face of the primary cell wall (Meshram and Palit 2013a). Jute fibre has typical physical properties in terms of tensile strength (30–50 g tex^{-1}), fibre fineness (2.0–5.0 tex) and fibre length:breadth ratio (110–140) (Hazra and Karmakar 2008). It is chemically characterized by cellulose (60–65 %), hemicellulose (14–16 %), lignin (15–18 %) and structural proteins, however, inherent with a high lignin content (Islam and Sarkanen 1993; Del Rio et al. 2009; Kundu et al. 2012), as compared with the other bast fibres, that detrimentally affects the yarn quality resulting in poor blending possibilities with other fine fibres, particularly cotton (Sarkar et al. 2010).

Yet, jute is one of nature's strongest vegetable fibres and is only next to cotton in terms of production (Mahapatra et al. 2012). It is perhaps the most spinnable of all natural fibres, 100 % biodegradable and recyclable (Palit and Meshram 2010). It has a high tensile strength, low extensibility, high insulating and antistatic properties, moderate moisture regain and low thermal conductivity (Rowell and Stout 2007). As the name implies, dark jute is more lustrous and golden colour, with softer and stronger fibre than white jute (Kundu et al. 1959), and thus, it is much valued for quality jute products (Kar et al. 2010). Although sacking and burlap (hessian) make up the bulk of manufactured jute products, jute yarn and twines are also used for manufacturing carpet backing clothes, curtains, chair coverings, rugs, linoleum backing, shopping and handbags, and decorative and household fabrics (Hazra et al. 2008). Increasing use of jute in composites and plastic reinforcement is opening up new possibilities for value-added diversified products (Teixeira-Pinto et al. 2009). Because jute is biodegradable and flexible, absorbs moisture and drains well, geosynthetics made from jute are used to prevent soil erosion and landslides, in addition to diversified uses in agriculture and civil engineering (Rowell and Stout 2007). Due to its unique versatility and durability, there is an increasing market for jute-blended yearns (Mahapatra et al. 2009). Whole jute with a similar strength of hardwood but with a low-lignin content and better lignin (syringyl monolignols) type is fast replacing wood in pulp and paper (Jahan et al. 2007).

With increasing uses in diversified products, jute requires, in addition to drastic yield enhancement, major qualitative changes in gross fibre architecture including texture and secondary cell wall constituents through a gain of control over inherent bast fibre characteristics (Sarkar et al. 2009). Since jute is distinguished by fibre (that matures under long day at 120 days) and seed (that matures under short day at 130 days) crops with contrasting ideotypes, it is not possible to harvest both fibre and seeds from the same plant (Kundu et al. 1959; Basak 1993). Conventional single-plant pedigree breeding that relies on indirect selection through correlated yield traits (e.g. plant height and stem-base diameter) has been historically used for yield improvement in jute (Palit et al. 1996; Palve and Sinha 2005; Karmakar et al. 2008; Kar et al. 2010; Sinha et al. 2011). However, bast fibre quality traits, such as tensile strength and fibre fineness, are complex under polygenic control, with low-to-medium heritability (Basak 1993). As a result, the genetic gain in selection for physical fibre quality traits has been frustratingly slow over the years, with the consequence that all present-day high-yielding jute varieties are crippled by average fibre quality values that render them unsuitable for being blended in finer fabrics. In terms of chemical fibre quality traits, reducing lignin content in jute fibres has been one of the long-standing research objectives that could not be realized through conventional breeding approaches due to lack of compatible breeding resources (Benor et al. 2012; Kundu et al. 2013), narrow genetic base vis-à-vis restricted breeding pool (Mir et al. 2008b) and strong sexual incompatibility, particularly between the two cultivated species (Patel and Datta 1960; Arangzeb 1994; Sinha et al. 2011). Modulating cellulose: hemicellulose ratio and/or interactions (Obembe et al. 2006) has also been proposed as an effective means to increase tensile strength of jute fibres (Sarkar et al. 2010). This makes jute an ideal crop, for which molecular breeding employing genome-wide or genomic selection (GS) together with reverse genetic approaches, such as TILLING and/or EcoTILLING, is almost indispensable.

Although the development and use of genomic resources in jute has lagged behind all other major crops including the allied bast fibre crops (i.e. flax, kenaf and ramie), recent devaluation of genome sizes in *Corchorus* spp. including its two cultivated species (Benor et al. 2011; Sarkar et al. 2011; Akashi et al. 2012) has resulted in a surge of efforts into genomics or high-throughput next-generation sequencing (NGS) approaches at affordable prices (Chakraborty et al. 2015; Kundu et al. 2015; Zhang et al. 2015b, c). Parallel with this development, a shoot tip-based stable genetic transformation system has been recently developed in *C. capsularis* (Saha et al. 2014a), which has proved to be frustratingly recalcitrant for regeneration in vitro from a variety of tissue explants. As a result, jute is increasingly being used as an attractive model for functional genomics, with a basic objective to study xylan-type bast fibre formation in relation to cell expansion, secondary cell wall deposition and cellulose and lignin biosynthesis (Chakraborty et al. 2015; Samanta et al. 2015; Zhang et al. 2015c). Here, in this chapter, we have described NGS-assisted recent advances in jute genomics, with a historical perspective on DNA markers development and utilization for QTL mapping of bast fibre yield and quality traits. We have also discussed the recent progress in jute transcriptomics, with a special reference to generation and characterization of bast transcriptomes. However, before so doing, we have briefly highlighted unique biological constrains that have limited the application of some of the recent techniques in jute, even though a few of those issues have been cursorily mentioned above.

9.2 Biological Constraints

Besides innate limitations with respect to bast fibre quality (Palit et al. 2006a; Sarkar et al. 2010), the jute crop has several biological constraints that not only limit its genetic improvement and production using conventional approaches, but also restrict the scope for deploying some of the latest molecular techniques. They are as follows: (1) narrow genetic base among the genotypes of both the cultivated jute species (i.e. restricted breeding pool) that represent only a small fraction of the existing genetic variation (Mir et al. 2008b, 2009; Benor et al. 2012; Kundu et al. 2013; Satya et al. 2014a); (2) complete reproductive isolation between the two cultivated species (Kundu 1956; Kundu et al. 1959; Islam and Rashid 1960; Patel and Datta 1960; Swaminathan et al. 1961; Haque 1987; Sinha et al. 2011) making it impossible to transfer desirable characters, particularly those of bast fibre quality-related traits, across them (Haque 1987); (3) varying degrees of sexual incompatibilities limiting the introgression of wild *Corchorus* gene pool and its effective utilization in conventional breeding programmes (Arangzeb 1994; Palve et al. 2003; Kar et al. 2009; Sinha et al. 2011); (4) contrasting photoperiod sensitivities to bast fibre development and seed production (Kundu 1956; Palit and Meshram 2008); (5) asynchronous vegetative-reproductive growth phases resulting in a prolonged interval between bast fibre and seed maturities (Kundu et al. 1959); (6) precocious reproductive behaviour restricting the vegetative growth vis-à-vis bast fibre formation (Sinha et al. 2011); (7) biological limitations on pyramiding agronomic and seed quality traits with bast fibre yield; (8) dearth of cytoplasmic male sterile (CMS) and compatible restorer lines in the entire breeding pool across the two cultivated species (Kar et al. 2009; Sinha et al. 2011); (9) post-harvest retting process to extract and process fibre strands, which continues to be a variable process across the environments depending on the quality of retting water (Basak 1993; Rowell and Stout 2007); and (10) susceptibility to various biotic (anthracnose, mosaic, stem rot, semilooper, stem weevil and yellow mite) and abiotic (cold, drought, salinity and waterlogging) stresses (Kundu 1956; Kundu et al. 1959). Lastly, it requires to be mentioned that jute is not amenable to standard and kit-based molecular biology work (e.g. laboratory extraction of macromolecules) and cell and tissue culture manipulations in vitro due to exceptionally high mucilage content in all tissues, from leaves to floral buds to anthers (Kundu et al. 1959), across developmental stages.

9.3 DNA Markers

Over the past one decade, a variety of DNA markers have been developed in jute (Table 9.1 and references therein), but most of the initial marker systems developed prior to 2006 were more of academic interest than of any applied importance. Major classes of DNA markers developed and validated in jute till date are amplified fragment length polymorphism (AFLP), CAAT box-derived polymorphism (CBDP), chloroplast-derived simple sequence repeat (cpSSR), expressed sequence tag-derived simple sequence repeat (EST-SSR), intersimple sequence repeat (ISSR), random amplified polymorphic DNA (RAPD), restriction-site-associated DNA (RAD), sequence characterized amplified region (SCAR), sequence-related amplified polymorphism (SRAP), sequence-tagged microsatellite site (STMS), (genomic) simple sequence repeat (SSR), single nucleotide polymorphism (SNP) and start codon-targeted (SCoT) polymorphism (see Satya and Chakraborti 2015). Isozyme markers developed in jute will not be discussed here, and readers are referred to Ali et al. (2012) and references therein. Readers are also referred to 'JuteMarkerdb', an integrated web portal (http://jutemakerdb.icar.gov.in; developed and maintained by Dipnarayan Saha, ICAR-CRIJAF, Barrackpore), to access all publicly available molecular markers in jute.

9.3.1 Dominant DNA Markers

Dominant markers, such as AFLP, ISSR, RAPD and SCoT, were mainly used for DNA fingerprinting and genetic diversity analyses, with a primary focus on identification of the level of variation between the two cultivated jute species

Table 9.1 DNA markers developed and validated in *Corchorus* spp. including the two cultivated jute species *C. capsularis* and *C. olitorius*

DNA marker[a]	Species typed	Assay objective	Reference
AFLP	*C. capsularis, C. olitorius*	DNA polymorphism, genetic diversity, population structure, species relationship	Hossain et al. (2003), Basu et al. (2004), Das et al. (2011), Benor (2011), Benor et al. (2012), Ghosh et al. (2014)
CBDP	*C. capsularis, C. olitorius*	Species relationship	Singh et al. (2014)
cpSSR	*C. aestuans, C. capsularis, C. depressus, C. fascicularis, C. olitorius, C. pseudocapsularis, C. pseudo-olitorius, C. tridens, C. trilocularis, C. urticifolius*	Chloroplast haplotyping, genetic diversity, species relationship	Basu et al. (2004), Kundu et al. (2013)
EST-SSR	*C. capsularis, C. olitorius*	DNA polymorphism	Zhang et al. (2014, 2015a, b, c)
GBD	*C. aestuans, C. capsularis, C. fascicularis, C. olitorius, C. pseudo-olitorius, C. siliquosus, C. tridens, C. trilocularis*	Genetic analysis, phylogenetic analysis, population structure	Satya et al. (2014a), Tanmoy et al. (2015)
RAD	*C. olitorius*	Comparative genomics, linkage mapping, QTL detection	Kundu et al. (2015)
RAPD	*C. aestuans, C. capsularis, C. incisifolus, C. olitorius, C. trilocularis*	DNA fingerprinting, genetic diversity, linkage mapping, species relationship	Hossain et al. (2002, 2003), Qi et al. (2003a, 2004), Roy et al. (2006), Haque et al. (2007, 2008), Ogunkami et al. (2010), Mir et al. (2011), Chen et al. (2014), Adeyemo and Abati (2015)
ISSR	*C. aestuans, C. capsularis, C. depressus, C. fascicularis, C. olitorius, C. pseudocapsularis, C. pseudo-olitorius, C.*	Genetic diversity, linkage mapping, species relationship	Qi et al. (2003b, 2004), Roy et al. (2006), Sultana et al. (2006), Tao et al. (2012), Chen et al. (2014), Saha et al. (2014b)

(continued)

Table 9.1 (continued)

DNA marker[a]	Species typed	Assay objective	Reference
	tridens, C. trilocularis, C. urticifolius		
SCAR	*C. capsularis*	Linkage mapping, Macrophomina phaseolina-resistance screening	Mir et al. (2011)
SCoT	*C. capsularis, C. olitorius*	Genetic diversity	Rana et al. (2012)
SNP	*C. olitorius*	Comparative genomics, linkage mapping, QTL detection	Kundu et al. (2015)
SRAP	*C. capsularis, C. olitorius*	DNA fingerprinting, genetic diversity, linkage mapping	Tao et al. (2012), Rana et al. (2013), Chen et al. (2014)
SSR	*C. aestuans, C. capsularis, C. fascicularis, C. olitorius, C. pseudocapsularis, C. pseudo-olitorius, C. tridens, C. trilocularis, C. urticifolius*	DNA fingerprinting, genetic diversity, linkage mapping, QTL detection, selection for mite resistance, population structure, species relationship	Akter et al. (2008), Keka et al. (2008), Mir et al. (2008a, b, 2009), Huq et al. (2009), Ghosh et al. (2010, 2014, 2015), Das et al. (2012a, b), Kundu et al. (2013), Topdar et al. (2013), Nag et al. (2014), Satya et al. (2014b)
STMS	*C. aestuans, C. capsularis, C. olitorius*	Genetic diversity	Roy et al. (2006)

[a]*AFLP* amplified fragment length polymorphism, *CBDP* CAAT box-derived polymorphism, *cpSSR* chloroplast-derived simple sequence repeat, *EST-SSR* expressed sequence tag-derived simple sequence repeat, *GBD* gene-based DNA, *ISSR* intersimple sequence repeat polymorphism, *RAD* restriction-site-associated DNA, *RAPD* random amplified polymorphic DNA, *SCAR* sequence characterized amplified region, *SCoT* start codon-targeted, *SNP* single nucleotide polymorphism, *SRAP* sequence-related amplified polymorphism, *SSR* simple sequence repeat, *STMS* sequence-tagged microsatellite site

(Hossain et al. 2002; Qi et al. 2003a, b; Basu et al. 2004; Roy et al. 2006; Das et al. 2011; Benor 2011; Benor et al. 2012; Rana et al. 2012; Ghosh et al. 2014; other references as in Table 9.1). The AFLPs were also used for the assessment of population structure, more exhaustively in *C. olitorius* than in *C. capsularis* (Benor 2011; Benor et al. 2012; Ghosh et al. 2014). All these dominant-marker studies revealed that (1) the level of variation between the two cultivated jute species is high, (2) genetic variability present at the intraspecific level is low, with extremely low levels of diversity in *C. capsularis* varieties, (3) the level of polymorphism is higher in *C. olitorius* than that in *C. capsularis*, (4) the germplasm accessions in both the cultivated species have considerably higher level of diversity, (5) there exists a scope for broadening the genetic variability at the intraspecific level by undertaking intraspecific cross-hybridization programmes, and (6) there is a lack of correspondence between molecular differentiation and geographic origin in jute genotypes, suggesting seed trade or germplasm exchange across boundaries. Using a set of ISSR, RAPD and STMS markers, Roy et al. (2006) evaluated genetic diversity in *C. capsularis* and *C. olitorius* in relation to their two wild extant relatives, viz. *C. aestuans* and *C. trilocularis*, and suggested a polyphyletic origin of the two cultivated jute species. The historic significance of this work perhaps lies in identifying, for the first time, the role of selection in the precise differentiation between the germplasm and commercial varieties of jute (Roy et al. 2006). Based on AFLP analysis, the two cultivated species were also found to be distantly related and thought to be allopatric, sharing certain common alleles (Basu et al. 2004). That the two cultivated jute species are distantly related were also inferred from AFLP (Ghosh et al. 2014) and ISSR data (Saha et al. 2014b). In a most comprehensive study, Benor et al. (2012) used AFLPs to investigate genetic diversity and relationships in a rather large population of *C. olitorius*, representing a total of 61 worldwide accessions. Based on an estimation of the overall amount of gene flow (N_m) among populations (1.8308), they confirmed that relatively high gene flow and little genetic differentiation occurred in *C. olitorius* populations. The overall diversity was relatively higher in African than in Asian populations. The Asian accessions were found to be phylogenetically nested within the African accessions, thereby supporting an African centre of origin of *C. olitorius* (Kundu 1951, 1956, 1959; Kundu et al. 2013). Though attempts were made to use dominant markers, such as ISSR and RAPD for genetic linkage mapping in jute (Sultana et al. 2006; Haque et al. 2008; Mir et al. 2011), very few markers, except for one recent report in *C. capsularis* (Chen et al. 2014), could be successfully mapped in either of the two cultivated species (see Sect. 9.4). The application of AFLP markers that indicated a higher level of polymorphism in *C. olitorius* than that in *C. capsularis* (Das et al. 2011) to genetic linkage mapping has also proved to be rather disappointing due to lack of desired polymorphisms between the parental lines of biparental mapping populations used so far (personal communication, PK Gupta).

9.3.2 Codominant DNA Markers

As early as in 2004, Basu et al. (2004) developed eight chloroplast microsatellite markers (cpSSRs; NTCP 8, 9, 10, 12, 28, 29, 37 and 48) from *Nicotiana tabacum* and used them for genetic diversity analyses in jute. To our knowledge, this was the first report on codominant SSR marker development in jute, which revealed that the two cultivated species have different maternal origin. Subsequent studies confirmed the usefulness of genomic SSR markers for DNA fingerprinting (Akter et al. 2008) and even selection of mite-tolerant jute varieties (Keka et al. 2008). In the Department of Biotechnology (DBT, India)-sponsored collaborative research programme between ICAR-Central Research Institute for Jute and Allied Fibres (CRIJAF, Kolkata) and Ch. Charan Singh University (CCU, Meerut), large-scale genomic SSRs ($\sim$2500) were developed in *C. olitorius* cvs O-4 and JRO-524 using SSR-enriched genomic libraries (Mir et al. 2008b, 2009). Nearly half of the total SSRs (51 %) comprised trinucleotide repeats followed by di- (39 %), tetra- (6 %) and pentanucleotide (4 %) repeats, with an overall density of 1 SSR/0.43 kb (Mir et al. 2009). These genomic SSR markers that showed high transferability across the two cultivated species (Mir et al. 2009) proved to be useful for the assessment of genetic diversity and population structure in jute (Banerjee et al. 2012). In agreement with AFLP-based results (Basu et al. 2004; Das et al. 2011), a relatively higher level of genetic diversity was also detected in *C. olitorius* than in *C. capsularis* (Mir et al. 2008b, 2009). Satya et al. (2014a) employed 15 of these *C. olitorius*-specific genomic SSR markers to assess the population structure in *Corchorus* spp. and could clearly differentiate between the African and Indian *C. olitorius* populations with low admixture, suggesting their spatial differentiation during the domestication process. However, the potential of these genomic SSRs for molecular mapping/QTL analysis and comparative genomics was not realized to the extent as expected because $\sim$21 % of these SSRs could only be mapped in an intraspecific jute mapping progeny (Das et al. 2012a, b; Topdar et al. 2013; see Sect. 9.4). This might be due to low polymorphism information content (PIC) of these SSR markers combined with an inherently narrow genetic base of jute (Mir et al. 2008b). Nevertheless, a set of 38 of these single-locus polymorphic MJM (Meerut jute markers) series of genomic SSRs were used to genotype nine Asian *Corchorus* species that led to reevaluation of the origins of the two cultivated jute species (Kundu et al. 2013). The study showed that both the cultivated jute species contained $\sim$70 % of nSSR diversity present in their wild extant relatives, and this reduction in neutral genetic diversity was suggested to be due to genetic drift in the form of domestication bottlenecks (Gepts 2004). These MJM series of SSRs were also used to assess genetic diversity in *C. olitorius* (Nag et al. 2014; Ghosh et al. 2015) and its mutant gene pool (Satya et al. 2014b). Ten consensus chloroplast microsatellite markers (enriched from *N. tabacum* cpDNA; Weising and Gardner 1999) were used to haplotype *Corchorus* spp. (Kundu et al. 2013), which revealed that either *C. aestuans* or *C. pseudo-olitorius* could be the maternal progenitor of *C. capsularis*. Together with nSSR and morphometric data, chloroplast microsatellite analysis concluded that both the cultivated jute species had their origin in equatorial region of East Africa, but were domesticated, most likely independently, in Asia (Kundu et al. 2013). Current results from organelle genetic diversity of 160 genotypes of both the cultivated jute species have also suggested a possible African origin of *C. capsularis* (Basu et al. 2016).

Zhang et al. (2014) used 838 jute EST sequences (from GenBank) and developed 66 EST-SSR markers, with predominant $(AT)_n$ or $(GC\text{-})_n$ repeat motifs. These EST-SSRs were used in jute for DNA fingerprinting/genetic diversity (Zhang et al. 2015a) and genetic structure and relationship analyses (Zhang et al. 2015d). Although population structure analysis divided an association mapping panel comprising 159 accessions of both the cultivated jute species into two groups and four subgroups (Zhang et al. 2015d), there was a lack of discrimination between genetic differentiation and geographic origin, in agreement with results obtained earlier using AFLPs and genomic SSRs (Ghosh et al. 2014). Recently, 1906

EST-SSR markers were developed from transcriptomic (Zhang et al. 2015c) unigenes (48,914 unigenes) of *C. capsularis* cv. Huangma 179 (Zhang et al. 2015b). In agreement with earlier results (Zhang et al. 2014), di-, tri- and tetra-nucleotide repeat motifs were the most abundant in these newly developed EST-SSRs, but AG- and GA-rich nucleotide repeats were found to be predominant. A total of 116 of these EST-SSRs (97 polymorphic) were located in genes that encode cellulose synthases A (CesAs) and transcription factors (TFs); the major TF families localized were WRKY, MYB, MYB-related, bHLH, AP2-EREBP, AUS/IAA, GRAS and SBP (Zhang et al. 2015b). Since some of these TFs, e.g. MYB, WRKY are involved in the regulation of secondary wall formation (Samanta et al. 2015) and CesA has been identified as the key protein for cellulose biosynthesis in jute fibres (Chakraborty et al. 2015; Zhang et al. 2015c), these EST-SSRs will prove to be useful as functional markers for the improvement of jute fibre yield and quality via marker-assisted backcrossing (MAB). However, in general, the PIC values of these EST-SSRs were lower than that of genomic SSRs (Zhang et al. 2014, 2015b), though the results were based on validation of only $\sim$6 % of the total EST-SSRs enriched (Zhang et al. 2015b). The efficacy of gene-based DNA markers based on peroxidase and phenylalanine ammonia-lyase (PAL) genes for genetic and population structure analyses in *Corchorus* spp. was reported to be comparable with results obtained by using genomic SSR markers (Satya et al. 2014a). Using next-generation sequencing of the whole genome of *C. olitorius* cv. Sudan Green based on Illumina NextSeq (2 $\times$ 150 bp chemistry), we have recently developed large number of genomic SSR (7259) and InDel (44) markers, of which a total of 956 SSRs [GenBank accessions: KT889381 to KT890272 (simple) and KU841709 to KU841771 (compound)] have been validated in both the cultivated jute species. It is expected that this set of single-locus SSR markers, 885 of which are functionally annotated, will be useful as functional markers across the two cultivated jute species including their wild relatives.

The lack of an abundance of informative DNA markers, especially AFLPs and SSRs, for the construction of moderately to highly dense linkage maps (see above) necessitated the development of second-generation genetic markers in jute, employing next-generation massively parallel and multiplexed DNA sequencing. Reduced-representation restriction-site-associated DNA sequencing (RAD-seq) that involves sequencing short genomic regions surrounding all restriction sites for a given restriction endonuclease was employed to discover codominant RAD markers de novo by calling single nucleotide polymorphisms (SNPs) in adjacent sequences flanking the restriction sites (Kundu et al. 2015). In this study, partially methylation-sensitive restriction endonuclease *ApeKI* (cut site GCWGC, where W = A or T) was used to generate fragments from low-copy genic regions of *C. olitorius* Sudan Green $\times$ *bast fibre-shy* (Kundu et al. 2012) F_2 progeny and its two founders (BioSamples SAMN02838329 to SAMN02838506) and RAD libraries were constructed, which were sequenced on an Illumina HiSeq 2000 (Illumina, San Diego, California) platform following 100-bp single-read sequencing chemistry. The quality-filtered demultiplexed RAD-seq data were analysed by using the software Stacks Version 1.09 (Catchen et al. 2011, 2013) to infer codominant RAD loci de novo and call SNPs (see Kundu et al. 2015). Effects of three different parameters of the *denovo_map.pl* pipeline (of the Stacks software) on RAD marker discovery vis-à-vis SNP calling are shown in Table 9.2. Maximum numbers of polymorphic codominant RAD markers were inferred de novo at 75 % (5635 RAD markers) and 90 % (3131 RAD markers) mappable progeny, using a minimum stack depth of 3 (-*m* 3) and maximum number of three nucleotide mismatches each allowed between stacks to merge into a locus within an individual (i.e. within-individual distance, -*M* 3) and between loci (i.e. between-individual distance, -*n* 3) while merging them into the Stacks catalog (see Table 9.2 for details). The between-individual distance parameter (-*n*) is a must for detecting RAD loci that are homozygous in individuals but polymorphic between individuals (see Catchen et al. 2011). No aa/bb-type RAD markers, which are only mapable in an intercross population (e.g. F_2) derived from two inbred lines, are detected when the -*n* parameter is set to zero (Table 9.2). The F_2-type aa/bb RAD markers were used to construct the first high-density framework linkage map in *C. olitorius* (see below). The detection of a large number of genome-wide SNPs employing RAD-seq has enabled the development of tools for marker-assisted selection (MAS) in jute.

9.4 Genetic Linkage Mapping

As has been pointed out earlier, inherently narrow genetic base continues to be one of the major limitations for genetic linkage mapping in either of the two cultivated jute species. Intraspecific mapping populations are characterized by extremely low levels of parental polymorphisms, assessed using a variety of DNA markers (see above). This is not unexpected because organelle-based genetic diversity analysis has traced the parentage of modern Indian jute cultivars to one or two landraces (Basu et al. 2016). Due to strong sexual incompatibility barrier between the two cultivated species (Patel and Datta 1960; Swaminathan et al. 1961; Haque 1987; Sinha et al. 2011), an interspecific mapping population could not be constructed yet. Wild *Corchorus* species that are endemic to Asia also exhibit varying levels of bidirectional sexual incompatibilities when hybridized with the cultivated species (Arangzeb 1994; Sinha et al. 2011). In rare interspecific hybrids, a skewed distribution of the recombinant types towards the female parent often occurs in F_2 and F_3 generations, suggesting the existence of some sorts of gametic and/or zygotic mechanisms that dictate gametic selection biased against the male parent (Swaminathan and Iyer 1961; Sinha et al. 2011). Notwithstanding, several intraspecific bi- and multiparental mapping populations have been constructed in jute, and some of them have been used to construct genetic linkage maps.

9.4.1 Synthetic Mapping Populations

A list of synthetic biparental mapping populations developed in both the cultivated jute species for genetic linkage mapping is shown in Table 9.3. The earliest

Table 9.2 Effects of three major parameters for running the Stacks pipeline de novo on calling codominant RAD markers from the cleaned 92-bp sequence reads of the jute Sudan Green × *bfs* intercross F_2 population and its two founder lines at two different levels of mappable progeny, for which a genotype could be inferred

Stack depth (-*m*)[a]	Within-individual distance (-*M*)[b]	Between-individual distance (-*n*)[c]	75 % mappable progeny			90 % mappable progeny		
			Total markers	aa/bb markers[d]	aa/bb markers with 1/2/3/4 SNPs	Total markers	aa/bb markers[d]	aa/bb markers with 1/2/3/4 SNPs
3	2	2	5486	1268	956/306/6/0	3092	650	480/167/3/0
	3	3	5635	1330	924/294/110/2	3131	664	461/156/46/1
	3	0[c]	4358	0	0	2502	0	0
5	2	2	2849	760	602/141/15/2	1917	390	328/60/2/0
	3	3	3062	792	588/137/63/4	2204	399	317/56/26/0
	3	0[c]	2282	0	0	1630	0	0

The Stacks Catalog of RAD loci was filtered employing recommended criteria as applicable for a diploid genetic cross, where homozygous and heterozygous loci contain one and two stacks, respectively

[a]The minimum number of identical cleaned sequence reads used to form a stack (set in the *ustacks* program; Catchen et al. 2011, 2013)

[b]The maximum number of nucleotide mismatches allowed between stacks before merging two or more stacks into a locus (set in the *ustacks* program)

[c]The maximum number of nucleotide mismatches allowed between loci while merging them into the catalog (-*n*; set in the *cstacks* program). This parameter allows to detect loci that are homozygous in individuals but polymorphic between individuals. A zero value of catalog mismatch together with defined -*m* and -*M* parameters has been used as a control; note that no mappable markers of the aa/bb grandparental configuration are inferred in the F_2 progeny

[d]The aa/bb marker configuration denotes that loci are homozygous within grandparents but heterozygous between grandparents. These markers are only mappable in an intercross F_2 population derived from two inbred founder lines

published F_2 mapping populations of *C. olitorius* comprised a very small number of individuals (22–35), derived from parental lines that differed in cold sensitivity (Sultana et al. 2006; Haque et al. 2008) or mite tolerance (Keka et al. 2008). These populations were used to construct preliminary low-resolution, incomplete linkage maps that consisted of few ISSR/RAPD/SSR markers tagged to cold or mite tolerance in jute; at the most, 25 RAPDs could be mapped on three linkage groups spanning a total distance of 463.7 cM, with an average marker interval of 19.6 cM (Haque et al. 2008). Similarly, in *C. capsularis*, an F_2 population, consisting of only 67 individuals, was constructed from parental lines differing in resistance to *Macrophomina phaseolina* (Mir et al. 2011). A total of nine RAPD and SCAR markers, tagged to *M. phaseolina* resistance, could be assigned to two linkage groups covering a total length of 628.4 cM, with an average marker interval of 28.0 cM. The first intraspecific RIL (recombinant inbred lines) mapping population was constructed in dark jute from a biparental cross between *C. olitorious* cv. JRO-524 (♀), a leading variety with ~80 % coverage in the total jute area and *C. olitorious* mt. PPO-4 (♂), a selection from *C. olitorius* exotic accession OIJ-154. As compared to the female parent JRO-524, the male parent PPO-4 was characterized by a phenotypic marker of red-tinted pale green stem that becomes crimson red at maturity and fine fibres (1.5 tex) with high tensile strength (19.7 g tex^{-1}) and low-lignin content (13.8 %). The RIL_6 generation of this population consisting of 120 individuals was used for linkage map construction and QTL analyses (see below). Recently, Chen et al. (2014) developed an F_2 population (185 individuals) in white jute from a cross between *C. capsularis* cvs Qiongyueqing (♀) and Xinxuan No. 1 (♂). The male parent used in the cross was a pure line collected from India that ensued maximum genetic divergence between the parental lines. Since 2011, in a project funded by National Agricultural Science Fund (NASF, ICAR, India), concerted efforts were made to develop a number of biparental mapping populations in *C. olitorius*. Some of those important *C. olitorius* mapping populations that have been advanced to RIL_8 are as follows (the first accession/cultivar in a cross represents the female parent): Sudan Green × *bast fibre-shy* (*bfs*), JRO-632 × *C. aestuans*, JRO-632 × *bfs* and JRO-632 × Sudan Green; *bfs* is a thermal neutron-induced mutant of *C. olitorius* cv. JRO-632, which has been identified as a secondary phloic mutant of jute defective in bast fibre development and is distinguished by diagnostic phenotype of dissected ribbon leaves, a trait controlled by a single recessive gene (Kundu et al. 2012). Till date, the F_2 population of Sudan Green × *bfs* that comprised 176 individuals have been used for linkage mapping and QTL analyses (see below). Since simple synthetic populations based on biparental crosses fail to capture the full genetic architecture of complex traits (Kover et al. 2009), such as fibre quality (fibre fineness and tensile strength) traits as in jute (see Introduction), the development of a multiparent advanced generation intercross (MAGIC) population (i.e. RILs from multiple parents whose genomes are mixed by several rounds of intermating prior to inbreeding to produce stable inbred lines) has been initiated in *C. olitorius*. The 19 parental lines (founders) selected from a dark jute association mapping panel (BioProject PRJNA207496/BioSamples SAMN03097738 to SAMN03097962) and *C. olitorius*

Table 9.3 Details of synthetic biparental mapping populations used for genetic linkage mapping and QTL analyses in jute

Species	Mapping population[a]	Generation	Size	Reference
C. capsularis	CIM-036 × JRC-412	F_2	67	Mir et al. (2011)
	Qiongyueqing × Xinxuan No. 1	F_2	185	Chen et al. (2014)
C. olitorius	O-9897 × No. 1805	F_2	22	Sultana et al. (2006), Haque et al. (2008)
	O-72 × O-7/95	F_2	35	Keka et al. (2008)
	JRO-524 × PPO4	RIL_6	120	Das et al. (2012a, b), Topdar et al. (2013)
	Sudan Green × *bfs*	F_2	176	Kundu et al. (2015)

RIL recombinant inbred line

[a]The first accession/cultivar in the cross represents the female parent; *bfs, bast fibre-shy* mutant

Table 9.4 Parental lines used to construct multiparent advanced generation intercross (MAGIC) population of dark jute (*Corchorus olitorius*)

Sample name	Cultivar	Country	Continent	Biosample accession[a]
CoAM-224	JRO-620	India	Asia	SAMN03097961
CoAM-225	Chinsurah Green	India	Asia	SAMN03097962
CoAM-012	Sudan Green	Sudan	Africa	SAMN03097749
CoAM-001	Tanganyika-1	Tanzania	Africa	SAMN03097738
CoAM-009	Australia via Brazil	Australia	Australia	SAMN03097746
CoAM-013	Russian Green	Russia	Europe	SAMN03097750
CoAM-057	BRA NONSOONG	Brazil	South America	SAMN03097794
CoAM-010	Germany	Germany	Europe	SAMN03097747
CoAM-005	Peaking	China	Asia	SAMN03097742
CoAM-004	Bangkok	Thailand	Asia	SAMN03097741
CoAM-011	Nigeria Ibaden	Nigeria	Africa	SAMN03097748
CoAM-007	Olit-3 Burma	Myanmar	Asia	SAMN03097744
CoAM-014	KEN/DS/015C	Kenya	Africa	SAMN03097751
CoAM-038	IDN/SU/053C	Indonesia	Asia	SAMN03097775
CoAM-043	NPL/JRC/550	Nepal	Asia	SAMN03097780
CoAM-070	Golden	Pakistan	Asia	SAMN03097807
CoAM-071	Vandarpur (Binpur)-1	India	Asia	SAMN03097808
CoAM-074	Olit. Deep Red	India	Asia	SAMN03097811
CoAM-075	Wild Olit. Dwarf	India	Asia	SAMN03097812
–	Bidhan Rupali	India	Asia	–

[a]The NCBI BioSample database (http://www.ncbi.nlm.nih.gov/biosample/)

cv. Bidhan Rupali were mixed by four rounds of intermating to generate 341 F_4 outcrossed families (Table 9.4). From each F_4 family, 2–3 MAGIC lines (MLs) will be produced by selfing an F_4 plant for six generations. The MLs will prove to be a valuable genomic resource for *C. olitorius*, for the detection QTLs for complex fibre quality traits.

9.4.2 Genetic Linkage Maps

Preliminary genetic linkage maps of jute based on ephemeral populations with an inadequate number of mapable progeny, as described above (Sultana et al. 2006; Haque et al. 2008; Keka et al. 2008; Mir et al. 2011), were mostly of academic interest. In the first microsatellite-based linkage map for the JRO-524 × PPO-4 cross in *C. olitorius* (see above), 36 SSR markers were mapped on six linkage groups (LGs) that covered a total genetic distance of 784.3 cM, with an average marker interval of 21.8 cM (Das et al. 2012a). However, this linkage map, with the smallest and longest LGs spanning to 53.1 and 216.3 cM respectively, was incomplete because it could not be resolved into seven LGs as expected on the basis of haploid chromosome number of jute (n = 7). Topdar et al. (2013) constructed the first complete (7 LGs) microsatellite-based linkage map of *C. olitorius* for the same JRO-524 × PPO-4 cross (RIL_6) used earlier by Das et al. (2012a), however, mapping a different set of microsatellite loci. This complete linkage map, with 82 MJM series of SSR markers (see above), covered a total genetic distance of 799.9 cM, with an average marker interval of 10.7 cM and 96.9 % of the genome within 20 cM to the nearest marker (Fig. 9.1). LG1 had the maximum (16) and LG7 the minimum (5) number of markers, whereas LG5 had the longest (258.1 cM) and LG7 the shortest (24.8 cM) genetic distance. However, there was no relationship between the number of markers mapped and the genetic distance covered. An inability to assign a large number of SSR markers to the linkage maps for the JRO-524 × PPO-4 cross could be attributed to the availability of relatively fewer alleles at each locus with unexpectedly low PIC that resulted in only ~21 % marker polymorphism between the two parental lines, though they were phenotypically distinct (Das et al. 2012a, b; Topdar et al. 2013). In addition, segregation distortion of the SSR loci was high (61–64 %), with majority of them skewed towards the female parent (76 %). This is not surprising because jute is characterized by a skewed distribution of the recombinant types towards the female parent in F_2 and F_3 generations (Swaminathan and Iyer 1961). The first genetic linkage map in *C. capsularis* was constructed using 57 ISSR, 18 RAPD and 44 SRAP markers (Chen et al. 2014). This linkage map for the Qiongyueqing × Xinxuan No. 1 cross (see above) was resolved into eight LGs, covering a total genetic distance of 2185.7 cM with an average marker interval of 18.7 cM; LG1 had the longest (573.9 cM) and LG8 the shortest (13.3 cM) genetic distance. Interestingly, the dominant markers mapped in this study showed a much lower proportion of segregation distortion (8.3 %) than that observed in *C. olitorius* JRO-524 × PPO-4 linkage map constructed using codominant SSR markers. However, 90 % of the

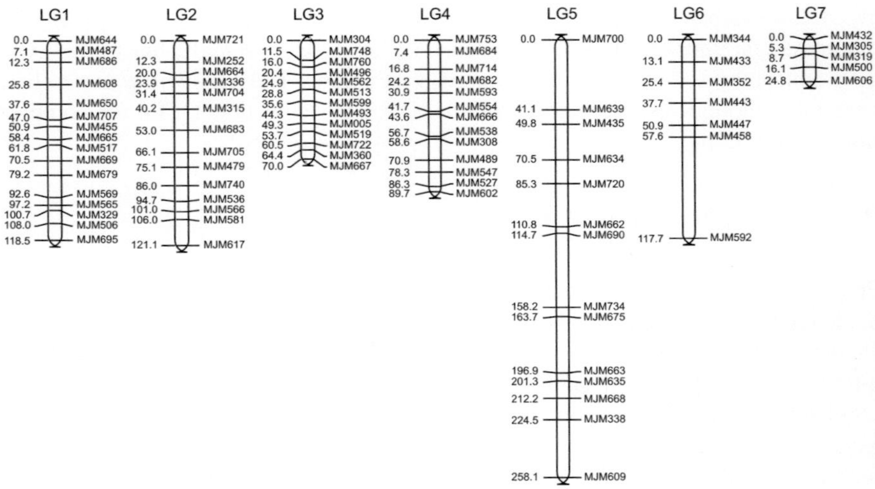

Fig. 9.1 A complete microsatellite-based genetic linkage map of dark jute (*Corchorus olitorius* L.), constructed using the JRO-524 × PPO-4 RI_6 progeny (*source* Topdar et al. 2013). Intermarker genetic distances are in Kosambi centiMorgan (cM). *LG* linkage group, *RI* recombinant inbred

distorted loci skewed towards the female parent Qiongyueqing, reiterating the hypothesis that some sorts of gametic selection may occur in jute in favour of the female gametes.

Recently, Kundu et al. (2015) developed a high-density RAD-SNP linkage map for the Sudan Green × *bfs* cross in *C. olitorius* that comprised a total of 638 SNPs (503 RAD markers) distributed over seven linkage groups spanning a genetic distance of 358.5 cM, with an average marker interval of 0.72 cM (Fig. 9.2). The genome length of this RAD-SNP linkage map was estimated to be 370.4 cM, suggesting that the linkage map represents 96.8 % of the genome length covering 87.0 % of the *C. olitorius* genome. This high-resolution linkage map of *C. olitorius* was in agreement with its karyotype (Saha et al. 2014b) and resolved with a density of 0.5 cM per SNP. The relative proportions of A/G–G/A and C/T–T/C SNPs were estimated to be 32.4 and 29.3 %, respectively. Genome-wide patterns of RAD marker segregation distortion are shown in Fig. 9.3. A total of 173 RAD markers (34.4 %) mapped on the Sudan Green × *bfs* linkage map displayed segregation distortion at $P \leq 0.05$. However, segregation distortion was found to be non-random across the linkage map, with a directional bias mostly towards the female parent for all linkage groups, except for LG3. In parallel with results obtained with microsatellite loci, a higher proportion of RAD loci (~62 %) were biased towards the female parent. Comparative mapping with other eudicot species revealed that jute (*C. olitorius*) had maximum syntenic relationships with *Theobroma cacao* (cocoa; 47.5 %) and *Gossypium raimondii* (diploid cotton; 29.2 %), both belonging to the family Malvaceae s. l. in which jute is classified. The mapped RAD loci on seven jute LGs had orthologous regions on nine out of ten cocoa chromosomes, but were not found to be colocalized to a single or two diploid cotton chromosomes (Fig. 9.4). Results implied that synteny was not conserved between jute and diploid cotton, whereas collinearity within syntenic chromosomes was not conserved between jute and cocoa (Kundu et al. 2015). Together with recent results from karyotype analysis and chromosomal evolution in Asian *Corchorus* species in relation to genome size variation (Saha et al. 2014b),

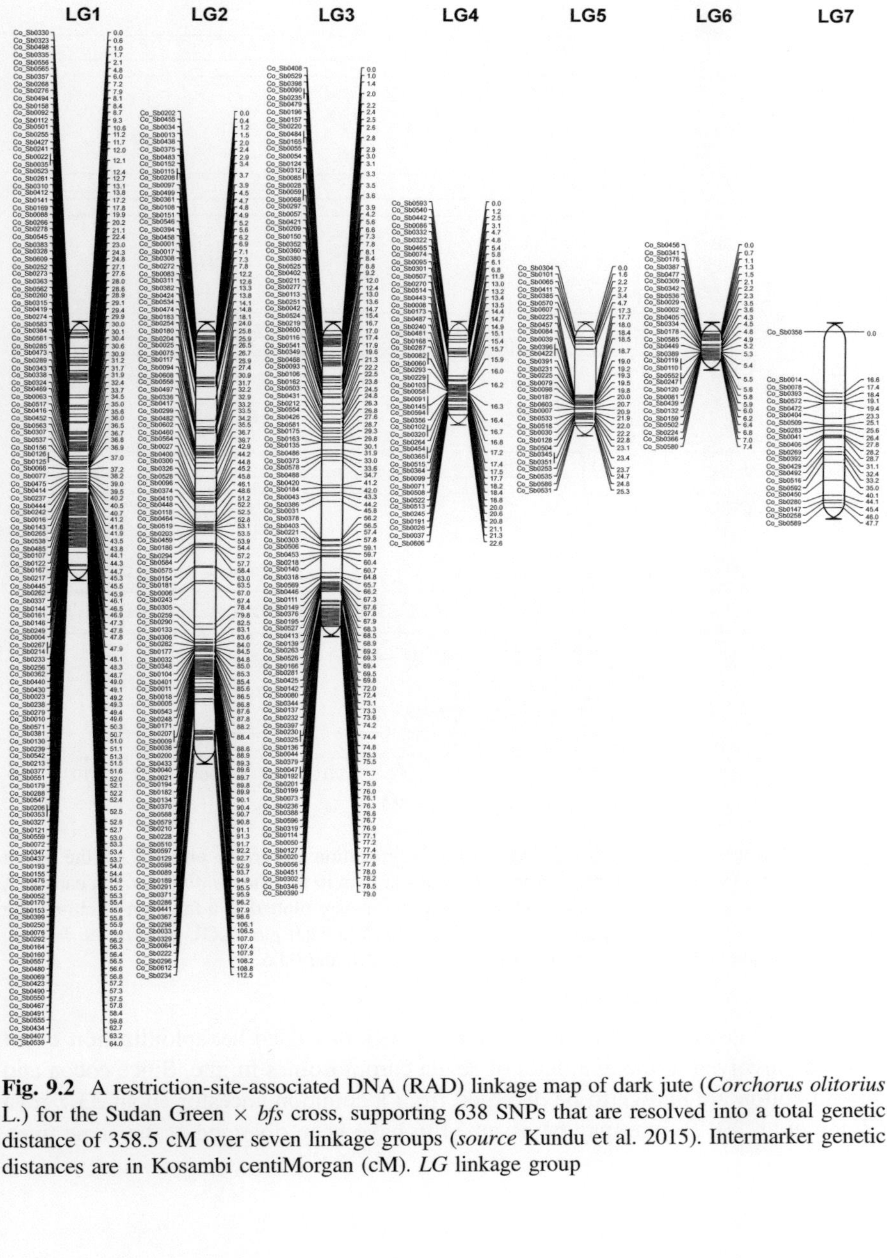

Fig. 9.2 A restriction-site-associated DNA (RAD) linkage map of dark jute (*Corchorus olitorius* L.) for the Sudan Green × *bfs* cross, supporting 638 SNPs that are resolved into a total genetic distance of 358.5 cM over seven linkage groups (*source* Kundu et al. 2015). Intermarker genetic distances are in Kosambi centiMorgan (cM). *LG* linkage group

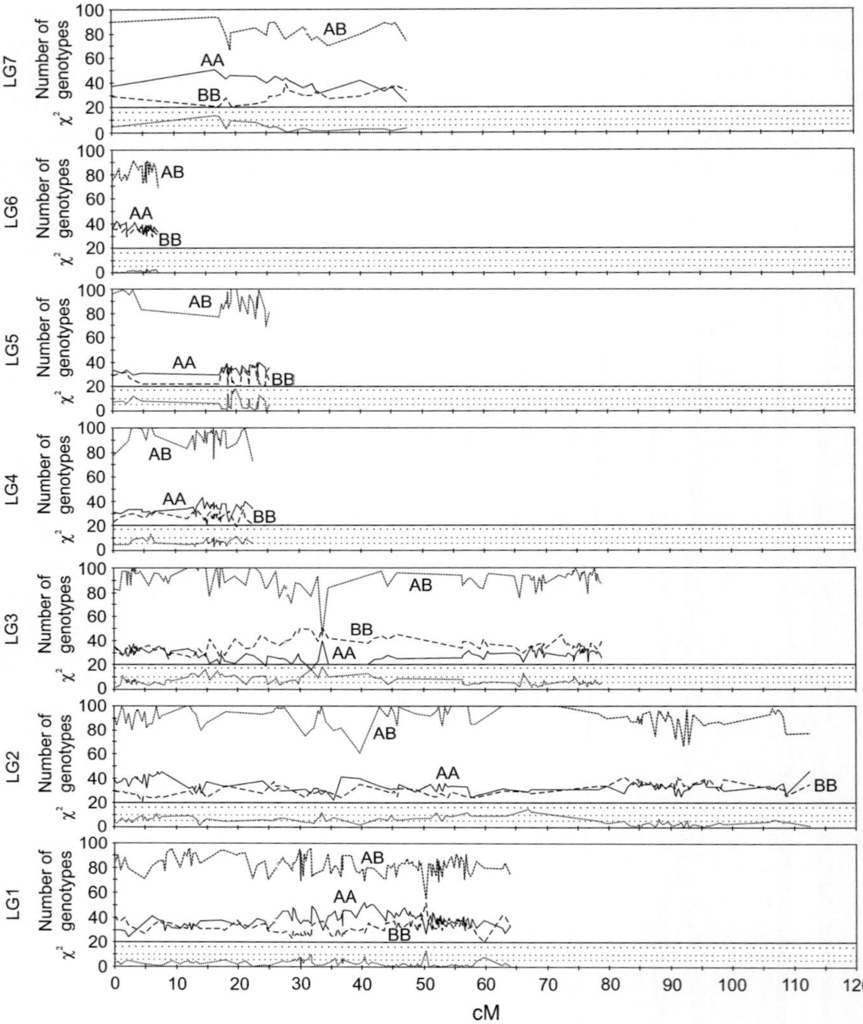

Fig. 9.3 Genome-wide patterns of RAD marker segregation distortion on each of the seven linkage groups (LGs) of the Sudan Green × *bfs* linkage map in *Corchorus olitorius*. On each LG, F_2 monogenic marker (AA, AB and BB) segregation ratios are plotted as a function of chi-square values against the marker position (cM). The *dotted lines* in each plot (LG), from *top* to *bottom*, indicate chi-square significance values at $P \leq 0.001$, 0.01 and 0.05

comparative genomics data suggest that an ancient palaeo-hexaploidization event might be involved in the evolution of seven chromosomes in jute. Since cocoa and diploid cotton are known to be diverged from a common ancestor about 33.7 mya (Wang et al. 2012), jute may be assumed to have been diverged at the same time.

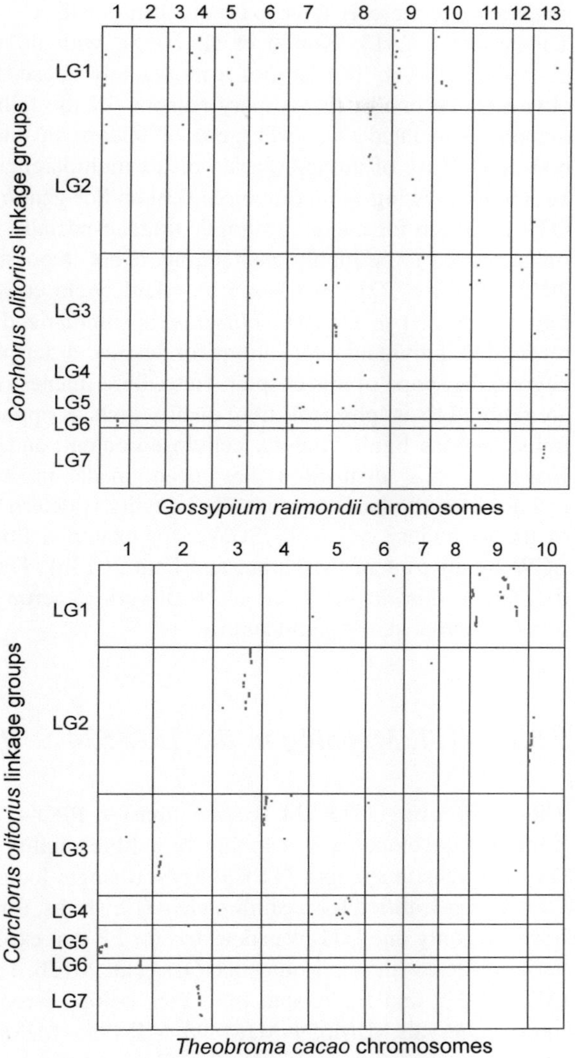

Fig. 9.4 The Oxford grids showing genome-wide comparisons of the dark jute genome (*Corchorus olitorius*) with those of cocoa (*Theobroma cacao*) and diploid cotton (*Gossypium raimondii*). Each *dot* represents the position of a homologous locus. The *x*-axis is proportional to physical length (chromosomes), while the *y*-axis is proportional to genetic distance (linkage groups) in Kosambi centiMorgan (cM)

9.5 QTL Mapping

Despite the availability of a rather large number of DNA markers, some of them even with high level of PIC and comparable cross-species transferability, over the past one decade or so (see Sect. 9.3 and Table 9.1), their use in the identification of quantitative trait loci (QTL) for major agronomic traits in jute has been limited, though attempts were made to use them for tagging traits of agronomic importance (Keka et al. 2008; Mir et al. 2008a, 2011). Till date at the time of the writing of this chapter, there are only three QTL publications for *C. olitorius* (i.e. Das et al. 2012b; Topdar et al. 2013; Kundu et al. 2015), with no reports on QTL mapping in *C. capsularis* yet. The lack of a moderately dense, framework linkage map may obviously be one of the primary reasons for the failure to identify and use DNA markers associated with QTL/gene of interest in jute (Das et al. 2012a), but the non-availability of an appropriate bi- or multiparental mapping population cannot be ignored. Owing to unique biological and/or genomic constraints (see Sect. 9.2), QTL detection for major agronomic traits, especially for complex bast fibre quality traits, proved to be highly challenging, if not impossible, in jute (see Topdar et al. 2013). A list of QTL for bast fibre yield, yield components and physical quality traits, identified so far in *C. olitorius*, is summarized in Table 9.5, but readers are referred to individual publications for detailed description of QTL effects, which are beyond the scope of this chapter. Here, fibre fineness and tensile (filament) strength are referred to as 'physical' fibre quality traits as opposed to 'chemical' fibre quality traits, such as lignin content, cellulose content, and cellulose:hemicellulose ratio. However, this distinction has been made in a literal sense without any techno-scientific basis because fibre quality is determined by chemical constituents of the secondary cell wall (SCW); for example, fibre fineness is conditioned by SCW lignin content (Meshram and Palit 2013b). Traits are abbreviated uniformly throughout this chapter, irrespective of various acronyms/conventions used in QTL nomenclatures across publications.

9.5.1 QTL Mapping in the JRO-524 × PPO-4 Progeny

The *C. olitorius* JRO-524 (coarse fibre) × PPO-4 (fine fibre) RI_6 progeny was primarily developed for mapping bast fibre quality traits, such as fibre fineness (FF) and tensile strength (TS). Based on single-locus QTL analyses, a total of 21 QTL were identified for bast fibre yield (FY) traits, for the cross JRO-524 × PPO-4; however, only one QTL was detected for FF that explained 8.3–10.6 % phenotypic variation across the environments (Das et al. 2012b; Table 9.5). Of these 21 QTL, 14 QTL for FY and FY-related traits (see below) were mapped on LG1 at the SSR marker intervals MJM659–MJM895, MJM895–MJM631 and MJM631–MJM1265, suggesting that these colocalized QTLs may either entail tightly linked QTL for different traits or individual pleiotropic QTL controlling a number of traits. This can only be resolved by QTL fine-mapping. Two-locus QTL analyses based on mixed-model composite interval mapping (MCIM) detected a total of 11 main-effect QTL (designated as M-QTL by the authors) for all FY-related traits, except for stem-top diameter (STD). M-QTL for plant height (PH), stem-base diameter (SBD), stem-mid diameter (SMD), number of nodes (NN), green biomass yield (GBY) and FY exhibited significant Q × E interactions. Only one M-QTL was detected for FF, but no QTL for TS could be identified either by single- or two-locus analysis (Das et al. 2012b). A total of 16 epistatic QTL (designated as E-QTL by the authors) that involved 9 QQ interactions were detected for six FY-related traits, such as FY,

Table 9.5 List of QTL identified for bast fibre yield, yield components and physical quality traits in *Corchorus olitorius*

Trait[a]	Cross[b]	QTL detected[c]	Linkage group[d]	LOD score[c]	R^2 (%)[e]	Reference
Plant height (PH)	JRO-524 × PPO-4	*QPh.ccsu-1.1, QPh.ccsu-1.3, qPH-l1, qPH-l2, qPH-l5, qPH-l7*	LG1, LG2, LG5, LG7	2.4–3.4	6.7–11.8	Das et al. (2012b), Topdar et al. (2013)
	Sudan Green × *bfs*	*qPH-l1, qPH-l2-1, qPH-l2-2*	LG1, LG2	3.6–4.0	6.3–9.5	Kundu et al. (2015)
Stem-base diameter (SBD)	JRO-524 × PPO-4	*QBd.ccsu-1.3, qSDB-l1, qSDB-l2, qSDB-l7*	LG1, LG2, LG7	2.7–4.4	7.3–14.8	Das et al. (2012b), Topdar et al. (2013)
	Sudan Green × *bfs*	*qSBD-l1, qSBD-l2-1, qSBD-l2-2*	LG1, LG2	3.7–5.8	6.6–10.3	Kundu et al. (2015)
Stem-mid diameter (SMD)	JRO-524 × PPO-4	*QMd.ccsu-2.3, QMd.ccsu-1.1, QMd.ccsu-1.3, QMd.ccsu-3.1, QMd.ccsu-3.3, QMd.ccsu-6.3, qSDM-l3, qSDM-l7*	LG1, LG2, LG3, LG6, LG7	2.0–4.2	8.5–36.5	Das et al. (2012b), Topdar et al. (2013)
Stem-top diameter (STD)	JRO-524 × PPO-4	*QTd.ccsu-2.4, qSDT-l1, qSDT-l3-1, qSDT-l3-2, qSDT-l5*	LG1, LG2, LG3, LG5	2.6–4.0	7.9–37.7	Das et al. (2012b), Topdar et al. (2013)
Number of nodes (NN)	JRO-524 × PPO-4	*QNn.ccsu-1.3, QNn.ccsu-1.2, QNn.ccsu-6.1, qNN-l1*	LG1, LG6	2.1–3.9	6.6–17.3	Das et al. (2012b), Topdar et al. (2013)

(continued)

Table 9.5 (continued)

Trait[a]	Cross[b]	QTL detected[c]	Linkage group[d]	LOD score[c]	R^2 (%)[e]	Reference
Fibre yield (FY)	JRO-524 × PPO-4	*QFw.ccsu-1.1, QFw.ccsu-1.2, qFY-l1, qFY-l4*	LG1, LG4	2.3–5.0	9.0–20.6	Das et al. (2012b), Topdar et al. (2013)
	Sudan Green × *bfs*	*qFY-l1*	LG1	3.6–3.9	7.5–8.1	Kundu et al. (2015)
Wood yield (WY)	JRO-524 × PPO-4	*QSw.ccsu-1.1, QSw.ccsu-1.2, QSw.ccsu-1.3, qWY-l1, qWY-l4, qWY-l5*	LG1, LG4, LG5	2.2–4.5	7.6–25.0	Das et al. (2012b), Topdar et al. (2013)
Green biomass yield (GBY)	JRO-524 × PPO-4	*QGw.ccsu-4.2, QGw.ccsu-1.1, QGw.ccsu-1.3, qGBY-l2, qGBY-l4*	LG1, LG2, LG4	2.0–3.9	6.3–13.5	Das et al. (2012b), Topdar et al. (2013)
Root weight (RW)	Sudan Green × *bfs*	*qRW-l1*	LG1	2.7–3.7	7.2–9.2	Kundu et al. (2015)
Fibre fineness (FF)	JRO-524 × PPO-4	*QFf.ccsu-5.3, qFF-l2, qFF-l3-1, qFF-l3-2, qFF-l5*	LG2, LG3, LG5	2.4–5.1	8.1–19.2	Das et al. (2012b), Topdar et al. (2013)
Tensile strength (TS)	JRO-524 × PPO-4	*qTS-l1*	LG1	3.0	11.0	Topdar et al. (2013)
Histological fibre content (FC)	Sudan Green × *bfs*	*qFC-l1*	LG1	4.5–4.7	10.2–10.6	Kundu et al. (2015)

[a]Trait acronyms presented here have been used in the text; however, traits are often abbreviated differently in QTL names depending on the convention employed

[b]Cross details are as in Table 9.3

[c]QTLs with a suffix '*ccsu*' were reported by Das et al. (2012b) and include both suggestive and definitive QTL, whereas those reported by Topdar et al. (2013) and Kundu et al. (2015) represent definitive QTL based on genome-wide LOD threshold values at $P \leq 0.05$, ignoring chromosome-wide significant LOD scores

[d]QTL detected in the JRO-524 × PPO-4 progeny were associated with microsatellite markers, whereas those in the Sudan Green × *bfs* progeny were associated with RAD-SNP markers

[e]The percentage of phenotypic variance explained by a QTL at the highest probability peak

PH, SBD, SMD, STD and NN. For FF and TS, four and six E-QTL involving two and three QQ interactions were identified. Negative QQ interactions for NN, SBD, STD, FY and FF indicated that recombinants (derived form the parents) had desirable epistatic effects on these traits. However, it must be noted here that Das et al. (2012b) used an incomplete linkage map (6 LGs; Das et al. 2012a) with only 36 mapped SSR loci (in their QTL analyses) and considered both suggestive and definitive QTL using less restrictive LOD thresholds (Table 9.5).

By comparison, Topdar et al. (2013) used a complete linkage map (82 mapped SSR loci) and detected a total of 26 definitive QTL, distributed over six LGs at genomic regions specified by 15 SSR markers, for bast fibre quality traits, FY and FY-related traits in the JRO-524 × PPO-4 progeny of *C. olitorius* (Table 9.5). Seven QTL for FY and its components including TS were colocalized on LG1 at genomic regions bracketed by three SSR markers, viz. MJM644, MJM650 and MJM679 (Figs. 9.5, 9.6, 9.7), providing a genetic basis for indirect selection for bast fibre yield based on FY components, such as PH and SBD (Kundu 1956; Kundu et al. 1959; Palit et al. 1996; Basak 1993; Palve and Sinha 2005). A QTL for FY that accounted for 9 % of the phenotypic variation was colocalized on LG4 with those for wood yield (WY) and GBY (Figs. 9.6 and 9.7). Bast fibre and wood (secondary xylem) formations are developmentally regulated by cambium function

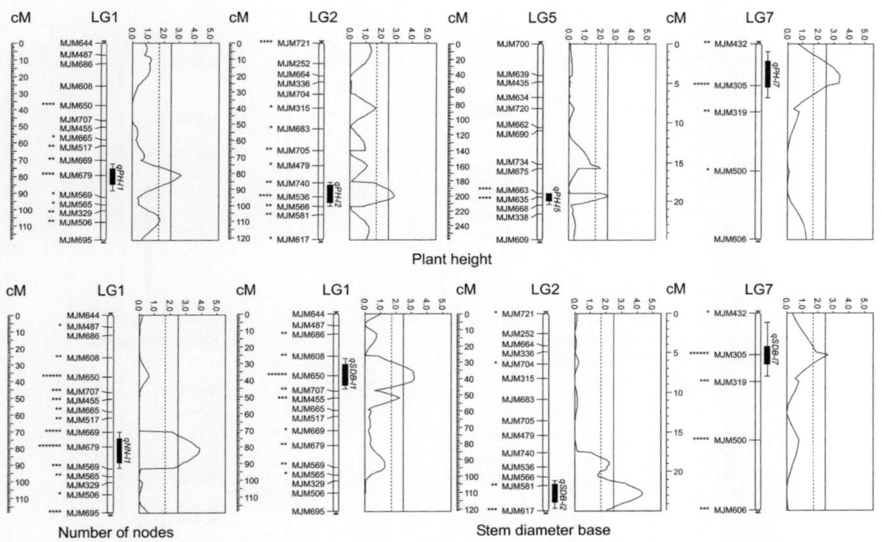

Fig. 9.5 QTL for major bast fibre yield components identified in the JRO-524 × PPO4 recombinant inbred (F_6) progeny of *Corchorus olitorius*, with graphs showing the corresponding LOD curves from multiple-QTL model (MQM) mapping. *Filled bars* and *whiskers* represent the 1-LOD and 2-LOD confidence intervals, while *dashed* and *solid lines* indicate chromosome-wide and genome-wide LOD significance thresholds, respectively. Microsatellite loci with significant *K* (coefficient of Kruskal–Wallis rank-sum test) values (nonparametric QTL mapping) are indicated by asterisks; *, **, ***, ****, *****, ****** and ******* significant at $P \leq 0.1$, 0.05, 0.01, 0.005, 0.001, 0.0005 and 0.0001, respectively

(Hazra and Karmakar 2008; Maiti and Satya 2010; Kundu et al. 2012), and therefore, the identification of a QTL for FY coincident with that for WY linked to the SSR marker MJM602 may be useful in MAS to accelerate yield improvement in jute. The female parent JRO-524, which was released in 1977 and extensively used in dark jute varietal improvement programme (Karmakar et al. 2008), was found to contribute all favourable QTL for FY and its components mapped on LG1 and LG4 genomic clusters. Unexpectedly, three out of four favourable QTLs for FF were derived from the female parent JRO-524 characterized by coarse fibre, not from the male parent PPO-4, a mutant that produces fine fibre. This might have accrued due to the creation of new gene combinations in the RI_6 progeny derived from the cross JRO-524 × PPO-4 (Paterson et al. 2003). However, the QTL identified in the JRO-524 × PPO-4 progeny needs to be validated in additional populations across environments.

9.5.2 QTL Mapping in the Sudan Green × bfs Progeny

The development of a high-density RAD-SNP linkage map for the Sudan Green × *bfs* F_2 cross has allowed to precisely map QTL for FY and its major components in *C.*

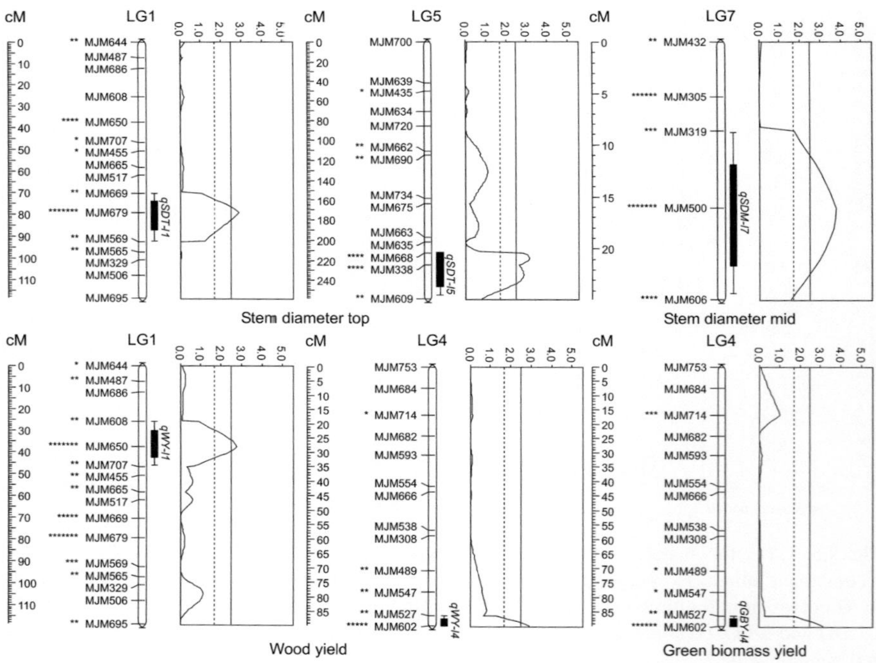

Fig. 9.6 QTL for major bast fibre yield components identified in the JRO-524 × PPO4 recombinant inbred (F_6) progeny of *Corchorus olitorius*. The other details are the same as in Fig. 5

olitorius (Kundu et al. 2015). Most importantly, the use of an F_2 mapping population has provided information, for the first time, on dominance effects of QTL for these traits in jute. In jute, retted FY is a highly variable character across the environments because the quality of retting water significantly affects the extraction of bast fibre from the stem core. To accurately measure FY prior to retting, a unique trait of histological fibre content (FC) has been identified in jute, which estimates the total number of fibre cell bundles (FCBs) in an entire stem transversal section at 90 days after sowing (Kundu et al. 2012). FC that predicates development-specific cambium activity leading to the formation of bast fibre is not only a reliable measure of the retted FY ($r_s = 0.92$, $P \leq 0.01$), but also positively correlated with PH ($r_s = 0.82$, $P \leq 0.01$) and SBD ($r_s = 0.77$, $P \leq 0.01$), the two most important FY components in jute (see above; Basak 1993; Palit et al. 1996; Mir et al. 2008b; Das et al. 2012b; Topdar et al. 2013). Genome-wide QTL scan in the Sudan Green × *bfs* progeny that detected a total of nine definitive QTL across the two environments (Table 9.5; Kundu et al. 2015) identified a single QTL for FC on LG1 on top of a single SNP marker (C/T) at 40.2 cM (Fig. 9.8). This FC QTL explained 10.2–10.6 % of the phenotypic variation and was colocalized with one QTL each for FY, PH, SBD and root weight (RW). The extremely narrow 2-LOD support intervals (0.8–4.2 cM) of this colocalized QTL cluster on LG1 indicated that MAS for FC may be an effective strategy for yield improvement in jute. Since a large population can be readily screened for FC using simple histological staining of freehand stem cross sections, implementation of FC-based selection for FY will also prove to be cost-effective and less labour-intensive in a conventional breeding programme.

All five QTL in the FC-related QTL cluster on LG1 (Fig. 9.9) had positive additive effects, implying that the female parent Sudan Green, an exotic cultivar developed in Sudan and extensively used in dark jute varietal improvement programme in India since 1950s (Karmakar et al. 2008), contributed favourable alleles to increase the trait values. However, they also showed varying degrees of partial dominance, with positive dominance always for the FC QTL but negative dominance for the other four QTL in the cluster (Kundu et al. 2015). It must, however, be noted here that QTL for FY, PH and SBD mapped on LG4, LG5 and LG7 in the JRO-524 × PPO-4 progeny (see Sect. 9.5.1) could not be detected in the Sudan Green × *bfs* progeny. This may be attributed to differences in the genetic backgrounds of the two mapping populations (Ross et al. 2006). Interestingly, two QTL linked in repulsion one each for PH (*qPH-l2-1* and *qPH-l2-2*) and SBD (*qSBD-l2-1* and *qSBD-l2-2*), with varying degrees of overdominance, were detected on LG2. For these two linked QTL, the expression of negative dominance at one of the two linked loci suggested that heterozygosity is not always favourable for PH and SBD, and higher order multilocus epistasis may be involved in the inheritance of these two complex traits, as reported earlier from classical genetical studies (Basak 1993; Palit et al. 1996). The use of a sequence-based RAD-SNP linkage map has enabled the identification of candidate genes underlying the QTL for FC- and FY-related traits (Table 9.6). Candidate genes encoding laccase 17, nitrate excretion transporter1 and Myb domain protein 36/ribosomal protein L11 family protein were associated with the FC-related QTL cluster with high fidelity. The genomic region

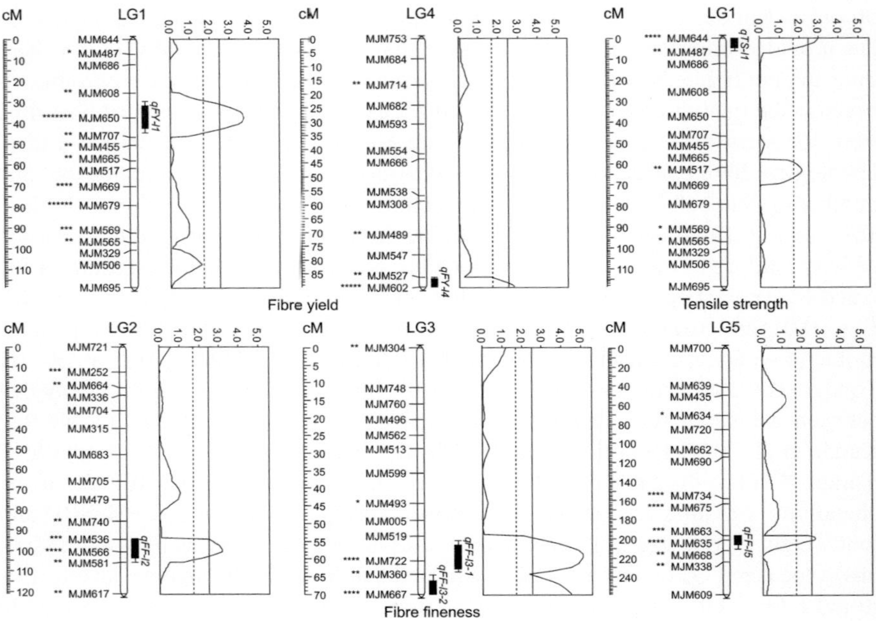

Fig. 9.7 QTL for bast fibre yield and physical quality traits identified in the JRO-524 × PPO4 recombinant inbred (F_6) progeny of *Corchorus olitorius*. The other details are the same as in Fig. 5

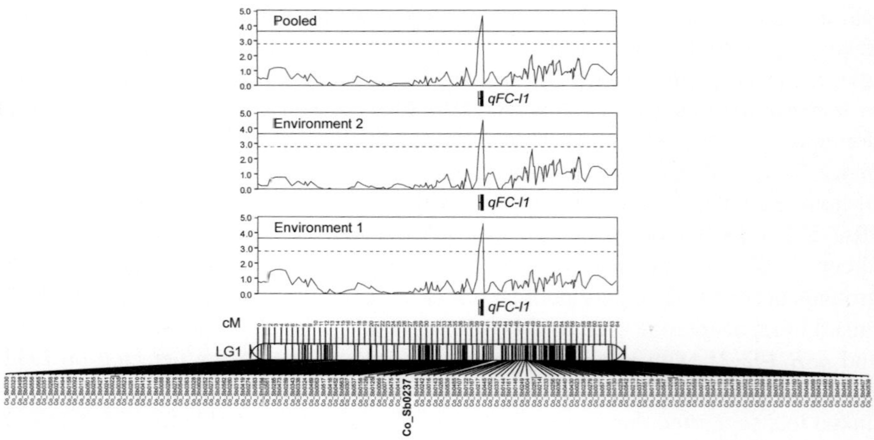

Fig. 9.8 QTL for histological fibre content (FC) identified in the Sudan Green × *bfs* F_2 progeny of *Corchorus olitorius* and mapped on top of a single SNP (C/T) RAD marker Co_Sb0237 at 40.2 cM on LG1 across the two environments and the pooled data, with 2-LOD support intervals at QTL peak positions. *Solid* and *dashed lines* indicate genome- and chromosome-wide significant LOD threshold values at $P \leq 0.05$, respectively

enclosing the two linked QTL for PH and SBD on LG2 were linked to candidate genes that encode tubulin beta 8, chromosome transmission fidelity 8-like protein isoform 1, AGC kinase family protein, low psii accumulation2 (lpa2) and EXORDIUM like 3 proteins. With the availability of whole-genome sequence of *C. olitorius*, more candidate genes are expected to be identified within the QTL detected in the Sudan Green × *bfs* progeny.

9.6 Gene Discovery and Transcriptomics

Parallel with the development and use of DNA-based markers (see Sect. 9.3), generation and characterization of expressed sequence tags (ESTs) were undertaken in jute, especially in *C. olitorius*, by way of constructing full-length cDNA libraries and screening for differentially expressed genes (Islam et al. 2005; Taliaferro et al. 2008; Ahmed et al. 2009). However, those molecular studies were mostly preliminary in nature, without any sustained efforts towards developing an EST database and characterizing those ESTs for gene discovery and downstream functional analyses. As of January 2009, ~1200 nucleotide sequences of jute were deposited in GenBank (Ahmed et al. 2009). A set of those GenBank ESTs were later utilized to develop EST-SSR markers followed by their validation in jute (Zhang et al. 2014; see above). Recently, Tao et al. (2015) constructed a full-length cDNA library of *C. capsularis*, which was analysed to obtain a total of 219 high-quality ESTs, of which 203 were functionally categorized and 61 were identified as unique sequences. Instead of a whole-genome- or transcriptome-based approach, molecular studies largely focused on identification and characterization of specific gene families that are involved in selected response pathways, e.g. stress-responsive leucine-rich repeat receptor-like kinase (LRR-RLK) (Alam et al. 2010) and a novel dehydration-responsive transcript that shows reduced expression under water deficit stress (Sharmin et al. 2011). Zhang et al. (2013) isolated the full-length cDNA of UDP-glucose pyrophosphorylase gene (*CcUGPase*) by homologous cloning and overexpressed it in *C. capsularis* to obtain an increase in cellulose content. At the genome level, isolation and characterization of the reverse transcriptase domains of long terminal repeat (LTR) retrotransposons led to identification of transcriptionally active retrotransposons in jute leaves, opening up the possibility of the development of LTR retrotransposon-based DNA markers in jute (Ahmed et al. 2011). About 31,000 Ty1-*copia* and 3600 Ty3-*gypsy* retrotransposons were reported in the jute (*C. olitorius*) genome (Ahmed et al. 2011), but Begum et al. (2013) later argued that these estimates were incorrect because of the limited specificity of the degenerate primers used, the unknown size of full-length retrotransposons reported and the inflated genome sizes of both *C. capsularis* (1000 Mbp) and *C. olitorius* (1250 Mbp) that were taken into account for computational bioinformatics. In this context, it must be noted here that most of the molecular studies including the so-called whole-genome sequencing of dark jute (by Bangladesh) reported prior to the devaluation of jute genome size (Benor et al. 2011; Sarkar et al. 2011; Akashi et al. 2012) were based on highly inflated genome sizes of

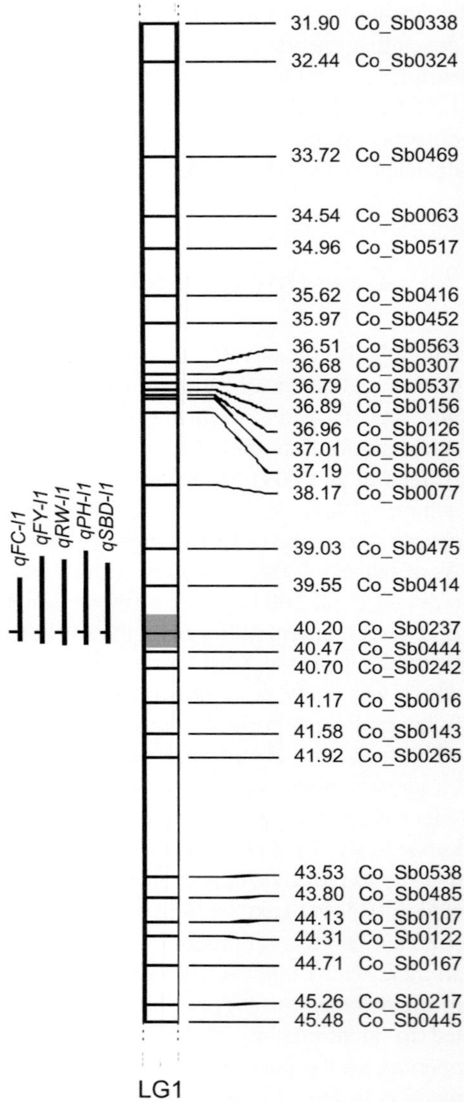

Fig. 9.9 A QTL cluster for bast fibre yield and yield-related traits on LG1 detected in the Sudan Green × *bfs* F_2 progeny of *Corchorus olitorius*, consisting of QTL for histological fibre content (FC), fibre yield (FY), plant height (PH), stem-base diameter (SBD) and root weight (RW)

both the cultivated jute species (Samad et al. 1992) that often, if not always, resulted in such incorrect estimates and interpretations therefrom.

Using differential display (DD) and reverse northern hybridization, Ray et al. (2011) analysed the transcriptomes of jute (*C. capsularis* cv. JRC-412) leaves following infection by *Macrophomina phaseolina* that causes stem rot disease, with or without treatment with β-amino butyric acid (BABA), an elicitor that activates defence responses in plants. They reported three classes of differentially expressed

Table 9.6 Candidate genes linked to QTL mapped in the Sudan Green × *bfs* progeny of *Corchorus olitorius*, using a RAD-SNP linkage map

RAD locus	Position (cM)	Candidate gene	E value
LG1[a]			
Co_Sb0066	37.2	11-oxo-beta-amyrin 30-oxidase-like/inactive protein kinase SELMODRAFT_444075-like [*Oryza brachyantha*]	1e−08
Co_Sb0077	38.2	Polymerase/histidinol phosphatase-like, putative isoform 1 [*Theobroma cacao*]	7e−06
Co_Sb0475	39.0	Laccase 17 [*Theobroma cacao*]	7e−19
Co_Sb0237	40.2	Nitrate excretion transporter1, putative [*Theobroma cacao*]	2e−13
Co_Sb0444	40.5	Nitrate excretion transporter1, putative [*Theobroma cacao*]	7e−19
Co_Sb0242	40.7	AP2.7 isoform 3/AP2.7 isoform 2 [*Theobroma cacao*]	2e−07
Co_Sb0016	41.2	11-oxo-beta-amyrin 30-oxidase-like/inactive protein kinase SELMODRAFT_444075-like [*Oryza brachyantha*]	4e−09
Co_Sb0538	43.5	Myb domain protein 36, putative/ribosomal protein L11 family protein [*Theobroma cacao*]	6e−26
Co_Sb0485	43.8	Myb domain protein 36, putative/ribosomal protein L11 family protein [*Theobroma cacao*]	3e−24
LG2[b]			
Co_Sb0254	24.0	11-oxo-beta-amyrin 30-oxidase-like/inactive protein kinase SELMODRAFT_444075-like [*Oryza brachyantha*]	4e−09
Co_Sb0180	25.8	Tubulin beta 8 [*Theobroma cacao*]	2e−33
Co_Sb0204	25.9	Tubulin beta 8 [*Theobroma cacao*]	2e−35
Co_Sb0025	26.5	Chromosome transmission fidelity 8-like protein isoform 1 [*Theobroma cacao*]	2e−19
Co_Sb0075	26.7	Methyl-CPG-binding domain protein 13, putative isoform 2/isoform 1 [*Theobroma cacao*]	4e−09
Co_Sb0117	26.9	Chromosome transmission fidelity 8-like protein isoform 1 [*Theobroma cacao*]	2e−19
Co_Sb0608	30.9	Hypothetical protein [*Citrus clementine*]	4e−15
Co_Sb0336	32.9	AGC (cAMP-dependent, cGMP-dependent and protein kinase C) kinase family protein [*Theobroma cacao*]	2e−25
Co_Sb0417	33.2	Low psii accumulation2, putative [*Theobroma cacao*]	1e−27
Co_Sb0299	33.5	EXORDIUM like 3/uncharacterized protein [*Theobroma cacao*]	2e−13

[a]RAD loci associated with the QTL cluster that comprises QTL for histological fibre content (*qFC-l1*), fibre yield (*qFY-l1*), plant height (*qPH-l1*), stem-base diameter (*qSBD-l1*) and root weight (*qRW-l1*)

[b]RAD loci associated with the two linked QTL for plant height (*qPH-l2-1* and *qPH-l2-2*) and stem-base diameter (*qSBD-l2-1* and *qSBD-l2-2*)

transcripts related to plant defence mechanism, signal transduction/gene expression and energy metabolism/other functions/unknown functions and could identify few SAR (systemic acquired resistance) genes in jute by studying the overlap between *M. phaseolina*- and BABA-induced gene expression patterns. Recently, using NGS-based transcript and microRNA (miRNA) analyses, SAR against *M. phaseolina* in *C. capsularis* has been shown to be dependent on genes involved in salicylic acid (SA), jasmonic acid (JA) and abscisic acid (ABA) pathways (Biswas et al. 2014). In the first systematic analysis of genes involved in bast fibre formation in jute, suppression subtractive hybridization (SSH) was applied, using the subtracted library of wild-type *C. capsularis* cv. JRC-321 as a tester and its fibre-deficient soft-stem mutant as a driver (Samanta et al. 2015). A total of 2685 ESTs representing the differentially expressed genes between the two genotypes were assembled into 456 unigenes, 377 of which were classified into 15 different functional categories and the rest remained functionally unaccounted for. The transcription factor WRKY was identified to have a predominant role as a regulatory element in controlling genes of different pathways that are involved in synthesizing constituents required for secondary cell wall deposition vis-à-vis bast fibre formation (Samanta et al. 2015). Despite the elegance of this SSH experiment, there are several ambiguities in the experimental approach, the most important being the isolation of total RNA from the stem tissue (that includes the core tissue representing the secondary xylem) instead of the bast tissue (stem outer layer consisting of epidermis, cortex and phloem FCBs). The CDNA library was enriched in sequences derived from genes that are upregulated at the fibre-forming stage (as opposed to non-fibre-forming stage) in the wild type, but it is not unlikely that genes differentially expressed are also related to cell wall deposition in secondary xylem tissue, especially when most of the soft-stem jute mutants are known to have thicker xylem elements as compared to their wild-types (Sengupta and Palit 2004).

For a crop such as jute without reference genome sequence, recent availability of high-throughput RNA-seq approach based on next-generation sequencing (NGS) has accelerated the pace of inexpensive development of whole or tissue-specific transcriptomes de novo. From an elite *C. capsularis* cv. Huangma 179, Zhang et al. (2015c) assembled the whole transcriptome comprising a total of 48,914 unigene sequences (accession number SRP060467 in the NCBI SRA database) and identified major bast-related genes involved in cellulose biosynthesis. Although this transcriptomic resource is not available in the public domain yet, it was analysed to develop and characterize 1906 EST-SSR markers (Zhang et al. 2015b; see Sect. 9.3.2). A total of 227 known microRNAs (miRNAs) and 17 novel miRNA candidates were identified in *C. olitorius* by deep sequencing of small RNA based on Illumina HiSeq platform (Islam et al. 2015). Of the 79 identified miRNA families, 116 potential target genes were predicted for 39 families, with miR397 family (designated as col-miR397a in *C. olitorius*, with the sequence UCAUUGAGUGCAGCGUUGAUG) having the highest number of target genes, all of which encode laccase, an ubiquitous cell wall-bound enzyme involved in lignin biosynthesis (polymerization) in higher plants (Hao and Mohnen 2014) including jute (Chakraborty et al. 2015). Incidentally, *LAC17* that encodes laccase 17 was one of the candidate genes linked to the FC QTL (*qFC-l1*) mapped on LG1 in the Sudan Green × *bfs* progeny of *C. olitorius* (Kundu et al. 2015; Table 9.6; see Sect. 9.5.2). Thus, overexpressing miR397 could be an attractive option, as also reiterated by Islam et al. (2015), for developing low-lignin jute fibres using the transgenic approach. Among the novel miRNAs reported in jute, col-miRN7 had maximum number of target genes, most of which encode NB-ARC domain containing protein, which is involved in pathogen recognition and activation of innate immune responses in plants (van Ooijen et al. 2008).

9.6.1 Bast Transcriptomes

The first reference bast transcriptomes of jute (*C. capsularis*) were generated by Illumina (Illumina HiSeq 2000 platform) paired-end sequencing and de novo assembly (Chakraborty et al. 2015). A total of 34,163 and 29,463 unigenes, with average lengths of 1442 (N50: 1999 bp) and 1136 (N50: 1528 bp) bp were assembled de novo in wild-type *C. capsularis* cv. JRC-212 (GenBank accession GBSD00000000.1) and its mutant *dlpf* (deficient lignified phloem fibre) [GenBank accession GBSE00000000.1], respectively. The size distributions of the coding sequences (CDS) of the unigenes and predicted proteins therefrom are shown in Fig. 9.10. The bast transcriptome of the wild-type jute was characterized by the longest average length and highest proportion of unigenes ≥ 1 kb, assembled till date in plants using Illumina paired-end transcriptome sequencing. About 77–79 % unigenes were functionally annotated, assigned to COG (clusters of orthologous groups) and GO (gene ontology) classifications (Fig. 9.11) and mapped to 189 KEGG (Kyoto Encyclopedia of Genes and Genomes) pathways. Unigenes representing the bast transcriptomes showed high sequence similarities with *Vitis vinifera* (grape), *Ricinus communis* (castor) and *Populus trichocarpa* (poplar), in agreement with that reported for EST-derived unigene sequences obtained by characterizing a full-length normalized cDNA library constructed from leaf tissues in *C. capsularis* (Tao et al. 2015). The KEGG-pathways mapping showed that metabolic processes underlying bast fibre formation mainly involve carbohydrate and amino acid metabolism and transport, with an active role of a combination of hormonal signals transduced via multiple hormone signal transduction pathways. Comparative analysis of the bast transcriptomes between the wild-type and its mutant identified major genes and their isoforms, such as cellulose synthase A (*CcCesA1* to *CcCesA12*), cellulose synthase-like (*CcCsl1*), sucrose synthase (*CcSuSy1* to *CcSuSy5*), β-galactosidase (*CcBGAL1* to *CcBGAL8*) and fasciclin-like arabinogalactan (*CcFLA1* to *CcFLA19*), involved in synthesizing components required for secondary cell wall formation in jute fibres (Chakraborty et al. 2015). Differential expression analyses based on qRT-PCR revealed that *CcCesA7* is secondary cell wall-specific in bast tissues; *CcFLA6* is involved in coordinating the deposition of S-layers in the xylan-type secondary cell walls of jute fibres, while *CcFLA15*, an ortholog of *AtFLA17* and *AtFLA18* of *Arabidopsis thaliana*, is involved in coordinating the developmental transition of phloem fibres from elongation to secondary cell wall deposition. Further, bast transcriptomics allowed the identification of a large number of bast-related transcription factors (TFs) and transcriptional regulatory families, particularly members of zinc fingers, MADS,

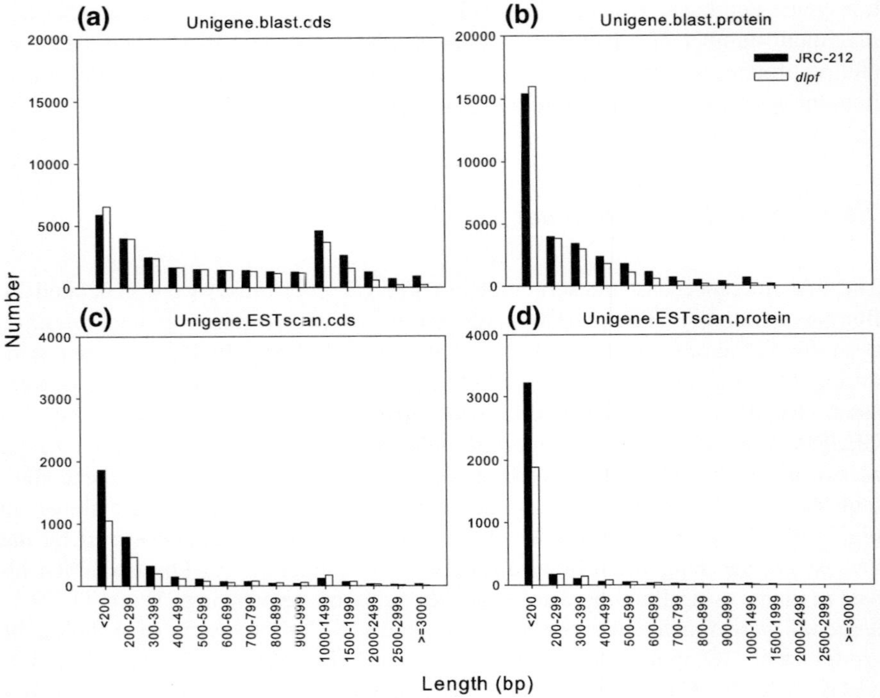

Fig. 9.10 An overview of the Trinity (version r2013-08-14; Grabherr et al. 2011)-assembled bast transcriptomes of *Corchorus capsularis* cv. JRC-212 and its mutant *dlpf*, in terms of the size distributions of coding sequences (CDS) (**a**), CDS-predicted proteins (**b**), ESTScan-derived ESTs (**c**) and ESTScan-derived proteins (**d**). The CDS were extracted from unigene sequences by searching against Nr, SwissProt, KEGG (Kanehisa et al. 2012) and COG (Tatusov et al. 2003) databases in priority order using BLASTx (E value < 10^{-5}), and then, proteins were predicted from the CDS. For unigenes without any BLASTx hits in either of the protein database, the CDS were predicted by ESTScan v3.0 (Iseli et al. 1999) and then translated into peptide sequences

FAR1, WRKY, NAC and MYB-related, which are known to regulate secondary cell wall formation and lignin biosynthesis (see below). C3H-type zinc finger and MADS were the most abundant types of TFs expressed in jute bast tissues. A total of 236 wild-type unigenes had high sequence similarity with WRKY, which was documented to have a predominant role in bast fibre formation in jute (Samanta et al. 2015).

Perhaps, the most significant application of bast transcriptomics was the discovery of genes and pathways associated with lignin biosynthesis in jute fibres. A total of 81 isoforms of 26 genes involved in lignin biosynthesis were identified in bast fibres (Chakraborty et al. 2015). Metabolic pathway mapping showed that lignin biosynthesis occurs in jute fibres via well-conserved upstream shikimate-aromatic amino acid and downstream monolignol (Fig. 9.12) biosynthetic pathways, which

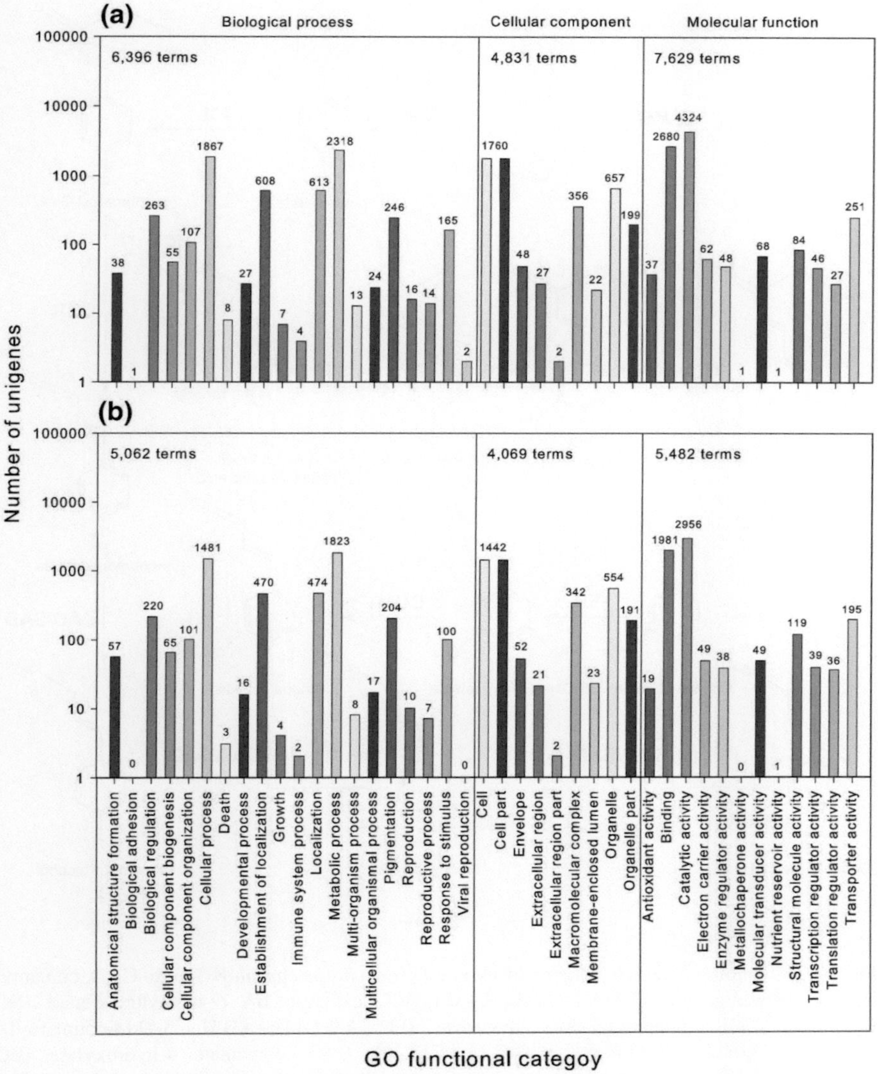

Fig. 9.11 Gene ontology (GO) annotations of the unigenes by Blast2GO (Conesa et al. 2005), representing the bast transcriptomes of *Corchorus capsularis* cv. JRC-212 (**a**) and its mutant *dlpf* (**b**)

are characterized by three important metabolic branch points at the metabolite levels of chorismate, phenylalanine (Phe) and *p*-coumaroyl CoA. In plants, Phe, a gate-keeper metabolite in the monolignol biosynthetic pathway, is essentially synthesized from prephenate via the arogenate route (Maeda and Dudareva 2012). However, a large number of PPY-AT (phenylpyruvate aminotransferase) isoforms that convert

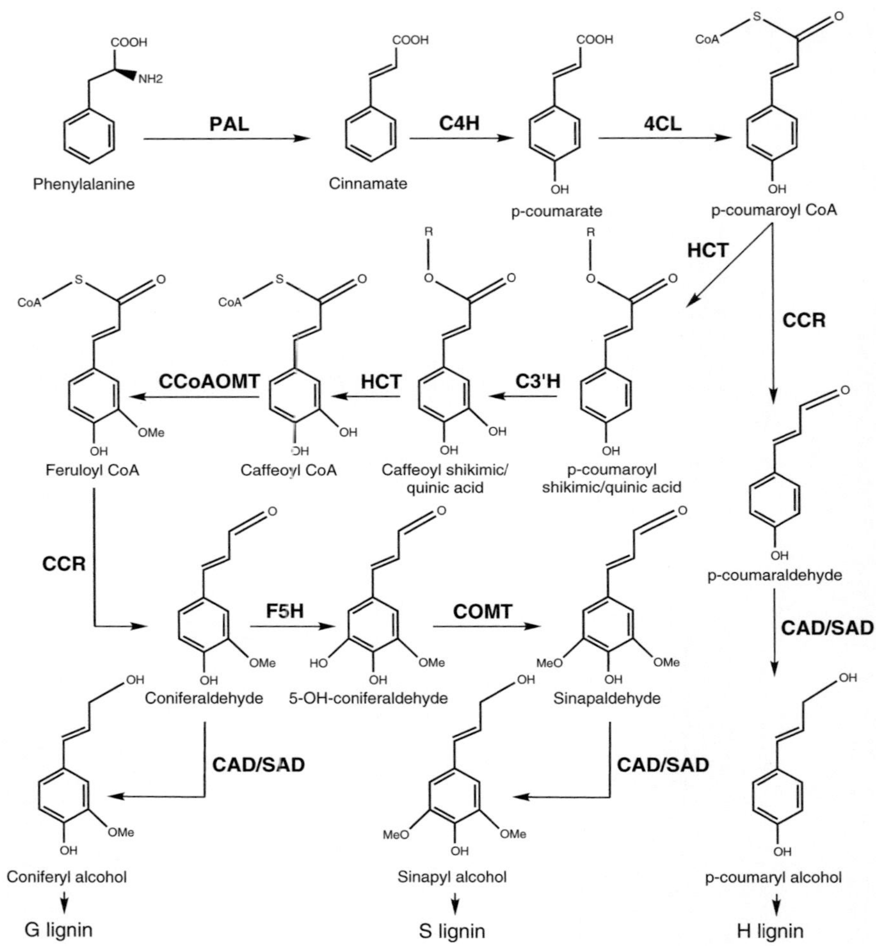

Fig. 9.12 Monolignol biosynthetic pathway in jute (*Corchorus capsularis*) fibre. CAD, cinnamyl alcohol dehydrogenase (EC 1.1.1.195); CCoAOMT, caffeoyl-CoA *O*-methyltransferase (EC 2.1.1.104); CCR, cinnamoyl-CoA reductase (EC 1.2.1.44); C3′H, 5-*O*-(4-coumaroyl)-D-quinate/shikimate 3′-hydroxylase (EC 1.14.13.36); C4H, cinnamate 4-hydroxylase (EC 1.14.13.11); 4CL, 4-coumarate:CoA ligase (EC 6.2.1.12); COMT, caffeic acid *O*-methyltransferase (EC 2.1.1.68); F5H, ferulate 5-hydroxylase (EC 1.14.-.-); HCT, shikimate *O*-hydroxycinnamoyltransferase (EC 2.3.1.133); PAL, phenylalanine ammonia-lyase (EC 4.3.1.24); SAD, sinapyl alcohol dehydrogenase (EC 1.1.1.195)

PPY to Phe were discovered, indicating that Phe is also synthesized in jute fibres via the PPY route. A larger number of HCT (shikimate *O*-hydroxycinnamoyltransferase) isoforms that catalyse the conversion of *p*-coumaroyl CoA to *p*-coumaroyl shikimic/quinic acid in the monolignol biosynthetic pathway (Fig. 9.12) were expressed in bast tissues, implying that the metabolic flux from the *p*-coumaroyl

CoA branch point is mostly routed to the monolignol instead of flavonoid biosynthetic pathway. These two metabolic characteristics of lignin biosynthetic pathways may perhaps account for increased accumulation of monolignols in jute fibres. Of all the monolignol genes, the highest number of homologous isoforms (*CcCAD1* to *CcCAD10*) was identified for CAD (cinnamyl alcohol dehydrogenase) that converts cinnamaldehydes to their corresponding alcohol derivatives. Since lignin deficiency in mutant bast fibre was associated with CAD (*CcCAD7*) disruption irrespective of developmental stages, *CAD* has been identified as a promising target for developing low-lignin jute fibres using the transgenic approach.

9.7 Conclusions and Future Prospects

As compared to major selfing crops, jute is characterized by an extremely narrow genetic base. And this has been one of the primary constraints that had limited the use of DNA markers and implementation of MAS/MAB in jute improvement. For example, in a large diversity panel comprising 292 genotypes of both *C. capsularis* and *C. olitorius*, the average number of microsatellite alleles per locus and PIC were 3.46 and 0.198–0.203, respectively (Banerjee et al. 2012), which are relatively much lower than that reported in other major selfing crops (Gupta and Varshney 2000; Kalia et al. 2011). Likewise, the AFLP analysis in a total of 61 *C. olitorius* accessions consisting of wild populations and cultivars from both Africa and Asia revealed very low genetic diversity within populations, with a mean Nei's genetic diversity (H_e) of 0.0763 (Benor et al. 2012). The frequency of transcriptomic unigenes containing at least one EST-SSR was reported to be relatively low (3.9 %), but the PIC values of these newly developed genic SSRs were found to be even lower than that of genomic SSRs (Zhang et al. 2015a, b). It is thus not possible at this stage to predict the potential usefulness of these functional microsatellite markers in genomic-assisted breeding approaches in jute that fundamentally require the deployment of high-density markers (Desta and Ortiz 2014). The recent discovery of biallelic SNP markers and their use in developing a dense linkage map and QTL mapping of bast fibre yield-related traits (Kundu et al. 2015) would enable the development of tools for MAS/MAB in jute. However, the challenge lies in saturating the jute genome, though it is relatively smaller than that of other major crops including natural fibre crops, with more SNPs using multiparental populations, e.g. MAGIC lines instead of relying on RI lines from biparental intraspecific crosses. It is well known that multiallelic SSRs are more informative than biallelic SNPs when performing diversity and relatedness analyses (Hamblin et al. 2007). SNP haplotypes that combine information from several SNPs within the same locus have been shown to provide a partial solution to this inherent disadvantage of biallelic SNP markers (Yan et al. 2010). RAD-seq allows the recovery of SNP haplotypes with multiple, up to 4, SNPs (Catchen et al. 2011, 2013). About 21 % RAD loci mapped in *C. olitorius* had more than one SNP per locus (Kundu et al. 2015). Further enrichment, preferably employing DD (double-digest)-RAD-seq, and the use of RAD markers with multiple SNPs per locus are likely to bolster the diversity and relatedness analyses in jute. No physical maps of jute are available yet; however, FISH (fluorescent in situ hybridization) karyotypes of both the cultivated species based on major satellite repeats (*Co*Sat I) and LTR retrotransposons (*Co*Retro I and *Co*Retro II) have been developed (Begum et al. 2013). Recently, a set of EST loci have been physically mapped on *C. olitorius* chromosomes by using FISH (Joshi et al. 2014). For integrating genetic linkage and chromosomal maps, a set of chromosome arm-specific RAD markers from the most distal position of each LG can be localized by multicolour FISH followed by the development of TRAP (target region amplification polymorphism) markers (Hu and Vick 2003) to terminally close the RAD-based genetic linkage map (Paesold et al. 2012).

Despite concerted efforts over the past several decades, no significant improvement (i.e. genetic gain in selection for) in physical fibre quality traits such as fibre fineness and tensile strength could be realized in jute. These two polygenic traits are not only complex with low-to-medium heritability (see Basak 1993), but also detrimentally influenced by the post-harvest biological retting process used to extract fibres from the jute stem (Rowell and Stout 2007). There are no histo-anatomical methods that can be used to assess the quality parameters in jute, for germplasm evaluation *en masse* (Maiti and Satya 2009; Maiti et al. 2011). The spinnable units of jute fibres are not simple cells such as in cotton and ramie, but pieces of filaments composed of ultimate fibre cells (see Sect. 9.1). Fibre quality is negatively affected by meshiness (i.e. network of fibre strands) of fibre bundles (Hazra and Karmakar 2008), which is thought to be under genetic control (Chen et al. 1990). Several of these interrelated factors impose serious constraints on genetic dissection of jute fibre quality traits and their improvement via phenotypic selection (PS). No major QTL for fibre fineness and tensile strength could be detected in the JRO-524 × PPO-4 progeny, the sole intraspecific RIL population suitable for mapping fibre quality traits in *C. olitoriu* till date (Das et al. 2012b; Topdar et al. 2013). Therefore, genome-wide association mapping has been initiated to dissect the complex architecture of fibre quality traits in jute (BioProject PRJNA207496). A structured association population consisting of 225 diverse accessions of *C. olitorius* (BioSamples SAMN03097738 to SAMN03097962) has been developed and genotyped by RAD-seq (SRA Accession SRP064554) to obtain a set of 1115 SNPs, with a call rate of >0.95 and minor allele frequency (MAF) of >5 %. Expectedly, this association study would enable the dissection of complex bast fibre quality traits in jute. Emerging results indicate that bast fibre quality traits are controlled by 'minor' genes of small effects, and therefore, genomic selection (GS) based on the prediction of the highest genomic estimated breeding values (GEBVs) will be a promising tool for augmenting fibre quality improvement in jute. However, the GEBV prediction accuracy is significantly influenced by trait heritability; high-heritable traits are positively correlated with higher GEBV prediction (Desta and Ortiz 2014). And this is one of the challenges for implementing GS in jute because of low-to-medium heritability (23–57 %) of physical fibre quality traits (Singh and Gupta 1985; see Basak 1993; Topdar et al. 2013). In jute, phenotyping of fibre fineness and tensile strength is performed using archaic methods based on airflow fineness (Sinha and Bandopadhyay 1968) and bundle strength (Bandopadhyay and Mukhopadhyay 1964) testers, respectively, which are not only cumbersome but also highly labour-intensive and time-consuming. Since reliable and rapid high-throughput phenotyping is required to improve the efficiency of GS (Jannink et al. 2010; Dhondt et al. 2013), there is a need to upgrade the existing phenotyping method for tensile strength and to employ high-throughput image-based technique for measuring fibre fineness in jute. Lignin estimation in retted fibre by chemical analysis (Sengupta et al. 1958) or analytical pyrolysis (Del Rio et al. 2009) is cumbersome and time-consuming, while FibreCap™ technique of the Fibretec™ 2021 system (FOSS Analytical A/S, Hilleroed, Denmark) that determines acid detergent lignin (ADL) is exorbitantly expensive, though it provides a reliable option for high-throughput screening. After harvest, phenotyping of fibre quality traits in jute is usually delayed, on an average, by 30 days due to biological retting and other associated post-harvest processes (see above). In the absence of high-throughput phenotyping, breeding cycles will be further extended (Cabrera-Bosquet et al. 2012) resulting in a lower rate of annual genetic gain per unit time and cost in GS (Desta and Ortiz 2014). These are some of the pressing issues that need to be addressed prior to implementing biparental or multifamily GS for fibre quality improvement in jute, with a compromise between marker density and training population size.

Recent results indicate a negative relationship between fineness and lignin content of jute fibre (Meshram and Palit 2013b). Though it has long been recognized that reducing the amount of lignin in bast fibre would allow more diversification of jute usage (Palit 1999; Palit et al. 2001), little attention has been paid by the geneticists/breeders to dissect the genetics of fibre lignin content and to develop low-lignin jute. This may perhaps be due to costly and cumbersome chemical method(s) associated with lignin estimation for large-scale germplasm evaluation and an inability to transfer low-lignin trait from *C. capsularis* to *C. olitorius* by interspecific hybridization. The lack of polymorphic parental lines vis-à-vis an appropriate biparental mapping population limited the scope for QTL mapping of fibre lignin content in jute. A biparental mapping population of *C. capsularis* has been recently constructed using the lignin-deficient *dlpf* mutant as a pollen donor. The usefulness of this biparental population for QTL mapping of fibre lignin content, however, seems to be limited due to disappointingly low levels of genomic SSR polymorphims between the two founder lines. The mapping potential of next-generation EST-SSRs or biallelic RAD-SNPs may be explored, but that would be too expensive to afford, especially considering the added cost that involves lignin phenotyping (see above) of bast fibres in jute. Thus, interspecific hybridization between the two cultivated species still remains an attractive option, considering the potential of rather high dividends associated with such an effort. Recently identified monolignol genes (Chakraborty et al. 2015) are promising targets for developing low-lignin jute using the transgenic approach, but it is feasible at this stage only in *C. capsularis* due to the availability of a high-throughput transgenic platform (Saha et al. 2014a). However, no reproducible high-throughput tissue culture-dependent transgenic platform exists for *C. olitorius* that occupies more than 80 % of jute growing area in the world. Reverse genetic technique such as TILLING (targeting induced local lesions in genomes) that does not require tissue culture or transgenic manipulations provides a better targeted option for developing elite breeding lines and/or varieties of *C. olitorius* with low-lignin quality fibres, with an added advantage that allelic series of mutations including hypomorphic alleles can be used for genetic analyses. With the availability of the reference draft genome of dark jute (BioProject PRJNA278717/BioSample SAMN04160039), rare lignin mutations can be discovered in *C. olitorius* populations by implementing TILLING by sequencing (Tsai et al. 2011). It is expected that NGS-driven genomic-assisted targeted breeding approaches will be more precisely applied to jute improvement once the dark jute whole-genome sequence is available.

Acknowledgments We thank Department of Biotechnology (DBT), Ministry of Science and Technology, Government of India (Grant ID: BT/PR/7143/AGR/16/672/2006), National Agricultural Science Fund (NASF), Indian Council of Agricultural Research (ICAR) (Grant ID: NFBSFARA/GB-2018/2011-12), ICAR Network Project on Transgenics in Crops (ICAR-NPTC), ICAR (Grant IDs: ICAR/NRCPB/NPTC-3052 and 3070) for financial supports. Authors who cannot be cited here due to space limitation, but whose works have significantly contributed to our present understanding of jute genomics are sincerely acknowledged.

References

Adeyemo O, Abati A (2015) Genetic diversity in *Corchorus olitorius* L. grown in south-west Nigeria inferred from RAPD data. Jordan J Agric Sci 11:329–337

Ahmed S, Nabi MZ, Alam MM, Islam MS, Samira R, Moosa MM, Khan H (2009) A computational and experimental approach for developing jute ESTs from genomic clones. Aust J Crop Sci 3:322–328

Ahmed S, Shafiuddin MD, Azam MS, Islam MS, Ghosh A, Khan H (2011) Identification and characterization of jute LTR retrotransposons: their abundance, heterogeneity and transcriptional activity. Mob Genet Elem 1:18–28. doi:10.4161/mge.1.1.16433

Akashi R, Fancy NN, Tanmoy AM, Khan H (2012) Estimation of genome size of jute (*Corchorus capsularis* (L.) var. CVL-1 using flow cytometry. Plant Tissue Cult Biotechnol 22:83–86. doi:10.3329/ptcb.v22i1.11264

Akter J, Islam MS, Sajib AA, Ashraf N, Haque S, Khan H (2008) Microsatellite markers for determining genetic identities and genetic diversity among jute cultivars. Aust J Crop Sci 1:97–107

Alam M, Sharmin S, Nabi Z, Mondal S, Islam M, Nayeem S, Shoyaib M, Khan H (2010) A putative leucine-rich repeat receptor-like kinase of jute involved in stress response. Plant Mol Biol Rep 28:394–402. doi:10.1007/s11105-009-0166-4

Ali MN, Ghosh A, Sasmal BG, Sarkar HK, Das PK (2012) Isozyme diversity in selected leaf mutants of 'tossa' jute (*Corchorus olitorius* L.). Indian J Biotechnol 11:333–336

Arangzeb S (1994) Cross compatibility of eight wild species of jute with cultivars and among themselves. Dissertation, Dhaka University, Dhaka

Bandopadhyay SB, Mukhopadhyay SK (1964) Assessment of jute fibre bundle strength. Jute Bull 27:193–199

Banerjee S, Das M, Mir RR, Kundu A, Topdar N, Sarkar D, Sinha MK, Balyan HS, Gupta PK (2012) Assessment of genetic diversity and population structure in a selected germplasm collection of 292 jute genotypes by microsatellite (SSR) markers. Mol Plant Breed 3:11–25. doi:10.5376/mpb.2012.03.0002

Basak SL (1993) Quantitative genetics of fibre yield and its components. In: Denton IR (ed) Review on the genetics and breeding of jute. International Jute Organization, Dhaka, pp 51–95

Basu A, Ghosh M, Meyer R, Powell W, Basak SL, Sen SK (2004) Analysis of genetic diversity in cultivated jute determined by means of SSR markers and AFLP profiling. Crop Sci 44:678–685. doi:10.2135/cropsci2004.6780

Basu T, Satya P, Sarkar D, Kar CS, Mitra J, Karmakar PG (2016) Organelle genetic diversity in a global collection of jute (*Corchorus capsularis* and *C. olitorius*, Malvaceae). S Afr J Bot 103:54–60. doi:10.1016/j.sajb.2015.09.016

Begum R, Zakrzewski F, Menzel G, Weber B, Alam SS, Schmidt T (2013) Comparative molecular cytogenetic analyses of a major tandemly repeated DNA family and retrotransposon sequences in cultivated jute *Corchorus* species (Malvaceae). Ann Bot 112:123–134. doi:10.1093/aob/mct103

Benor S (2011) Phylogeney of the genus *Corchorus* (Malvaceae s. l.) and diversity analyses in selected species: evidence from morphology, flow cytometry, and molecular data. Dissertation, Universität Kassel, Kassel

Benor S, Fuchs J, Blattner FR (2011) Genome size variation in *Corchorus olitorius* (Malvaceae s. l.) and its correlation with elevation and phenotypic traits. Genome 54:575–585. doi:10.1139/g11-021

Benor S, Demissew S, Hammer K, Blattner FR (2012) Genetic diversity and relationships in *Corchorus olitorius* (Malvaceae s. l.) inferred from molecular and morphological data. Genet Resour Crop Evol 59:1125–1146. doi:10.1007/s10722-011-9748-8

Biswas C, Dey P, Karmakar PG, Satpathy S (2014) Next-generation sequencing and micro RNAs analysis reveal SA/MeJA1/ABA pathway genes mediated systemic acquired resistance (SAR) and its master regulation via production of phased, trans-acting siRNAs against stem rot pathogen *Macrophomina phaseolina* in a RIL population of jute (*Corchorus capsularis*). Physiol Mol Plant Pathol 87:76–85. doi:10.1016/j.pmpp.2014.07.003

Cabrera-Bosquet L, Crossa J, von Zitzewitz J, Serret MD, Luis Araus J (2012) High-throughput phenotyping and genomic selection: the frontiers of crop breeding converge. J Integr Plant Biol 54:312–320. doi:10.1111/j.1744-7909.2012.01116.x

Catchen JM, Amores A, Hohenlohe P, Cresko W, Postlethwait JH (2011) Stacks: building and genotyping loci de novo from short-read sequences. G3 1:171–182. doi:10.1534/g3.111.000240

Catchen J, Hohenlohe PA, Bassham S, Amores A, Cresko WA (2013) Stacks: an analysis tool set for population genomics. Mol Ecol 22:3124–3140. doi:10.1111/mec.12354

Chakraborty A, Sarkar D, Satya P, Karmakar PG, Singh NK (2015) Pathways associated with lignin biosynthesis in lignomaniac jute fibres. Mol Genet Genom 290:1523–1542. doi:10.1007/s00438-015-1013-y

Chen SH, Lu HR, Zheng YY (1990) The genetic relationship between anatomical characters and fibre yield and quality in jute. J Fujian Agric Coll 20:378–384

Chen Y, Zhang L, Qi J, Chen H, Tao A, Xu J, Lin L, Fan P (2014) Genetic linkage map construction for white jute (*Corchorus capsularis* L.) using SRAP, ISSR and RAPD markers. Plant Breed 133:777–781. doi:10.1111/pbr.12205

Conesa A, Götz S, García-Gómez JM, Terol J, Talón M, Robles M (2005) Blast2GO: a universal tool for annotation, visualization and analysis in functional genomics research. Bioinformatics 21:3674–3676. doi:10.1093/bioinformatics/bti610

Das M, Banerjee S, Topdar N, Kundu A, Sarkar D, Sinha MK, Balyan HS, Gupta PK (2011) Development of large-scale AFLP markers in jute. J Plant Biochem Biotechnol 20:270–275. doi:10.1007/s13562-011-0058-1

Das M, Banerjee S, Dhariwal R, Mir RR, Vyas S, Topdar N, Kundu A, Khurana JP, Tyagi AK, Sarkar D, Sinha MK, Balyan HS, Gupta PK (2012a) Development of SSR markers and construction of a linkage map in jute. J Genet 91:21–31. doi:10.1007/s12041-012-0151-9

Das M, Banerjee S, Topdar N, Kundu A, Mir RR, Sarkar D, Sinha MK, Balyan HS, Gupta PK (2012b) QTL identification for molecular breeding of fibre yield and fibre quality traits in jute. Euphytica 187:175–189. doi:10.1007/s10681-011-0603-y

Del Rio JC, Rencoret J, Matques G, Li J, Gellerstedt G, Jiménez-Barbero J, Martínez AT, Gutiérrez A (2009) Structural characterization of the lignin from jute (*Corchorus capsularis*) fibers. J Agric Food Chem 57:10271–10281. doi:10.1021/jf900815x

Desta ZA, Ortiz R (2014) Genomic selection: genome-wide prediction in plant improvement. Trends Plant Sci 19:592–601. doi:10.1016/j.tplants.2014.05.006

Dhondt S, Wuyts N, Inzé D (2013) Cell to whole-plant phenotyping: the best is yet to come. Trends Plant Sci 18:428–439. doi:10.1016/j.tplants.2013.04.008

Edmonds JM (1990) Herbarium survey of African *Corchorus* L. species. International Board for Plant Genetic Resources, Rome

Gepts P (2004) Crop domestication as a long-term selection experiments. Plant Breed Rev 24:1–43

Ghosh A, Sharmin S, Islam S, Pahloan MU, Islam S, Khan H (2010) SSR markers linked to mite (*Polyphagotarsonemus latus* Banks) resistance in jute (*Corchorus olitorius* L.). Czech J Genet Plant Breed 46:64–74

Ghosh RK, Wongkaew A, Sreewongchai T, Nakasathien S, Phumichai C (2014) Assessment of genetic diversity and population structure in jute (*Corchorus* spp.) using simple sequence repeat (SSR) and amplifed fragment length polymorphism (AFLP) markers. Kasetsart J Nat Sci 48:83–94

Ghosh S, Meena K, Sinha MK, Karmakar PG (2015) Genetic diversity in *Corchorus olitorius* genotypes using jute SSRs. Proc Natl Acad Sci India Sect B Biol Sci. doi:10.1007/s40011-015-0652-4

Grabherr MG, Haas BJ, Yassour M, Levin JZ, Thompson DA, Amit I, Adiconis X, Fan L, Raychowdhury R, Zeng Q, Chen Z, Mauceli E, Hacohen N, Gnirke A, Rhind N, di Palma F, Birren BW, Nusbaum C, Lindblad-Toh K, Friedman N, Regev A (2011) Full-length transcriptome assembly from RNA-Seq data without a reference genome. Nat Biotechnol 29:644–652. doi:10.1038/nbt.1883

Gupta PK, Varshney RK (2000) The development and use of microsatellite markers for genetic analysis and plant breeding with emphasis on bread wheat. Euphytica 113:163–185. doi:10.1023/A:1003910819967

Hamblin MT, Warburton ML, Buckler ES (2007) Empirical comparison of simple sequence repeats and single nucleotide polymorphisms in assessment of maize diversity and relatedness. PLoS ONE 2:e1367. doi:10.1371/journal.pone.0001367

Hao Z, Mohnen D (2014) A review of xylan and lignin biosynthesis: foundation for studying Arabidopsis *irregular xylem* mutants with pleiotropic phenotypes. Crit Rev Biochem Mol Biol 49:212–241. doi:10.3109/10409238.2014.889651

Haque I (1987) Analysis of progenies of the cross, *Corchorus olitorius* × *C. capsularis* through tissue culture and biochemical methods. Dissertation, University of Dhaka, Dhaka

Haque S, Begum S, Sarker RH, Khan H (2007) Determining genetic diversity of some jute varieties and accessions using RAPD markers. Plant Tissue Cult Biotechnol 17:183–191. doi:10.3329/ptcb.v17i2.3239

Haque S, Ashraf N, Begum S, Sarkar RH, Khan H (2008) Construction of genetic map of jute (*Corchorus olitorius* L.) based on RAPD markers. Plant Tissue Cult Biotechnol 18:165–172. doi:10.3329/ptcb.v18i2.3647

Hazra SK, Karmakar PG (2008) Anatomical parameters of bast fibres for yield and quality improvement. In: Karmakar PG, Hazra SK, Ramasubramanian T, Mandal RK, Sinha MK, Sen HS (eds) Jute and allied fibre updates. Central Research Institute for Jute and Allied Fibres, Kolkata, pp 38–45

Hazra SK, Saha A, Karmakar PG (2008) Unexploited potentials of jute and allied fibre crops: whole plant, plant parts, stick and waste fibre. In: Karmakar PG, Hazra SK, Ramasubramanian T, Mandal RK, Sinha MK, Sen HS (eds) Jute and allied fibre updates. Central Research Institute for Jute and Allied Fibres, Kolkata, pp 297–319

Hossain MB, Haque S, Khan H (2002) DNA fingerprinting of jute germplasm by RAPD. J Biochem Mol Biol 35:414–419

Hossain MB, Awal A, Rahman MA, Haque S, Khan H (2003) Distinction between cold-sensitive and -tolerant jute by DNA polymorphism. J Biochem Mol Biol 36:427–432

Hu J, Vick BA (2003) Target region amplification polymorphism: a novel marker technique for plant genotyping. Plant Mol Biol Rep 21:289–294. doi:10.1007/BF02772804

Huq S, Islam MS, Sajib AA, Ashraf N, Haque S, Khan H (2009) Genetic diversity and relationships in jute (*Corchorus* spp.) revealed by SSR markers. Bangladesh J Bot 38:153–161. doi:10.3329/bjb.v38i2.5140

Iseli C, Jongeneel CV, Bucher P (1999) ESTScan: a program for detecting, evaluating, and reconstructing potential coding regions in EST sequences. In: Lengauer T, Schneider R, Bork P, Brutlag DL, Glasgow JI, Mewes H-W, Zimmer R (eds) Proceedings of the seventh international conference on intelligent systems for molecular biology. AAAI Press, Palo Alto, pp 138–158

Islam AS, Rashid A (1960) First successful hybrid between the two jute-yielding species, *Corchorus olitorius* L. (Tossa) × *C. capsularis* L. (White). Nature 185:258–259. doi:10.1038/185258b0

Islam AS, Sarkanen K (1993) The isolation and characterization of lignins of jute (*Corchorus capsularis*). Holzforschung 47:123–132

Islam AS, Taliaferro MJ, Lee CT, Ingram CI, Montalvo RJ, van der Ende G, Alam S, Siddiqui J, Sathasivan K (2005) Preliminary progress in jute (*Corchorus* species) genome analysis. Plant Tissue Cult Biotechnol 15:145–156

Islam MT, Ferdous AS, Najnin RA, Sarker SK, Khan H (2015) High-throughput sequencing reveals diverse sets of conserved, nonconserved, and species-specific miRNAs in jute. Int J Genom. doi:10.1155/2015/125048

Jahan MS, Al-Maruf A, Quaiyyum MA (2007) Comparative studies of pulping of jute fiber, jute cutting and jute caddis. Bangladesh J Sci Ind Res 42:425–434

Jannink J-L, Lorenz AJ, Iwata H (2010) Genomic selection in plant breeding: from theory to practice. Brief Funct Genom 9:166–177. doi:10.1093/bfgp/elq001

Joshi A, Das SK, Samanta P, Paria P, Sen SK, Basu A (2014) Chromosome-specific physical localisation of expressed sequence tag loci in *Corchorus olitorius* L. Plant Biol 16:1133–1139. doi:10.1111/plb.12158

Kalia RK, Rai MK, Kalia S, Singh R, Dhawan AK (2011) Microsatellite markers: an overview of the recent progress in plants. Euphytica 177:309–334. doi:10.1007/s10681-010-0286-9

Kanehisa M, Goto S, Sato Y, Furumichi M, Tanabe M (2012) KEGG for integration and interpretation of large-scale molecular data sets. Nucleic Acids Res 40:D109–D114. doi:10.1093/nar/gkr988

Kar CS, Kundu A, Sarkar D, Sinha MK, Mahapatra BS (2009) Genetic diversity in jute (*Corchorus* spp.) and its utilization: a review. Indian J Agric Sci 79:575–586

Kar CS, Satya P, Mitra J, Sarkar D, Sinha MK, Kundu A, Mahapatra BS (2010) Varietal development of jute and allied fibres in India. Indian Farming 60:5–9

Karmakar PG, Hazra SK, Sinha MK, Chaudhury SK (2008) Breeding for quantitative traits and varietal development in jute and allied fibre crops. In: Karmakar PG, Hazra SK, Ramasubramanian T, Mandal RK, Sinha MK, Sen HS (eds) Jute and allied fibre updates: production and technology. Central Research Institute for Jute and Allied Fibres, Barrackpore, pp 57–75

Keka SI, Samsuzzaman M, Pahloan MU, Pervin SP, Rahman MM, Khan H (2008) Identifying simple sequence repeat (SSR) marker linked to mite tolerance in jute species. Bangladesh J Bot 37:161–171. doi:10.3329/bjb.v37i2.1725

Kover PX, Valdar W, Trakalo J, Scarcelli N, Ehrenreich IM, Purugganan MD, Durrant C, Mott R (2009) A multiparent advanced generation inter-cross to fine-map quantitative traits in *Arabidopsis thaliana*. PLoS Genet 5:e1000551. doi:10.1371/journal.pgen.1000551

Kundu BC (1951) Origin of jute. Indian J Genet Plant Breed 11:95–99

Kundu BC (1956) Jute: world's foremost bast fibre. I. Botany, agronomy, diseases and pests. Econ Bot 10:103–133. doi:10.1007/BF02985322

Kundu BC, Rao NS (1975) Fine structure of jute fibre. Indian Bot Soc 54:85–94

Kundu BC, Basak KC, Sarkar PB (1959) Jute in India. The Indian Central Jute Committee, Calcutta

Kundu A, Sarkar D, Mandal NA, Sinha MK, Mahapatra BS (2012) A secondary phloic (bast) fibre-shy (*bfs*) mutant of dark jute (*Corchorus olitorius* L.) develops lignified fibre cells but is defective in cambial activity. Plant Growth Regul 67:45–55. doi:10.1007/s10725-012-9660-z

Kundu A, Topdar N, Sarkar D, Sinha MK, Ghosh A, Banerjee S, Das M, Balyan HS, Mahapatra BS, Gupta PK (2013) Origins of white (*Corchorus capsularis* L.) and dark (*C. olitorius* L.) jute: a reevaluation based on nuclear and chloroplast microsatellites. J Plant Biochem Biotechnol 22:372–381. doi:10.1007/s13562-012-0165-7

Kundu A, Chakraborty A, Mandal NA, Das D, Karmakar PG, Singh NK, Sarkar D (2015) A restriction-site-associated DNA (RAD) linkage map, comparative genomics and identification of QTL for histological fibre content coincident with those for retted bast fibre yield and its major components in jute (*Corchorus olitorius* L., Malvaceae s. l.). Mol Breed 35:19. doi:10.1007/s11032-015-0249-x

Maeda H, Dudareva N (2012) The shikimate pathway and aromatic amino acid biosynthesis in plants. Annu Rev Plant Biol 63:73–105. doi:10.1146/annurev-arplant-042811-105439

Mahapatra BS, Mitra S, Ramasubramanian T, Sinha MK (2009) Research on jute (*Corchorus olitorius* and *C. capsularis*) and kenaf (*Hibiscus cannabinus* and *H. sabdariffa*): present status and future prospects. Indian J Agric Sci 79:951–967

Mahapatra BS, Mitra S, Kumar M, Ghorai AK, Sarkar SK, Kar CS, Kundu DK, Karmakar PG (2012) An overview of research and development in jute and allied fibre crops in India. Indian J Agron 57:72–82

Maiti RK, Satya P (2009) Fibre bundle anatomy determines the yield potentials, and fibre quality of bast fibre (long fibre). Int J Agric Env Biotechnol 2:1–6

Maiti RK, Satya P (2010) Fibre bundle anatomy determines the yield potential of bast fibre (long fibre): a hypothesis. Environ Biotechnol 2:A1–A6

Maiti RK, Rodriguez HG, Satya P (2011) Horizon of world plant fibres: an insight. Pushpa Publishing House, Kolkata

Meshram JH, Palit P (2013a) Biology of industrial bast fibers with reference to quality. J Nat Fibers 10:176–196. doi:10.1080/15440478.2013.765669

Meshram JH, Palit P (2013b) On the role of cell wall lignin in determining the fineness of jute fibre. Acta Physiol Plant 35:1565–1578. doi:10.1007/s11738-012-1198-1

Mir JI, Karmakar PG, Chattopadhyay S, Chaudhury SK, Ghosh SK, Roy A (2008a) SSR and RAPD profile based grouping of selected jute germplasm with respect to fibre fineness trait. J Plant Biochem Biotechnol 17:29–35

Mir RR, Rustgi S, Sharma S, Singh R, Goyal A, Kumar J, Gaur A, Tyagi AK, Khan H, Sinha MK, Balyan HS, Gupta PK (2008b) A preliminary genetic analysis of fibre traits and the use of new genomic SSRs for genetic diversity in jute. Euphytica 161:413–427. doi:10.1007/s10681-007-9597-x

Mir RR, Banerjee S, Das M, Gupta V, Tyagi AK, Sinha MK, Balyan HS, Gupta PK (2009) Development and characterization of large-scale simple sequence repeats in jute. Crop Sci 49:1687–1694. doi:10.2135/cropsci2008.10.0599

Mir JI, Roy A, Ghosh SK, Karmakar PG (2011) Development of linkage map in F_2 population of selected parents with respect to *Macrophomina phaseolina* resistance trait using screened polymorphic RAPD and developed SCAR markers of jute. Arch Phytopathol Plant Protect 44:671–683. doi:10.1080/03235400903308883

Nag S, Mitra J, Satya P, Kar CS, Karan M, Ali N (2014) A comparison of efficiency parameters of SSR primers and genetic diversity analysis in jute (*Corchorus* spp.). Trends Biosci 7:2882–2885

Obembe OO, Jacobsen E, Visser RGF, Vincken J-P (2006) Cellulose-hemicellulose networks as target for *in planta* modification of the properties of natural fibres. Biotechnol Mol Biol Rev 1:76–86

Ogunkami LA, Okunowo WO, Oyelakin OO, Oboh BO, Adesina OO, Adekoya KO, Ogundipe OT (2010) Assessment of genetic relationships between two species of jute plants using phenotypic and RAPD markers. Int J Bot 6:7–11

Paesold S, Borchardt D, Schmidt T, Dechyeva D (2012) A sugar beet (*Beta vulgaris* L.) reference FISH karyotype for chromosome and chromosome-arm identification, integration of genetic linkage groups and analysis of major repeat family distribution. Plant J 72:600–611. doi:10.1111/j.1365-313X.2012.05102.x

Palit P (1999) Jute. In: Smith DL, Hamel C (eds) Crop yield, physiology and processes. Springer, Berlin, pp 271–286

Palit P, Meshram JH (2008) Physiology of jute yield and quality. In: Karmakar PG, Hazra SK, Ramasubramanian T, Mandal RK, Sinha MK, Sen HS (eds) Jute and allied fibre updates. Central Research Institute for Jute and Allied Fibres, Kolkata, pp 112–124

Palit P, Meshram JH (2010) Production and utilization of jute and allied fibre: potentialities and problems. In: Palit P, Sinha MK, Meshram JH, Mitra S, Laha S, Saha AR, Mahapatra BS (eds) Jute and allied fibre production, utilization and marketing. Indian Fibre Society (Eastern Region), Kolkata, pp 22–26

Palit P, Sasmal BC, Bhattacharryya AC (1996) Germplasm diversity and estimate of genetic advance of four morpho-physiological traits in a world collection of jute. Euphytica 90:49–58. doi:10.1007/bf00025159

Palit P, Sengupta G, Datta P, Meshram JH (2001) Lignin and lignification with special reference to its down regulation for the improvement of wood and bast fibre quality. Indian J Plant Physiol 6:217–228

Palit D, Meshram JH, Palit P (2006a) Biology of jute fibre quality. Sci Cult 72:379–382

Palit D, Meshram JH, Palit P (2006b) Genotypic variation in the characteristics of secondary phloem fibre cells of jute in relation to its yield and quality. J Bot Soc Bengal 60:32–37

Palve SM, Sinha MK (2005) Genetic variation and interrelationships among fibre yield attributes in secondary gene pool of *Corchorus* spp. SABRAO J Breed Genet 37:1–11

Palve SM, Sinha MK, Mandal RK (2003) Preliminary evaluation of wild species of jute (*Corchorus* species). Plant Genet Resour Newsl 134:41–44

Patel GI, Datta RM (1960) Interspecific hybridization between *Corchorus olitorius* Linn. and *C. capsularis* Linn. and the cytogenetical basis of incompatibility between them. Euphytica 9:89–110. doi:10.1007/BF00023259

Paterson AH, Saranga Y, Menz M, Jiang C-X, Wright RJ (2003) QTL analysis of genotype × environment interactions affecting cotton fibre quality. Theor Appl Genet 106:384–396. doi:10.1007/s00122-002-1025-y

Qi JM, Zhou D, Wu W, Lin L, Fang P, Wu J (2003a) The application of RAPD technology in genetic diversity detection of jute. Yi Chuan Xue Bao 30:926–932

Qi JM, Zhou D, Wu W, Lin L, Wu J, Fang P (2003b) Application of ISSR technology in genetic diversity detection of jute. Ying Yong Sheng Tai Xue Bao 14:1473–1477

Qi JM, Zhou D, Wu WR, Wu W, Lin L, Fang P, Wu JM, Wu J (2004) A comparison between RAPD and ISSR technology in detection of genetic diversity of jute. Sci Agric Sin 37:2006–2011

Rana MK, Singh S, Singh AK, Kak A (2012) Genetic diversity assessment in Indian jute (*Corchorus* spp) cultivars using gene-targeted molecular markers. India J Agric Sci 82:660–666

Rana MK, Arora K, Singh S, Singh AK (2013) Multi-locus DNA fingerprinting and genetic diversity in jute (*Corchorus* spp.) based on sequence-related amplified polymorphism. J Plant Biochem Biotechnol 22:1–8. doi:10.1007/s13562-012-0104-7

Ray R, Ghosh A, Bera A, Dutta N, Chattopadhyay DJ, Chakrabarti K (2011) Analysis of differentially expressed transcripts in jute upon fungal infection and beta-amino butyric acid treatment. Physiol Mol Plant Pathol 76:59–66. doi:10.1016/j.pmpp.2011.05.001

Ross AJ, Hallauer AR, Lee M (2006) Genetic analysis of traits correlated with maize ear length. Maydica 51:301–313

Rowell RM, Stout HP (2007) Jute and kenaf. In: Lewin M (ed) Handbook of fibre chemistry, 3rd edn. CRC Press, Boca Raton, pp 405–452

Roy A, Bandyopadhyay A, Mahapatra AK, Ghosh SK, Singh NK, Bansal KC, Koundal KR, Mohapatra T (2006) Evaluation of genetic diversity in jute (*Corchorus* species) using STMS, ISSR and RAPD markers. Plant Breed 125:292–297. doi:10.1111/j.1439-0523.2006.01208.x

Saha P, Datta K, Majumder S, Sarkar C, China SP, Sarkar SN, Sarkar D, Datta SK (2014a) *Agrobacterium* mediated genetic transformation of commercial jute cultivar *Corchorus capsularis* cv. JRC 321 using shoot tip explants. Plant Cell, Tissue Organ Cult 118:313–326. doi:10.1007/s11240-014-0484-6

Saha P, Sarkar D, Kundu A, Majumder S, Datta SK, Datta K (2014b) Karyotype analysis and chromosomal evolution in Asian species of *Corchorus* (Malvaceae s. l.). Genet Resour Crop Evol 61:1173–1188. doi:10.1007/s10722-014-0099-0

Samad MA, Kabir G, Islam AS (1992) Interphase nuclear structure and heterochromatin in two species of *Corchorus* and their F_1 hybrid. Cytologia 57:21–25. doi:10.1508/cytologia.57.21

Samanta P, Sadhukhan S, Basu A (2015) Identification of differentially expressed transcripts associated with bast fibre development in *Corchorus capsularis* by suppression subtractive hybridization. Planta 241:371–385. doi:10.1007/s00425-014-2187-y

Sarkar D, Kharbikar LL, Roy A, Kar CS, Kundu A, Sinha MK, Mahapatra BS (2009) Jute biotechnology roadmap: present status and emerging trends. In: Anonymous (ed) Proceedings of the international conference on emerging trends in production, processing and utilisation of natural fibres, vol 1. Indian Society for Cotton Improvement and Indian Fibre Society, Mumbai, pp 116–125

Sarkar D, Sinha MK, Kundu A, Kar CS, Saha A, Kharbikar LL, Mahapatra BS (2010) Why is ramie the strongest yet stiffest of bast fibres? Curr Sci 98:1570–1572

Sarkar D, Kundu A, Saha A, Mondal NA, Sinha MK, Mahapatra BS (2011) First nuclear DNA amounts in diploid ($2n = 2x = 14$) *Corchorus* spp. by flow cytometry: genome sizes in the cultivated jute species (*C. capsularis* L. and *C. olitorius* L.) are ~300 % smaller than the reported estimate of 1100–1350 Mb. Caryologia 64:147–153. doi:10.1080/00087114.2002.10589776

Satya P, Chakraborti M (2015) Development and utilization of DNA markers for genetic improvement of bast fibre crops. In: Tashki K (ed) Applications of molecular markers in plant genome analysis and breeding. Research Signpost, Trivandrum, pp 119–142

Satya P, Banerjee R, Biswas C, Karan M, Ghosh S, Ali N (2014a) Genetic analysis of population structure using peroxidase gene and phenylalanine ammonia-lyase gene-based DNA markers: a case study in jute (spp.). Crop Sci 54:1609–1620. doi:10.2135/cropsci2013.08.0518

Satya P, Banerjee R, Ghosh S, Karmakar PG (2014b) Morpho-anatomical and SSR diversity in mutant gene pool of jute (*Corchorus olitorius* L.). Indian J Genet 74:416–422. doi:10.5958/0975-6906.2014.00873.6

Sengupta G, Palit P (2004) Characterization of a lignified secondary phloem fibre-deficient mutant of jute (*Corchorus capsularis*). Ann Bot 93:211–220. doi:10.1093/aob/mch029

Sengupta AB, Mazumdar SK, Macmillan NG (1958) Isolation of jute holocellulose by the action of sodium chlorite. Indian J Appl Chem 25:105–110

Sharmin S, Moosa MM, Islam MS, Kabir I, Akter A, Khan H (2011) Identification of a novel dehydration responsive transcript from tossa jute (*Corchorus olitorius* L.). J Cell Mol Biol 9:21–29

Singh DP, Gupta D (1985) Genetics of fibre strength in jute (*Corchorus capsularis* L.). Bangladesh J Bot 14:82–83

Singh A, Rana MK, Singh S, Kumar S, Kumar R, Singh R (2014) CAAT box-derived polymorphism (CBDP): a novel promoter-targeted molecular marker for plants. J Plant Biochem Biotechnol 23:175–183. doi:10.1007/s13562-013-0199-5

Sinha NG, Bandopadhyay SB (1968) An air-flow method for the determination of the fibre fineness of jute and mesta. J Text Inst 59:148–156

Sinha MK, Kar CS, Ramasibramanian T, Kundu A, Mahapatra BS (2011) *Corchorus*. In: Kole C (ed) Wild crop relatives: genomic and breeding resources, industrial crops. Springer, Berlin, pp 29–61. doi:10.1007/978-3-642-21102-7_2

Sultana N, Khan H, Ashraf N, Sharkar MTK (2006) Construction of an interspecific linkage map of jute. Asian J Plant Sci 5:758–762

Swaminathan MS, Iyer RD (1961) Skewed recombination in a rare interspecific jute hybrid. Nature 192:893–894. doi:10.1038/192893b0

Swaminathan MS, Iyer RD, Sulbha K (1961) Morphology, cytology and breeding behaviour of hybrids between *Corchorus olitorius* and *C. capsularis*. Curr Sci 30:67–68

Taliaferro MJ, Islam AS, Sathasivan K (2008) Expressed sequence tags (ESTs) from a jute (*Corchorus olitorius*) cDNA library. Plant Tissue Cult Biotechnol 16:95–104. doi:10.3329/ptcb.v16i2.1110

Tanmoy AM, Alam MM, Moosa MM, Ghosh A, Quarni W, Ahmed F, Zaman NR, Sharmin S, Islam MT, Islam MS, Hossain K, Ahmed R, Khan H (2015) *Corchorus* L. and *Hibiscus* L.: molecular phylogeny helps to understand their relative evolution and dispersal routes. Biores Commun 1:1–10

Tao A, Qi J, Li M, Fang P, Lin L, Xu J (2012) Origin and evolution of jute analysed by SRAP and ISSR methods. Sci Agric Sin 45:16–35. doi:10.3864/j.issn.0578-1752.2012.01.003

Tao A, Li X, Qi J, Fang P, Lin L, Xu J, Zhang L, Wu J, Lin P (2015) Construction of a full-length cDNA library and analysis of expressed sequence tags in white jute (*Corchorus capsularis* L.). Afr J Biotechnol 14:1928–1935. doi:10.5897/AJB2015.14619

Tatusov R, Fedorova N, Jackson J, Jacobs A, Kiryutin B, Koonin E, Krylov D, Mazumder R, Mekhedov S, Nikolskaya A, Rao BS, Smirnov S, Sverdlov A, Vasudevan S, Wolf Y, Yin J, Natale D (2003) The COG database: an updated version includes eukaryotes. BMC Bioinform 4:41. doi:10.1186/1471-2105-4-41

Teixeira-Pinto A, Varela B, Shrotri K, Panandiker RSP, Lawson J (2009) Geopolymer-jute composite: a novel environmentally friendly composite with fire resistant properties. In: Brito M, Case E, Kriven WM, Salem J, Zhu D (eds) Developments in porous, biological and geopolymer ceramics: ceramic engineering and science proceedings, vol 28. Wiley, Hoboken, New Jersey, pp 337–346. doi:10.1002/9780470339749.ch30

The Angiosperm Phylogeny Group (2009) An update of the Angiosperm Phylogeny Group classification for the orders and families of flowering plants: APG III. Bot J Linn Soc 161:105–121. doi:10.1111/j.1095-8339.2009.00996.x

Topdar N, Kundu A, Sinha MK, Sarkar D, Das M, Banerjee S, Kar CS, Satya P, Balyan HS, Mahapatra BS, Gupta PK (2013) A complete genetic linkage map and QTL analyses for bast fibre quality traits, yield and yield components in jute (*Corchorus olitorius* L.). Cytol Genet 47:129–137. doi:10.3103/s0095452713030092

Tsai H, Howell T, Nitcher R, Missirian V, Watson B, Ngo KJ, Lieberman M, Fass J, Uauy C, Tran RK, Khan AA, Filkov V, Tai TH, Dubcovsky J, Comai L (2011) Discovery of rare mutations in populations: TILLING by sequencing. Plant Physiol 156:1257–1268. doi:10.1104/pp.110.169748

van Ooijen G, Mayr G, Kasiem MMA, Albrecht M, Cornelissen BJC, Takken FLW (2008) Structure–function analysis of the NB-ARC domain of plant disease resistance proteins. J Exp Bot 59:1383–1397. doi:10.1093/jxb/ern045

Wang K, Wang Z, Li F, Ye W, Wang J, Song G, Yue Z, Cong L, Shang H, Zhu S, Zou C, Li Q, Yuan Y, Lu C, Wei H, Gou C, Zheng Z, Yin Y, Zhang X, Liu K, Wang B, Song C, Shi N, Kohel RJ, Percy RG, Yu JZ, Zhu Y-X, Wang J, Yu S (2012) The draft genome of a diploid cotton *Gossypium raimondii*. Nat Genet 44:1098–1103. doi:10.1038/ng.2371

Weising K, Gardner RC (1999) A set of conserved PCR primers for the analysis of simple sequence repeat polymorphisms in chloroplast genomes of dicotyledonous angiosperms. Genome 42:9–19. doi:10.1139/g98-104

Yan J, Yang X, Shah T, Sánchez-Villeda H, Li J, Warburton M, Zhou Y, Crouch JH, Xu Y (2010) High-throughput SNP genotyping with the GoldenGate assay in maize. Mol Breed 25:441–451. doi:10.1007/s11032-009-9343-2

Zhang G, Qi J, Xu J, Niu X, Zhang Y, Tao A, Zhang L, Fang P, Lin L (2013) Overexpression of UDP-glucose pyrophosphorylase gene could increase cellulose content in jute (*Corchorus capsularis* L.). Biochem Biophys Res Comm 442:153–158. doi:10.1016/j.bbrc.2013.11.053

Zhang L, Yuan M, He X, Liu X, Fang P, Lin L, Tao A, Xu J, Qi J (2014) Development and universality evaluation of EST-SSR markers in jute (*Corchorus* spp.) from GenBank database. Acta Agron Sin 40:1213–1219. doi:10.3724/SP.J.1006.2014.01213

Zhang L, Cai R, Yuan M, Tao A, Xu J, Lin L, Fang P, Qi J (2015a) Genetic diversity and DNA fingerprinting in jute (*Corchorus* spp.) based on SSR markers. Crop J 3:416–422. doi:10.1016/j.cj.2015.05.005

Zhang L, Li Y, Tao A, Fang P, Qi J (2015b) Development and characterization of 1,906 EST-SSR markers from unigenes in jute (*Corchorus* spp.). PLoS ONE 10:e0140861. doi:10.1371/journal.pone.0140861

Zhang L, Ming R, Zhang J, Tao A, Fang P, Qi J (2015c) De novo transcriptome sequence and identification of major bast-related genes involved in cellulose biosynthesis in jute (*Corchorus capsularis L.*). BMC Genomics 16:1–13. doi:10.1186/s12864-015-2256-z

Zhang L, Yuan M, Tao A, Xu J, Lin L, Fang P, Qi J (2015d) Genetic structure and relationship analysis of an association population in jute (*Corchorus* spp.) evaluated by SSR markers. PLoS ONE 10:e0128195. doi:10.1371/journal.pone.0128195

Chapter 10
Transgenic Cotton for Agronomical Useful Traits

Chandrakanth Emani

Abstract Cotton (*Gossypium* spp.) is an important crop known for its commercial significance both as a fiber and oil yielding cultivar grown in over eighty countries around the world. The crop's economic importance underlined by its significance in textile industry makes it all the more important for agronomists and crop researchers to strategize novel approaches to overcome challenges in the form of abiotic and biotic stresses such as pests, pathogens, weeds, and more recently environmental challenges in the form of climate change. Conventional breeding technology made significant in roads into developing novel varieties with improved fiber, enhanced heat tolerance and high yields, but significant challenges remain in the areas of pathogen and insect resistance. Transgenesis or genetic engineering was an ideal solution to these challenges with its ability to introduce diverse agronomical important genes from various biological species, and over 80 % of cotton grown currently employs this technology. The present review is a comprehensive account of the historical progress made in the area of transgenic cotton both in terms of the evolution of the methodologies of tissue culture and genetic transformation, and lessons learned from conventional breeding in identifying agronomical important genes to improve cotton production.

Keywords Cotton · *Gossypium* sp. · Transgenesis · Biotechnology · Agronomic genes

10.1 Introduction

Cotton (*Gossypium hirsutum* L.) is one of the world's significant economic crops that serves as a fiber, feed, and edible oil crop. Grown in more than eighty countries, it ranks as the sixth most planted crop and in the beginning of the twenty first century

---

C. Emani (✉)
Department of Biology, Western Kentucky University-Owensboro, 4821 New Hartford Road, Owensboro, KY 42303, USA
e-mail: chandrakanth.emani@wku.edu

K.G. Ramawat and M.R. Ahuja (eds.), *Fiber Plants*, Sustainable Development and Biodiversity 13, DOI 10.1007/978-3-319-44570-0_10

33 million hectares of agricultural fields bore 52 million metric tons of cotton seeds and 18 million metric tons of cotton lint (FAO statistics). The annual cotton yield around the world contributes to $500 billion with livelihood contributions to more than 180 million people (Rahman et al. 2012). Of the 51 species classified as *Gossypium* spp., 46 are diploid ($2n = 2x = 26$) and five species are tetraploid ($2n = 2x = 52$) (Fryxell 1992). Of these, the cultivated species are the diploid *G. arboretum* L. and *G. herbaceum* L., and the tetraploid *G. hirsutum* L. and *G. barbadense* L. (Sun et al. 2006). The upland amphidiploid tetraploid *Gossypium hirsutum* L. at 90 % commercial production holds the lion's share of cotton cultivation throughout the world as the premium source of natural fiber and seed oil followed by *Gossypium barbadense* L., the other amphidiploid tetraploid Egyptian cotton contributing to 8 % of cotton cultivation (Wendel et al. 1992). Cotton and its by-products find diverse utility in products such as mattress padding, paper, plastics, toothpaste, salad dressings, fertilizers (Wilkins et al. 2000), and more recently the substance gossypol in cotton tissues is being as a phytopharmaceutical to treat cancer (Moon et al. 2008) and as a male contraceptive (Coutinho 2002).

The main challenge to cotton agriculture is in the form of biotic and abiotic stress factors, with biotic stress factors such as insects, weeds, and pathogens resulting in 85 % of annual losses (Oerke 2006). Extensive research that also translated into successful field level applications by conventional plant breeders achieved significant successes in generating cotton hybrids with improved yield (Lu and Zeiger 1994), heat tolerance (Garay and Barrow 1988), fiber quality (Dutt et al. 2004; Muthusamy and Jayabalan 2011), fungal resistance (Ganesan and Jayabalan 2006), herbicide tolerance (Rajasekaran et al. 1996), and early maturity (Kandhro et al. 2002). Similar levels of success were unachievable in generating improved biotic stress tolerant cotton varieties as conventional breeding is limited by evolutionary bottlenecks (Wilkins et al. 2000), and the fact that decades-long breeding with a limited number of genotypes severely narrowed down the gene pool for improvement of cotton varieties (Gingle et al. 2006). The annual losses of 36.8 % due to animal pests and 35.9 % to weeds (Oerke 2006) can be attributed not only to the limitations of conventional breeding as mentioned earlier, but also the prolonged and indiscriminate use of broad spectrum chemical pesticides, insecticides, and weedicides that causes both environmental and economic problems. The economic problems are in the form of increased costs of cotton cultivation that does not justify the loss in crops, and the ecological or environmental problems arise due to development of resistance especially by insect pests that led to unwanted pest outbreaks (Kranthi et al. 2002). The environmental awareness among the agriculturists that guided the current research goals of the global cotton community naturally gravitated toward adopting transgenic technology as the most effective alternative to generate improved cotton varieties with desired traits. The effective tissue culture technology that resulted in the first reports of transgenic cotton by somatic embryogenesis in the Coker 210 and 312 varieties (Firoozabady et al. 1987; Umbeck et al. 1987) led to extensive applications covering diverse genetic traits, finally resulting in the current cultivation area of transgenic cotton worldwide to be

24.3 Mha which is nearly 81 % of the total cotton cultivation (James 2012). In USA, insect-resistant *Bt* transgenic cotton varieties resulted in a net gain of $250 per hectare and more significantly reduced the insecticide application costs by 50 % (James 2012). In the global cotton cultivation perspective, India occupies the primary position in terms of transgenic cotton production with 10.6 million hectares accounting for 88 % of total cotton cultivation areas followed by USA with 4 million hectares, China with 3.9 million hectares, and Pakistan with 2.6 million hectares (James 2012).

There are many comprehensive reviews in literature on cotton biotechnology. Earlier reviews focused on tissue culture and emphasized focus on insect resistant, herbicide tolerant, and improved fiber quality (Wilkins et al. 2000; Kumria et al. 2003; Showalter et al. 2009). More recent reviews focused on the role of somatic embryogenesis in transgenesis (Obembe et al. 2011), focus on specific countries such as China (Zhu et al. 2011), and two recent reviews gave a comprehensive account on the current status of tissue culture and genetic engineering in cotton (Chakravarthy et al. 2014; Juturu et al. 2014). The present review is an extensive account of the chronological progress made in transgenic cotton research summarizing the factors associated with different transgenic methodologies with a breakdown of the diverse transgenes introduced along with the future roadmap that extends the application of transgenic technology both in basic and applied research in cotton.

10.2 Methods of Genetic Transformation

The advent of tissue culture methodology in cotton through somatic embryogenesis and organogenesis was beset with challenges in terms of limited number of cultivars and efforts by researchers to increase the range of cultivars was left with very limited success (Sakhanokho et al. 2004). Additionally, prolonged culture periods introduced somaclonal variation, abnormal embryos, and low conversion of somatic embryos into plantlets (Stelly et al. 1989; Sun et al. 2006), thus limiting the production of true to type transgenic cotton. Further, the high genotype dependency further restricted the diversity in application of transgenic technology in cotton. These challenges notwithstanding, *Agrobacterium*-mediated remained the method of choice followed by particle bombardment. More recently, a non-tissue culture-based in planta transformation approach is providing an effective alternative method to reduce genotype dependency as well as avoid a robust as well as prolonged tissue culture process. This section outlines the various transgenic methodologies as an evolving continual technology as researchers had continuously strived to develop effective methods to generate transgenic cotton in ideal in vitro conditions minimizing to a large extent the pitfalls arising out of prolonged tissue culture while maintaining desirable and stable transgene expression.

10.2.1 Particle Bombardment

Particle bombardment or biolistics, also known as direct gene transfer utilizes high velocity gold or tungsten particles in a vacuum-driven gene gun apparatus was the method of choice for researchers that wanted to overcome the challenge of genotype dependency in genetically engineering cotton. The first report of particle bombardment (Finer and McMullen 1990) employed suspension cultures derived from embryogenic calli of Coker 310 to introduce *aph*IV (hygromycin phosphotransferase) gene. The subsequent decade saw the particle bombardment introduce genes such as *pha*B (acetoacety CoA reductase), *pha*C (polyhydroxyalkanoate synthase), *gus*A (β-glucuronidase) in DP50 embryos (John and Keller 1996), *uid*A (β-glucuronidase) and *npt*II (kanamycin resistance) in Acala B1654 and Coker 315 embryogenic suspension cultures (Rajasekaran et al. 2000), *aph*A (acid phosphatase) through plastid transformation of embryogenic calli of Coker 310FR (Kumar et al. 2004), *Ahas* (acetohydroxy acid synthetase) in embryogenic axes of 7MH, Antares, CD 401, and ITA 94 (Aragao et al. 2005). More recently, *phy*A (phytochrome) gene was engineered into embryo tips of Giza 88 and Giza 90 and the *Aspergillus ficuumphy* A (phytase) gene for enhanced phosphorus usage was engineered into shoot apices of ND94-7 (Liu et al. 2011). From these reports, it is evident that particle bombardment successfully overcomes the challenge of genotype dependency in generating transgenic cotton as it embraces diverse genotypes. The transformation frequencies are also higher up to 4 % (Rajasekaran et al. 2000). However, the main drawback of particle bombardment comes in the form of transgene silencing or co-suppression due to multi-copy or fragmented gene insertions (Depicker and Montagu 1997). The shoot apex method also results in chimeras and high frequency of epidermal transformants (DeBlock 1993; Wilkins et al. 2000).

10.2.2 *Agrobacterium*-Mediated Transformation

The successful reports of gene transfer utilizing the natural genetic engineer *Agrobacterium tumefaciens* into tobacco (Barton et al. 1983) and into diverse plant species found a natural extension into cotton transgenesis (John 1997; Wilkins et al. 2000). The earliest reports of *Agrobacterium*-mediated cotton transformation utilized hypocotyl and cotyledon explants through somatic embryogenesis to engineer *npt*II gene into Coker 210 (Firoozabady et al. 1987; Umbeck et al. 1987). Subsequently, insect-resistant transgenic cotton plants were generated by engineering the *cry1A* (*b*) and *cry1A* (*c*) genes of *Bacillus thuringenesis* into Coker 312 (Perlak et al. 1990). The introduction of *cry* genes to generate insect-resistant transgenic cotton utilized diverse explants such as hypocotyl (Kumar et al. 2009), cotyledon (Singh et al. 2004), and embryogenic calli (Wu et al. 2005). The main challenge of *Agrobacterium*-mediated transformation was recalcitrance in terms of tissue culture of commercially important cotton varieties. The prolonged tissue culture that runs almost a year was another challenge. *Agrobacterium*-mediated through shoot apex explants (Gould and Cedeno 1998; Zapata et al. 1999) and apical meristems (Lv et al. 2008; Nandeshwar et al. 2009) was a strategy to include diverse cotton varieties. The significant advantage of *Agrobacterium*-mediated method is the introduction of low copy number of genes that negates the transgene silencing making it the method of choice in generating transgenic cotton. A comprehensive and systematic transformation optimization study by Sunilkumar and Rathore (2001) examined various aspects of *Agrobacterium*-mediated cotton transformation and the regeneration processes by utilizing the *gfp* (jellyfish green fluorescent protein) gene and were successful in developing an effective protocol that cut down the time from a year of prolonged tissue culture to less than 6 months.

10.2.3 In Planta-Mediated Transformation

In Planta transformation involves engineering genes into intact plants or seeds through physical or biological methods and recovering whole plants by non-tissue culture protocols. The methodology involves three broad protocols, namely the pollen tube pathway, the pollen-mediated method, and meristem transformation (Juturu et al. 2014). Pollen tube pathway involves direct introduction of exogenous DNA into embryos of self-pollinated cotton flowers (Zhou et al. 1983). Successful adaption of this methodology resulted in engineering of the *gfp* gene (Haung et al. 1999) and the whole cellulose synthesizing *ayacs* operon (Lu et al. 2002). Direct inoculation of *Agrobacterium* culture with a herbicide-tolerant gene on the pistils of pollinated cotton flowers (Tian et al. 2010), manual pollination of cotton flowers with pollen co-cultivated with *Agrobacterium* harboring cellulose synthesizing genes along with *hpt* (hygromycin phosphotransferase) genes (Li et al. 2004), and fertilization of flowers with pollen bombarded with *Arabidopsis thaliana* 3′-hydroxymethy glutaryl coenzyme A reductase (*hmgr*) gene are some applications which show successful integration of other transformation methods described earlier into a non-tissue culture-based platform. Meristem transformation method involved co-culturing 1–2 day old germinating cotton seedlings with *Agrobacterium* culture and transferring them directly into greenhouse for hardening (Keshamma et al. 2008; Kumar et al. 2013). The main challenge with these methods is the tendency to generate chimeras and the need to screen large number of primary transformants to discard the extensive numbers of transgenic escapes.

10.2.4 Miscellaneous Transformation Methods

Chemical-induced transgenic cotton plants were generated by exposing cotyledon-derived suspension cultures in tissue culture medium supplemented with spermidine and polybrene with *hpt* and *gus*A genes (Sawahel 2001).

Cotyledon-derived embryogenic calli agitated in a solution containing mannitol and sorbitol together with silicon carbide microfibers (WHISKERS™) coated with the *bar* (basta resistance) gene (Beringer et al. 2004), *AVP1* (*Arabidopsis* vacuolar pyrophosphatase proton pump), and *uid*A gene (Asad et al. 2008; Arshad et al. 2013) provided a direct gene transfer method to generate transgenic cotton. These methods were replete with problems of non-reproducibility and the need for extensive screening of a large numbers of transformed plants to identify very few stable transformants and then the failure of transmission of transgenes to subsequent generations (Showalter et al. 2009).

10.3 Agronomical Traits Targeted for Cotton Improvement and the Donor Genes

A robust transgenic technology for a specific crop enables the introduction or stacking of desirable genes for diverse agronomically important traits overcoming the challenges encountered in conventional breeding techniques. The stable and desired level of expression with proper Mendelian transmission across generations is another achievable goal of this technology and cotton is no exception. This section is a comprehensive overview of the traits targeted in cotton transgenesis with the engineered donor genes.

10.3.1 Abiotic Stress

Salinity, drought, water logging, erratic light, extreme temperatures, and mineral imbalance are some of the significant abiotic stress factors that affect cotton cultivation (Bohnert and Jensen 1996; Chen and Murata 2002; Gong et al. 2012). Potential donor genes conferring abiotic stress tolerance in plants included single action genes involved in signal and regulatory pathways such as those coding for antioxidants, osmoprotectants, late embryogenesis abundant proteins (LEAs), heat shock proteins (HSPs), and membrane transporters or multi-functional genes such as transcription factors and protein kinases (Allen 2010). Some of the genes engineered in cotton to tide over diverse abiotic stresses include the *Arabidopsis thaliana* sodium–hydrogen antiporter exchanger (*AtNHX1*) that confers improved salt tolerance as well as plant biomass (He et al. 2005), tobacco Mn superoxide dismutase (Mn-SOD) for chilling and high light intensity tolerance (Kornyeyev et al. 2003), *E. coli* choline dehydrogenase (*bet*A) for salt tolerance and increased yield (Zhang et al. 2009), *ipt* (isopentenyl transferase) gene for improved fiber quality under saline conditions, rice *SNAC1* gene for increased root development and reduced transpiration rates (Liu et al. 2014), and *VP1* (*Arabidopsis* vascular protein) gene for improved salt and drought along with improved fiber yield.

10.3.2 Insect Resistance

The inherent nature of cotton to have indeterminate growth traits render it as food and shelter for about 130 diverse species of insect pests. Donor genes for insect resistance have been identified from diverse sources such as bacteria (Perlak et al. 1990), insects (Thomas et al. 1995), and plants (Wu et al. 2006).

The pioneer shifting discovery that had a major role in generating insect-resistant transgenic cotton that had major agricultural impact across the world was from the donor genes encoding crystalline (Cry) proteins of *Bacillus thuringenesis,* more specifically, the Cry δ-endotoxin engineered transgenic cotton provided resistance against lepidopteran pests such as bollworm *Helicoverpa zea*, tobacco budworm *Heliothis virescens* (Siebert et al. 2008), pink bollworm *Pectinophora gossypiella* (Tabashnik et al. 2002), and fall armyworm *Spodoptera furgiperda* (Greenplate et al. 2003). A synthetic Cry 1EC protein fused to a tobacco pathogenesis-related promoter showed 100 % mortality against *S. litura* larvae with enhanced expression upon insect bite and salicylic acid treatment, thus showing a broad spectrum integrated pest control (Kumar et al. 2009). Ecological and evolutionary consequences soon took over the Bt transgenic strategy and the insects such as *H. viscerens*, *P. gossypiella*, *H. armigera*, and *H. zea* were found to develop resistance against Cry proteins (Bravo and Soberon 2008). To tide over these challenges, transgenic researchers employed genes encoding vegetative insecticidal proteins (Vips) that express during the pests' vegetative growth phase starting at mid-log phase and sporulation unlike the regular δ-endotoxins that express only during sporulation (Estruch et al. 1996). This increased the range and diversity of action against lepidopteran pests (Zhu et al. 2006) and coleopteran pests (Wu et al. 2011).

Insect derived toxins such as *Androctonus austrais* hector insect toxin (AaHIT) conferred resistance to cotton bollworm (Wu et al. 2008) and *Manduca sexta* derived protease inhibitors against whitefly *Bemesia tabaci* (Thomas et al. 1995). More recently, RNAi-mediated suppression of the cotton bollworm P450 monoxygenase gene (*CYP6AE14*) in transgenic cotton significantly retarded the larva growth of *H. armigera* (Mao et al. 2011).

Plant derived insecticidal proteins such as proteases and lectins, specifically, the cowpea (*Vigna unguiculata*) trypsin inhibitor gave protection against cotton bollworm (Li et al. 1998), lectin derivatives *A. caudatus* agglutinin (*aca*) gene conferred resistance against aphids (Wu et al. 2006) and *A. sativum* agglutinin (*asa*) gene conferred resistance against sap-sucking insects (Balogun et al. 2011).

10.3.3 Fungal and Viral Resistance

Mycoparasitic fungi proved to be rich sources of antifungal genes that were engineered into cotton and in a first report of its kind, a single gene introduced into cotton, specifically, the chitinase gene from the biocontrol agent *Trichoderma virens* conferred resistance against to major fungi viz., the soil-borne pathogen *Rhizoctonia solani* and a foliar pathogen *Alternaria alternata* (Emani et al. 2003). Rice chitinase (*ChiII*) gene conferred resistance against *Fusarium oxysporum* and *Alternaria macrospora* (Ganesan et al. 2009). Overexpression of *Gastrodiaelata* antifungal protein (*gafp*) (Wang et al. 2004) and baculovirus apoptosis inhibitor genes (*p35* and *op-iap*) (Tian et al. 2010) conferred resistance against Verticillium wilt. Other examples of fungal resistance include the synthetic antimicrobial peptide (D4E1) against *Fusarium verticillioides* and *Verticillium dahlia* (Rajasekaran et al. 2005), *Talaromyces flavus* glucose oxidase gene against root pathogen *V. dahliae* (Murray et al. 1999), *Xanthomonas oryzae* hairpin-encoding gene (*hpa* 1_{Xoo}) against *Verticillium* wilt, and AtNPR1 gene against a variety of pathogens such as *V. dahliae*, *F. oxysporum*, *R. solani*, and *A. alternata* (Parkhi et al. 2010).

Viral diseases especially the leaf curl disease (CLCuD) was tackled by engineering cotton plants with the leaf curl virus (CLCuV), antisense movement protein (*AV2*) (Sanjaya et al. 2005), and antisense coat protein (*ACP*) (Amudha et al. 2011). Introduction of a truncated version of cotton leaf curl Khokran virus replicase (*tAC1*) resulted in decreased symptoms of CLCuV infection (Hashmi et al. 2011).

10.3.4 Fiber Quality

The competition of cotton fiber with synthetic yarns due to its poor characteristics (Karaca et al. 2002) gave rise to competitive goals for transgenic cotton researchers to focus on improving traits such as fiber length, strength, and uniformity employing donor genes driven by fiber specific promoters. Chapter 11 of this book (*Cotton fiber biotechnology: Transgenic manipulation of elongation and cell wall thickening*) has a detailed exposition of this aspect.

10.3.5 Herbicide Resistance

Weeds in cotton fields compete with the crop for water, nutrients, and sunlight, and also they are responsible for increasing the trash content of harvested cotton fibers (Nida et al. 1996). Indiscriminate use of various herbicides posed an environmental problem and ineffectiveness in field also extended to the non-specific killing of plants owing to the inability of herbicides to distinguish between crop plants and weeds (Lyon et al. 1993). Engineering of herbicide-tolerant genes such as the *Agrobacterium tumefaciens* strain CP4 derived 5-enopyruvilshikimate-3-phosphate synthase (CP4-EPSPS) gave the cotton farmers the first commercialized glyphosate-tolerant Roundup Ready® transgenic cotton (Nida et al. 1996) that effectively resisted glyphosate applications up to the four-leaf stage (May et al. 2004). Beyond the four-leaf stage, applications caused damage to reproductive organs that led to substantial yield losses. This challenge was successfully overcome by developing the Roundup Ready Flex® cotton by adding an additional copy of the gene driven by Arabidopsis thaliana-derived elongation factor-1α (P-EF1α) promoter along with the figwort mosaic virus (FMV) promoter-driven gene originally used. The resulting transgenic cotton plants survived glyphosate applications well beyond the four-leaf stage with more than 97 % pollen viability (Chen et al. 2006). Other effective strategies of developing transgenic herbicide-tolerant cotton involved the *Streptomyces hygroscopicus* basta-resistant (*bar*) gene engineered bialaphos-resistant transgenic cotton (Keller et al. 1997), engineering of chimeric *aroA-M1* gene fused to chloroplast transit peptide of *A. thaliana* 5-enopyruvyl-3-phosphoshikimate synthase (*ASP*) gene for glyphosate tolerance (Zhao et al. 2006), a double mutated version of *Zea mays* derived EPSPS of Bayer's GlyTol™ (Wallace et al. 2011), and the commercially developed Liberty Link® of Bayer resistant to gluphosinate that was albeit less effective than GlyTol™, but did not result in producing any resistant weed species as the 12 weed species produced by the glyphosate-resistant transgenics (Dill et al. 2008).

10.3.6 Increased Yield

Transgenic cotton with increased yield characteristics was generally a tangential benefit seen when engineered with abiotic stress-tolerant genes. Cotton plants engineered with vacuolar sodium antiporter *At-NHX1* gene from *A. thaliana* and *Brassica juncea* derived annexin (*AnnBj1*) exhibited improved photosynthetic ability, increased plant biomass, and fibers (He et al. 2005). Transgenic cotton engineered for salinity tolerance with H^+-pyrophosphatase genes of *Thellungiella halophila* (*Ts-VP*) and *A. thaliana* (*AVP1*) exhibited 40 % enhanced seed yield (Lv et al. 2008). Expression of *A. thaliana* drought tolerant (*AtLOS5*) produced enhanced biomass fater exposure to drought conditions (Yue et al. 2012). Transgenic cotton with overexpressed *A. tumefaciens* isopentenyl transferase (*ipt*) resulted in improved fiber quality in saline conditions (Liu et al. 2012).

10.3.7 Oil Quality and Nutritional Enhancement

Cotton's ranking as the fifth best oil producing plant in the world and the second best potential source of protein (Benbouza et al. 2010) sets the goal for further improvement of oil quality and nutritional enhancement. With its fatty acid profile of 55 % polyunsaturated fatty acids, 17 % monosaturated fatty acids, 17 % monounsaturated fatty acids, and 26 % saturated fatty acids (Lukonge et al. 2007), it is a desirable cooking oil, but the presence of cardio and hepato-toxic terpenoid substances such as gossypol limits its human consumption (Risco et al. 1997). Conventional breeding developed glandless cotton varieties that were gossypol free, but the lack of natural immunity jeopardized their commercial success due to high susceptibility to insect pests (Lusas and Jividen 1987). Tissue-specific reduction of gossypol in seeds up to 50 % while retaining gossypol in vegetative parts utilizing varieties with different gland densities was another attempt by conventional breeders (Romano and Scheffler 2008). Transgenic approach proved much more effective with anti-sense engineering of *G. arboreum*δ-(+) cadinene synthase (cdn1-Cl) resulted in gossypol reduction up to 70 % in seeds and 92.4 % in foliar tissues (Martin et al. 2003). RNA interference proved the most effective in terms of overcoming the undesirable reduction of gossypol in non-target tissue with transgenic plants produced with 0.1 μg/mg seed levels of gossypol (Sunilkumar et al. 2006).

10.4 The Regulatory Aspects of Transgenic Cotton

The rapidity with which the *Bt* transgenic cotton took over cotton cultivation across the word underlines the importance placed by the cotton farmers on tiding over abiotic and biotic stresses in the field to recover the substantial losses of yesteryears and the recognition of transgenesis as an effective tool (James 2012). But the ecological consequences of unwanted resistant pests resulting because of planting transgenics without properly monitored field trials became a debatable issue when commercialization of the *Bt* transgenics started spreading across countries. Ecologists underlined the importance of conserving natural enemies to avoid outbreak of secondary pests that are insensitive to genetically engineered insecticidal proteins, and the equally important aspect of minimizing the damage to non-target organisms as part of an effective integrated pest management (Kos et al. 2009). It has been established that continual high expression of transgenes in plant species creates an unwanted opportunity for non-target herbivores to accumulate toxins for higher trophic levels (Torres and Ruberson 2008). An effective monitoring and avoidance of such unwarranted ecological and environmental consequences that may jeopardize the full realization of transgenic technology as a valuable tool reducing it to unwanted prolonged debates can be through focused laboratory studies and closely monitored field trials before releasing transgenic cotton varieties into the market. Several such field studies with *Bt* cotton that established the fact that predator and parasitoid abundance are similar in *Bt* and non-*Bt* crops (Romeis et al. 2006) that was later extended to a 6-year focused study in *Bt* cotton fields examining the prevalence of non-target arthropods clearly indicated that *Bt* toxin exposure over multiple generations had no adverse effects (Naranjo 2005). Related studies in *Bt* maize where non-target spider species revealed no differences in mortality, weight, and life cycle across generations (Meissle and Romeis 2009). *Bt* cotton studies on non-target *A. gossypii* showed no negative effects on the aphid population (Zhang et al. 2012). A 3-year study on the effect of Bt cotton fields on soil organisms also did not show any negative results (Li et al. 2011). Globally, genetically modified crops in general are subjected to strict regulatory controls and lessons learned from studies have found their way as valuable source material for regulating transgenic cotton as well. Signatory countries of the Cartagena protocol mandate environmental risk assessment for regulatory approval of genetically modified organisms (Hilbeck et al. 2011).

10.5 Conclusion and Future Roadmap

The global population expected to reach 9 billion by the year 2050 mandates the goal of producing as much food as what was produced till date. Transgenic technology is now rapidly emerging as the only feasible alternative for meeting such a demand as agricultural and arable lands are dwindling by the day. Cotton with its status as a major fiber and oil crop and with the current research also christening it as a valuable protein source will have a natural significant role to play in such a scenario. Commercially available *Bt* cotton's significant contributions to agricultural economy underline the fact that transgenic cotton's role is already underway as a crop of the future. The continual identification of donor candidate genes combined with cutting-edge technologies such as RNAi, the "Omics" approach to understand the molecular biology of the crop will play a major role for future genetic enhancement of the crop. A further refinement of tissue culture techniques to avoid prolonged culture and also the improvement in methodologies of in planta transformation also demand greater attention. Lastly, encompassing more genotypes in the purview of the transgenic technology should also be a focus of research.

References

Allen RD (2010) Opportunities for engineering abiotic stresses. In: Zher UB (ed) Biotechnology advances in agriculture and forestry, 65th edn. Springer, Berlin, pp 127–148

Amudha J, Balasubramani G, Malathi VG, Monga D, Kranthi KR (2011) Cotton leaf curl virus resistance transgenics with antisense coat protein gene (*AV1*). Curr Sci 101:300–307

Aragao FJL, Vianna GR, Carvalheira SBRC, Rech EL (2005) Germ line genetic transformation in cotton (*Gossypium hirsutum* L.) by selection of transgenic meristematic cells with a herbicide molecule. Plant Sci 168:1227–1233

Arshad M, Zafar Y, Asad S (2013) Silicon carbide whisker-mediated transformation. In: Zhang B (ed) Transgenic cotton: methods and protocols, methods in molecular biology, vol 958. Springer, New York, pp 79–91

Asad S, Mukhtar Z, Nazir F, Hashmi JA, Mansoor S, Zafar Y, Arshad M (2008) Silicon carbide whisker-mediated embryogenic callus transformation of cotton (*Gossypium hirsutum* L.) and regeneration of salt tolerant plants. Mol Biotechnol 40:161–169

Balogun NB, Inuwa HM, Sani I, Ishiyaku MF et al (2011) Expression of mannose-binding insecticidal lectin gene in transgenic cotton (*Gossypium*) plant. Cotton Genomics Genet 2:1–7

Barton KA, Binns AN, Matzke AJ, Chilton MD (1983) Regeneration of intact tobacco plants containing full length copies of genetically engineered T-DNA, and transmission of T-DNA to R1 progeny. Cell 32:1033–1043

Benbouza H, Lacape JM, Jacquemin JM, Courtois B et al (2010) Introgression of the low-gossypol seed & high-gossypol plant trait in upland cotton: analysis of [(*Gossypium hirsutum* × *G. raimondii*) 2 × *G. sturtianum*] trispecific hybrid and selected derivatives using mapped SSRs. Mol Breed 25:273–286

Beringer J, Palta AM, Baker LW, Petolino JF (2004) Transgenic cotton via whiskers-mediated transformation. Recent Res Dev Crop Sci 1:335–347

Bohnert HJ, Jensen RG (1996) Strategies for engineering water-stress tolerance in plants. Trends Biotechnol 14:89–97

Bravo A, Soberon M (2008) How to cope with insect resistance to Bt toxins? Trends Biotechnol 26:573–579

Chakravarthy VS, Reddy TP, Reddy VD, Rao KV (2014) Current status of genetic engineering in cotton (*Gossypium hirsutum* L.): an assessment. Crit Rev Biotech 34:144–160

Chen YC, Hubmeier C, Tran M, Martens A et al (2006) Expression of CP4 EPSPS in microspores and tapetum cells of cotton (*Gossypium hirsutum*) is critical for male reproductive development in response to late-stage glyphosate applications. Plant Biotechnol J 4:477–487

Chen TH, Murata N (2002) Enhancement of tolerance of abiotic stress by metabolic engineering of betaines and other compatible solutes. Curr Opin Plant Biol 5:250–257

Coutinho EM (2002) Gossypol: a contraceptive for men. Contraception 65:259–263

DeBlock M (1993) The cell biology of plant transformation: current state, problems, prospects and the implications for plant breeding. Euphytica 71:1–14

Depicker A, Montagu MV (1997) Post-transcriptional gene silencing inplants. Curr Opin Cell Biol 9:373–382

Dill GM, Cajacob CA, Padgette SR (2008) Glyphosate-resistant crops: adoption, use and future considerations. Pest Manag Sci 64:326–331

Dutt Y, Wang XD, Zhu YZ, Li YY (2004) Breeding for high yield and fibre quality in colored cotton. Plant Breed 123:145–151

Emani C, Garcia JM, Lopata-Finch E, Pozo MJ et al (2003) Enhanced fungal resistance in transgenic cotton expressing an endochitinase gene from *Trichoderma virens*. Plant Biotechnol J 1:321–336

Estruch JJ, Warren GW, Mullins MA, Nye GJ et al (1996) Vip3A, a novel *Bacillus thuringiensis* vegetative insecticidal protein with a wide spectrum of activities against lepidopteran insects. Proc Natl Acad Sci USA 93:5389–5394

Finer JJ, McMullen MD (1990) Transformation of cotton (*Gossypium hirsutum* L.) via particle bombardment. Plant Cell Rep 8:886–889

Firoozabady E, DeBoer D, Merlo D, Halk E et al (1987) Transformation of cotton (*Gossypium hirsutum* L.) by *Agrobacterium tumefaciens* and regeneration of transgenic plants. Plant Mol Biol 10:105–116

Fryxell PA (1992) A revised taxonomic interpretation of *Gossypium* L. (Malvaceae). Rheedea 2:108–165

Ganesan M, Bhanomathi P, Kumari G, Prabah AL et al (2009) Transgenic Indian cotton (*Gossypium hirsutum* L.) harboring rice chitinase gene (chiII) confers resistance to two fungal pathogens. Am J Biochem Biotechnol 5:63–74

Ganesan M, Jayabalan N (2006) Isolation of disease-tolerant cotton (*Gossypium hirsutum* L. cv. SVPR 2) plants by screening somatic embryos with fungal culture filtrate. Plant Cell Tissue Organ Cult 87:273–284

Garay BR, Barrow JR (1988) Pollen selection for heat tolerance in cotton. Crop Sci 5:857–859

Gingle AR, Yang H, Chee PW, May OL et al (2006) An integrated web resource for cotton. Crop Sci 46:1998–2007

Gong SY, Huang GQ, Sun X, Li P et al (2012) GhAGP31, a cotton non-classical arabinogalactan protein, is involved in response to cold stress during early seedling development. Plant Biol 14:447–457

Gould JH, Cedeno M (1998) Adaptation of cotton shoot apex culture to *Agrobacterium*-mediated transformation. Plant Mol Biol Rep 16:283–285

Greenplate JT, Mullins JW, Penn SR, Dahm A et al (2003) Partial characterization of cotton plants expressing two toxin proteins from *Bacillus thuringiensis*: relative toxin contribution, toxin interaction, and resistance management. J Appl Entomol 127:340–347

Hashmi JA, Zafar Y, Arshad M, Mansoor S, Asad S (2011) Engineering cotton (*Gossypium hirsutum* L.) for resistance to cotton leaf curl disease using viral truncated AC1 DNA sequences. Virus Genes 42:286–296

Haung GC, Dong YM, Sun JS (1999) Introduction of exogenous DNA into cotton via the pollen-tube pathway with GFP as reporter. Chin Sci Bull 44:698–701

He C, Yan J, Shen G, Fu L et al (2005) Expression of an Arabidopsis vacuolar sodium/proton antiporter gene in cotton improves photosynthetic performance under salt conditions and increases fiber yield inthe field. Plant Cell Physiol 46:1848–1854

Hilbeck A, Meier M, Rombke J, Jansch S et al (2011) Environmental risk assessment of genetically modified plants-concepts and controversies. Environ Sci Eur 23:13

James C (2012) Global status of commercialized Biotech/GM crops. ISAAA brief no. 44, Ithaca, New York. http://www.isaaa.org/resources/publications/briefs/44/download/isaaa-brief-44–2012.pdf

John ME (1997) Cotton crop improvement through genetic engineering. Crit Rev Biotech 17:185–208

John ME, Keller G (1996) Metabolic pathway engineering in cotton: biosynthesis of polyhydroxybutyrate in fiber cells. Proc Natl Acad Sci USA 93:12768–12773

Juturu VN, Mekala GK, Kirti PB (2014) Current status of tissue culture and genetic transformation research in cotton (*Gossypium* spp.). Plant Cell Tiss Org Cult 120:813–839

Kandhro MM, Laghari S, Sial MA, Nizamani GS (2002) Performance of early maturing strains of cotton (*Gossypium hirsutum* L.) developed through induced mutation and hybridization. Asian J Pl Sci 1:581–582

Karaca M, Saha S, Jenkins JN, Zipf A, Kohel R, Stelly DM (2002) Simple sequence repeat (SSR) markers linked to the Ligon lintless (Li1) mutant in cotton. J Hered 93:221–224

Keller G, Spatola L, McCabe D, Martinell B, Swain W, John ME (1997) Transgenic cotton resistant to herbicide bialaphos. Transgen Res 6:385–392

Keshamma E, Rohini S, Rao KS, Madhusudhan B, Udayakumar M (2008) Tissue culture-independent in planta transformation strategy: an *Agrobacterium tumefaciens*-mediated gene transfer method to overcome recalcitrance in cotton (*Gossypium hirsutum* L.). J Cotton Sci 12:264–272

Kornyeyev D, Logan BA, Payton P, Allen RD, Holaday AS (2003) Effect of chloroplastic overproduction of ascorbate peroxidase on photosynthesis and photoprotection in cotton leaves subjected to low temperature photoinhibition. Funct Plant Biol 30:101–110

Kos M, van Loon JJ, Dicke M, Vet LE (2009) Transgenic plants as vital components of integrated pest management. Trends Biotechnol 27:621–627

Kranthi KR, Jadhav DR, Kranthi S, Wanjari RR, Ali SS, Russell DA (2002) Insecticide resistance in five major insect pests of cotton in India. Crop Prot 21:449–460

Kumar M, Shukla AK, Singh H, Verma PC, Singh PK (2013) A genotype-independent *Agrobacterium* mediated transformation of germinated embryo of cotton (*Gossypium hirsutum* L). Int J Biotechnol Res 3(1):81–90

Kumar M, Shukla AK, Singh H, Tuli R (2009) Development of insect resistant transgenic cotton lines expressing cry1EC gene froman insect bite and wound inducible promoter. J Biotechnol 140:143–148

Kumar S, Dhingra A, Daniell H (2004) Stable transformation of the cotton plastid genome and maternal inheritance of transgenes. Plant Mol Biol 56:203–216

Kumria R, Leelavathi S, Bhatnagar RK, Reddy VS (2003) Regeneration and genetic transformation of cotton: present status and future perspectives. Plant Tissue Cult 13:211–225

Li X, Liu B, Cui J, Liu D et al (2011) No evidence of persistent effects of continuously planted transgenic insect-resistant cotton on soil microorganisms. Plant Soil 339:247–257

Li X, Wang XD, Zhao X, Dutt Y (2004) Improvement of cotton fiber quality by transforming the acsA and acsB genes into *Gossypium hirsutum* L. by means of vacuum infiltration. Plant Cell Rep 22:691–697

Li YE, Zhu Z, Chen ZX, Wu X, Wang W, Li SJ (1998) Obtaining transgenic cotton plants with cowpea trypsin inhibitor. Acta Gossypii Sin 10:237–243

Liu GZ, Li XL, Jin SX, Liu XY, Zhu LF, Nie YC, Zhang XL (2014) Overexpression of rice NAC gene SNAC1 improves drought and salt tolerance by enhancing root development and reducing transpiration rate in transgenic cotton. PLoS One 9(1):e86895

Liu JF, Wang XF, Li QL, Li X, Zhang GY, Li MG, Ma ZY (2011) Biolistic transformation of cotton (*Gossypium hirsutum* L.) with the phyA gene from *Aspergillus ficuum*. Plant Cell Tissue Organ Cult 106:207–214

Liu YD, Yin ZJ, Yu JW, Li J, Wei HL, Han XL, Shen FF (2012) Improved salt tolerance and delayed leaf senescence in transgenic cotton expressing the Agrobacterium IPT gene. Biol Plant 56:237–246

Lu Z, Zeiger E (1994) Selection of higher yield and heat resistance in pima cotton has caused genetically determined changes in stomatal conductance. Physiol Plant 92:273–278

Lu YC, Wei G, Zhu YX (2002) Cloning whole cellulose-synthesizing operon (*ayacs* operon) from *Acetobacter xylenium* and transforming it into cultivated cotton plants. Acta Bot Sin 44:441–445

Lukonge E, Labuschagne MT, Hugo A (2007) The evaluation of oil and fatty acid composition in seed of cotton accessions from various countries. J Sci Food Agri 87:340–347

Lusas EW, Jividen GM (1987) Glandless cottonseed: a review of the first 25 years of processing and utilization research. J Am Oil Chem Soc 64:839–854

Lv S, Zhang K, Gao Q, Lian L, Song Y, Zhang J (2008) Overexpression of an H+-PPase gene from *Thellungiella halophila* in cotton enhances salt tolerance and improves growth and photosynthetic performance. Plant Cell Physiol 49:1150–1164

Lyon BR, Cousins YL, Llewellyn DJ, Dennis ES (1993) Cotton plants transformed with a bacterial degradation gene are protected from accidental spray drift damage by the herbicide 2,4-dichlorophenoxyacetic acid. Transgen Res 2:162–169

Mao YB, Tao XY, Xue XY, Wang LJ, Chen XY (2011) Cotton plants expressing CYP6AE14 double-stranded RNA show enhanced resistance to bollworms. Transgen Res 20:665–673

Martin GS, Liu J, Benedict CR, Stipanovic RD, Magill CW (2003) Reduced levels of cadinane sesquiterpenoids in cotton plants expressing antisense (+)-delta-cadinene synthase. Phytochemistry 62:31–38

May OL, Culpepper AS, Cerny RE, Coots CB et al (2004) Transgenic cotton with improved resistance to glyphosate herbicide. J Crop Sci 44:234–240

Meissle M, Romeis J (2009) The web-building spider *Theridion impressum* (Araneae: Theridiidae) is not adversely affected by Bt maize resistant to corn rootworms. Plant Biotechnol J 7:645–656

Moon DO, Kim MO, Lee JD, Kim GY (2008) Gossypol suppresses NF-kappaB activity and NF-kappaB-related gene expression in human leukemia U937 cells. Cancer Lett 264:192–200

Muthusamy A, Jayabalan N (2011) *In vitro* induction of mutation in cotton (*Gossypium hirsutum* L.) and isolation of mutants with improved yield and fiber characters. Acta Physiol Plant 33:1793–1801

Murray F, Llewellyn D, McFadden H, Last D et al (1999) Expression of the *Talaromyces flavus* glucose oxidase gene in cotton and tobacco reduces fungal infection, but is also phytotoxic. Mol Breed 5:219–232

Nandeshwar SB, Moghe S, Chakrabarty PK, Deshattiwar MK et al (2009) *Agrobacterium*-mediated transformation of cry1Ac gene into shoot-tip meristem of diploid cotton *Gossypium arboreum* cv. RG8 and regeneration of transgenic plants. Plant Mol Biol Rep 27:549–557

Naranjo SE (2005) Long-term assessment of the effects of transgenic Bt cotton on the abundance of nontarget arthropod natural enemies. Environ Entomol 34:1193–1210

Nida DL, Kolacz KH, Buehler RE, Deaton WR et al (1996) Glyphosate-tolerant cotton: genetic characterization and protein expression. J Agric Food Chem 44:1960–1966

Obembe OO, Khan T, Popoola JO (2011) Use of somatic embryogenesis as a vehicle for cotton transformation. J Med Plants Res 5:4009–4020

Oerke EC (2006) Crop losses to pests. J Agric Sci 144:31–43

Parkhi V, Kumar V, Campbell LM, Bell AA, Shah J, Rathore KS (2010) Resistance against various fungal pathogens and reniform nematode in transgenic cotton plants expressing Arabidopsis NPR1. Transgen Res 19:959–975

Perlak FJ, Deaton RW, Armstrong TA, Fuchs RL, Sims SR, Greenplate JT, Fischhoff DA (1990) Insect resistant cotton plants. Biotechnology 8:939–943

Rahman M, Shaheen T, Tabbasam N, Iqbal MA, Ashraf M, Zafar Y, Paterson AH (2012) Cotton genetic resources. A review. Agron Sustain Dev 32:419–432

Rajasekaran K, Cary JW, Jaynes JM, Cleveland TE (2005) Disease resistance conferred by the expression of a gene encoding asynthetic peptide in transgenic cotton (*Gossypium hirsutum* L.) plants. Plant Biotechnol J 3:545–554

Rajasekaran K, Hudspeth RL, Cary JW, Anderson DM, Cleveland TE (2000) High-frequency stable transformation of cotton (*Gossypium hirsutum* L.) by particle bombardment of embryogenic cell suspension cultures. Plant Cell Rep 19:539–545

Rajasekaran K, Grula JW, Anderson DM (1996) Selection and characterization of mutant cotton (*Gossypium hirsutum* L.) cell lines resistant to sulfonylurea and imidazolinone herbicides. Plant Sci 119:115–124

Risco CA, Chase CC, DMello JPF (eds) (1997) Handbook of plantand fungal toxicants. CRC Press, Boca Raton, pp 87–98

Romano GB, Scheffler JA (2008) Lowering seed gossypol content in glanded cotton (*Gossypium hirsutum* L.) lines. Plant Breed 127:619–624

Romeis J, Meissle M, Bigler F (2006) Transgenic crops expressing *Bacillus thuringiensis* toxins and biological control. Nat Biotechnol 24:63–;71

Sakhanokho HF, Ozias-Akins P, May OL, Chee PW (2004) Induction of somatic embryogenesis and plant regeneration in select Georgia and Pee Dee cotton lines. Crop Sci 44:2199–2205

Sanjaya, Satyavathi VV, Prasad V, Kirthi N et al (2005) Development of cotton transgenics with antisense AV2 gene for resistance against cotton leaf curl virus (CLCuD) via *Agrobacterium tumefaciens*. Plant Cell Tissue Organ Cult 81:55–63

Sawahel WA (2001) Stable genetic transformation of cotton plants using polybrene-spermidine treatment. Plant Mol Biol Rep 19:377a–377f

Siebert MW, Nolting S, Leonard BR, Braxton LB et al (2008) Efficacy of transgenic cotton expressing Cry1Ac and Cry1F insecticidal protein against heliothines (Lepidoptera: Noctuidae). J Econ Entomol 101:1950–1959

Singh PK, Kumar M, Chaturvedi CP, Yadav D, Tuli R (2004) Development of a hybrid delta-endotoxin and its expression in tobacco and cotton for control of a polyphagous pest *Spodoptera litura*. Transgen Res 13:397–410

Showalter AM, Heuberger S, Tabashnik BE, Carriere Y, Coates B (2009) A primer for using transgenic insecticidal cotton in developing countries. J Insect Sci 9:22

Stelly DM, Altman DW, Kohel RJ, Rangan TS, Commiskey E (1989) Cytogenetic abnormalities of cotton somaclones from callus cultures. Genome 32:762–770

Sun Y, Zhang X, Huang C, Guo X, Nie Y (2006) Somatic embryogenesis and plant regeneration from different wild diploid cotton (*Gossypium*) species. Plant Cell Rep 25:289–296

Sunilkumar G, Campbell LM, Puckhaber L, Stipanovic RD, Rathore KS (2006) Engineering cottonseed for use in human nutrition by tissue-specific reduction of toxic gossypol. Proc Natl Acad Sci USA 103:18054–18059

Sunilkumar G, Rathore KS (2001) Transgenic cotton: factors influencing *Agrobacterium*-mediated transformation and regeneration. Mol Breed 8:37–52

Tabashnik BE, Dennehy TJ, Sims MA, Larkin K et al (2002) Control of resistant pink bollworm (*Pectinophora gossypiella*) by transgenic cotton that produces *Bacillus thuringiensis* toxin Cry2Ab. Appl Environ Microbiol 68:3790–3794

Tian CZ, Wu SJ, Zhao J, Guo WZ, Zhang Z (2010) Pistil drip following pollination: a simple in planta *Agrobacterium*-mediated transformation in cotton. Biotechnol Lett 32:547–555

Thomas JC, Adams DG, Keppenne VD, Wasmann CC, Brown JK, Bohnert HJ, Kanost MR (1995) Protease inhibitors of *Manducasexta* expressed in transgenic cotton. Plant Cell Rep 14:758–762

Torres JB, Ruberson JR (2008) Interactions of *Bacillus thuringiensis* Cry1Ac toxin in genetically engineered cotton with predatory heteropterans. Transgen Res 17:345–354

Umbeck P, Johnson G, Barton K, Swain W (1987) Genetically transformed cotton (*Gossypium hirsutum* L.) plants. Biotechnology 5:263–266

Wallace RD, Sosnoskie LM, Culpepper AS, York AC et al (2011) Tolerance of GlyTol® and GlyTol® + LibertyLink® cotton toglyphosate and glufosinate in the southeastern US. J Cotton Sci 15:80–88

Wang YQ, Chen DJ, Wang DM, Huang QS et al (2004) Over-expression of gastrodia anti-fungal protein enhances Verticillium wilt resistance in coloured cotton. Plant Breed 123:454–459

Wendel JF, Brubaker CL, Edward PA (1992) Genetic diversity in *Gossypium hirsutum* L. and the origin of upland cotton. Am J Bot 79:1291–1310

Wilkins TA, Rajasekaran K, Anderson DM (2000) Cotton biotechnology. Crit Rev Plant Sci 19:511–550

Wu J, Luo X, Zhang X, Shi Y, Tian Y (2011) Development of insect resistant transgenic cotton with chimeric TVip3A* accumulating in chloroplasts. Transgen Res 20:963–973

Wu J, Luo X, Wang Z, Tian Y, Liang A, Sun Y (2008) Transgenic cotton expressing synthesized scorpion insect toxin AaHIT gene confers enhanced resistance to cotton bollworm (*Heliothis armigera*) larvae. Biotechnol Lett 30:547–554

Wu J, Luo X, Guo H, Xiao J, Tian Y (2006) Transgenic cotton, expressing Amaranthus caudatus agglutinin, confers enhanced resistance to aphids. Plant Breed 125:390–394

Wu J, Zhang X, Nie Y, Luo X (2005) High-efficiency transformation of *Gossypium hirsutum* L. embryogenic calli mediated by *Agrobacterium tumefaciens* and regeneration of insect-resistant plants. Plant Breed 124:142–146

Yue Y, Zhang M, Zhang J, Tian X, Duan L, Li Z (2012) Overexpression of the AtLOS5 gene increased abscisic acid level and drought tolerance in transgenic cotton. J Exp Bot 63:3741–3748

Zapata C, Park SH, El-Zik KM, Smith RH (1999) Transformation of a Texas cotton cultivar by using Agrobacterium and the shoot apex. Theor Appl Genet 98:252–256

Zhang JH, Guo JY, Xia JY, Wan FH (2012) Long-term effects of transgenic *Bacillus thuringiensis* cotton on the non-target *Aphis gossypii* (Homoptera: Aphididae) maintained for multiple generations. Afr J Biotechnol 11:9873–9880

Zhang H, Zhao F, Zhao Y, Guo C, Li C, Xiao K (2009) Establishment of transgenic cotton lines with high efficiency via pollen-tube pathway. Front Agric China 3(4):359–365

Zhao FY, Li YF, Xu P (2006) *Agrobacterium*-mediated transformation of cotton (*Gossypium hirsutum* L. cv. Zhongmian 35) using glyphosate as a selectable marker. Biotechnol Lett 28:1199–1207

Zhou GY, Weng J, Zeng Y et al (1983) Introduction of exogenous DNA into cotton embryos. Methods Enzymol 101:433–481

Zhu SJ, Li L, Chen JH, He QL, Fang XX, Ye CY, Yan SF, Huang ZR, Mei L (2011) Advance in research and utilization of cotton biotechnology in China. Plant Omics J 4:329–338

Zhu C, Ruan L, Peng D, Yu Z, Sun M (2006) Vegetative insecticidal protein enhancing the toxicity of *Bacillus thuringiensis* subsp kurstaki against *Spodoptera exigua*. Lett Appl Microbiol 42:109–114

Part III
Applications

Chapter 11
Modification of Cellulose Acetate Films

Francisco Rodríguez, María J. Galotto, Abel Guarda and Julio Bruna

Abstract In the last decades, the petroleum-based plastic materials have become essential materials for many industries. This kind of materials has excellent physical properties with a wide range of applications, and moreover, they are low cost and easy processing materials. However, despite the advantages offered by these materials, a massive accumulation of their wastes has produced a negative impact on the environment. So, the development of plastic materials based on bio-based polymers has been proposed as a solution to address this issue, because these materials are more environmentally friendly. A very interesting option to develop eco-friendly plastics is cellulose because this polysaccharide is the most abundant in the nature. New development based on cellulose is focused on its derivatives because they have better processability than pure cellulose. A good option is cellulose acetate because it is a biodegradable polysaccharide with good physical properties such as optical clarity and high toughness. Moreover, it is a very interesting material because its physical properties can be modified by mean of different strategies. In this context, this chapter focuses primarily to describe the use of nanofillers to produce reinforced cellulose acetate nanocomposites, and the use of active substances to generate active materials.

Keywords Cellulose acetate · Nanocomposites · Active films

F. Rodríguez (✉) · M.J. Galotto · A. Guarda · J. Bruna
Food Packaging Laboratory, Department of Food Science and Technology,
Faculty of Technology, University of Santiago de Chile, Obispo Manuel Umaña 050,
9170201 Estación Central, Santiago, Chile
e-mail: francisco.rodriguez.m@usach.cl

F. Rodríguez · M.J. Galotto · A. Guarda · J. Bruna
Center for the Development of Nanoscience and Nanotechnology (CEDENNA),
University of Santiago de Chile, Obispo Manuel Umaña 050, 9170201 Estación Central,
Santiago, Chile

K.G. Ramawat and M.R. Ahuja (eds.), *Fiber Plants*, Sustainable Development
and Biodiversity 13, DOI 10.1007/978-3-319-44570-0_11

11.1 Introduction

Cellulose is the most abundant polysaccharide which is the major structural component of the vegetable material. For example, wood has a cellulose content of around 40 % and cotton over 90 %. Photosynthesis in plants is responsible for the formation of 10^9–10^{12} ton range per year of cellulose (Klemm et al. 2005). So, cellulose is an eco-friendly material with a high availability that does not compete with the food chain (Ganster and Fink 2014). Structurally, cellulose consisting of d-glucose repeating units is linked by β-1,4 glycosidic bonds (Fig. 11.1). Therefore, cellulose is a linear polymer with an important number of free hydroxyl groups (Klemm et al. 2005), which allows to produce a stable system of hydrogen bonds. Because of this, cellulose has a negligible solubility in water and traditional organic solvents. Moreover, cellulose cannot melt processed (e.g., extrusion and injection molding), like other thermoplastic materials, because it decomposes before it undergoes melt flow (Edgar et al. 2001; Rodríguez et al. 2012c). To solve this, the structure of cellulose can be modified by chemical reactions (e.g., esterification) to produce derivative compounds which can be processing by melting and solubilized in different solvents. The presence of hydroxyl groups into anhydroglucose units with three reactive groups (one primary and two secondary) has allowed to prepare cellulose esters from various organic acids (Gupta 2012). Presence of this kind of functionality into the polymer structure reduces the formation of hydrogen bond into the structure of cellulose because the –OH groups are blocked by the presence of different substituents (Ganster and Fink 2014). From a commercial perspective, cellulose acetate (CA), cellulose acetate propionate (CAP), and cellulose acetate butyrate (CAB) are the most important cellulose derivatives. Cellulose acetate is well known by wide range of applications, such as fibers, plastics, films, separation membranes, and coatings (Wertz et al. 2010).

Cellulose acetates are traditionally produced by reaction of the polysaccharide with acetic anhydride, acetic acid, and sulfuric acid as catalysts. Different degrees of substitution (DS) can be obtained in cellulose which is according to the reaction conditions. However, the traditional cellulose acetate is a secondary acetate (Fig. 11.2) (Fischer et al. 2008). Because cellulose esters can transform by melt process (e.g., extrusion and injection), the commercial application requires the use of high content of plasticizer (15–35 wt%) (Ganster and Fink 2014). For example,

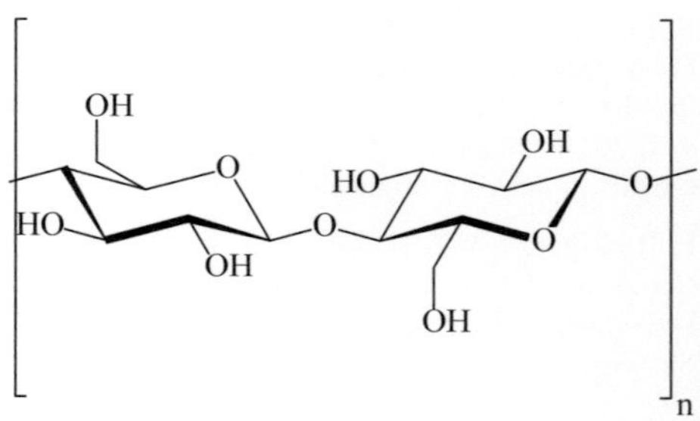

Fig. 11.1 General structure of cellulose

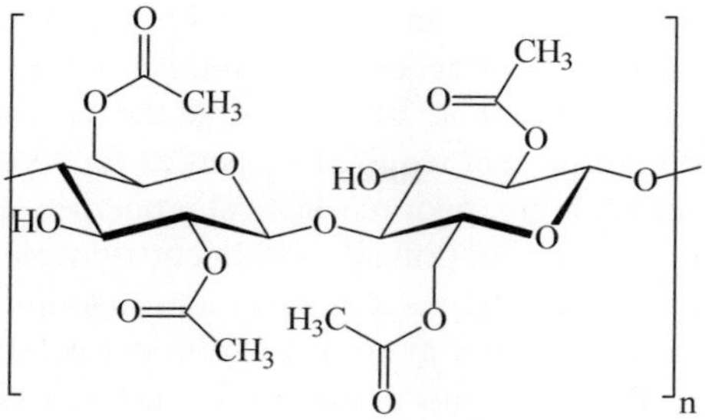

Fig. 11.2 General structure of cellulose acetate

use of eco-friendly plasticizer triethyl citrate (TEC) has allowed to process cellulose acetate at 170–180 °C which are much lower temperatures than the melting point of this polymer ($T_m \approx 233$ °C) (Rudnik 2008).

Cellulose acetate is used to produce a variety of products including synthetic fibers, cigarette filters, plastic films, and eyeglass frames. Regarding cellulose acetate films, these can be obtained by an extrusion process (Mohanty et al. 2003) and solvent cast with organic solvents such as acetone or chloroform (Meier et al. 2004). Cellulose acetate films are clear and strong, and have a high gas permeability (e.g., oxygen and water vapor). The last property favors their application in packaging films to bakery products and fresh produce. Therefore, the films have a high resistance to oils and fats (Mark 2004).

Recently, it is possible to find in literature two important strategies to modify the properties of CA films: use of nanofillers to produce nanocomposites and use of active compound to produce active films.

11.2 Nanocomposites of Cellulose Acetate

At present, a large number of industries are consumers of fossil-based plastic (e.g., polyolefins). These plastic materials have excellent physical properties with a wide range of applications, and moreover, they are low cost and easy processing materials. As a result, these materials have become indispensable materials to our society. Despite the advantages of these materials, the lack of separate, collection systems and processing sequences of municipal solid waste have led a massive accumulation, which has forced international organizations such as the European Union to seek solutions to decrease the environmental impact generated by plastic waste (Rodríguez et al. 2014). For example, directives associated with the management of plastic and plastic wastes have been oriented to prevent or minimize the generation of plastic waste, as well as promoting reuse, recycling, and other forms of recovering (Rudnik 2008). Different strategies have been proposed to reduce the environmental impact of waste plastics, such as reducing the weight of packaging and the use of biodegradable, recyclable, and/or compostable packaging materials. Bio-based polymers extracted from biomass or synthesized from monomers derived from biomass can be applied in food industry considering that most of them are obtained from renewable resources and in some cases they are also biodegradable (Rodríguez et al. 2014). Despite the environmental benefits of these materials, they do not present required properties for food packaging, mainly because of its high permeability, poor mechanical properties, and low melt viscosity, among others. To overcome this difficulty, different strategies have been proposed, and among them nanotechnology is a tool that allows to improve some of these properties becoming these polymers more like traditional polymer materials (Ray and Okamoto 2003).

Polymer nanocomposites are heterogeneous systems where a polymer matrix is reinforced a filler that has at least one dimension under 100 nm. These nanofillers can be one-dimensional (e.g., nanofibers), two-dimensional (e.g., clays), or three-dimensional (e.g., spherical particles) (Fig. 11.3).

Uniform dispersion of these nanosized fillers produces systems with a high aspect ratio. Presence of a significant interfacial area and the nanoscale of these fillers essentially differentiate polymer nanostructured composites from conventional filled plastics and composites. Consequently, the performance of nanocomposites cannot be understood by simple scaling rules that apply to traditional polymer composites (Koc 2006). So, the physical properties (barrier, thermal, optical, magnetic, and electric) of polymeric materials are significantly modified when low feed of nanofillers is used.

One of the advantages of using nanofillers combined with polymers instead of conventional fillers (e.g., glass fiber and talc) is that they produced important changes on the polymer properties with low filler feed (<5 wt%). In this regard, layered silicates (clays) have been the reinforcing materials most commonly used in this area. Structural unity, stability, easy modification, high performance, availability, low cost, and eco-friendly material are some features that have transformed

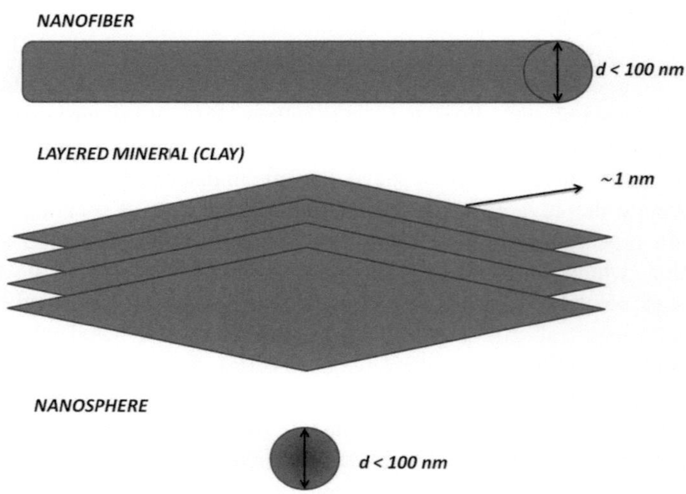

Fig. 11.3 Nanofiller structures

the clay into a perfect material for the nanocomposite development (Ray and Okamoto 2003).

Clays, also known as phyllosilicates, are crystalline minerals with a multilayer structure where each layer presents two types of structural sheets: tetrahedral and octahedral. Different arrangements of these two structures provide different classes of clay minerals (Uddin 2008). Normally, the clays used to prepare nanocomposites belong to 2:1 family. These clays are characterized by a crystal lattice, which is formed of two-dimensional layers where there is a central octahedral sheet of magnesia or alumina that is fused with two external silica tetrahedrons. The thickness of each layer is around 1 nm while the lateral dimension can vary from 30 nm to several microns. The organization of the structural layer over layer in clays is based on Van der Walls interactions. Because of the isomorphic substitution within the layers, the layers have permanent negative charge what are neutralized by cations situated in the interlayer zone (Alexandre and Dubois 2000). On the other hand, the interaction between layers is relatively weak; then, it is possible to modify the structure of the clay by means of the intercalation of some molecules. So, it is possible to modify the hydrophilic nature of interlayer by a hydrophobic one which produces an organoclay (Paiva et al. 2008). This modification has been decisive to the development of nanocomposites because the modification produces an increasing interlayer distance which facilitates the entry of the polymer into the clay structure (Fig. 11.4) (Rodríguez et al. 2015); in addition, the presence of organic groups within the clay promotes compatibility between the polymer and clay (Vazquez et al. 2008).

Depending on the polymer/organoclay interactions, two different structures of nanocomposites can be formed (Fig. 11.5). Intercalated structure in which polymer chains are intercalated between the layers of the clay also produces an increase of the distance between layers. This kind of nanocomposite is characterized by a morphology where polymer and silicate layers are arranged alternately. On the other hand, exfoliated structure shows the rupture of the layer over layer structure of

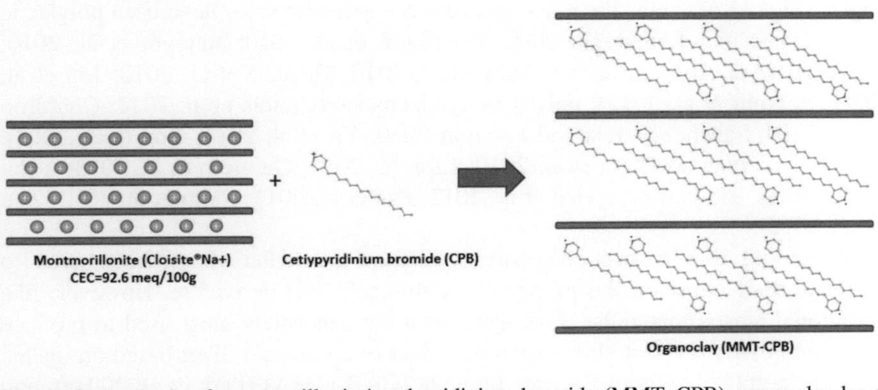

Fig. 11.4 Synthesis of montmorillonite/cetylpyridinium bromide (MMT–CPB) organoclay by a cationic interchange process (Rodríguez et al. 2015)

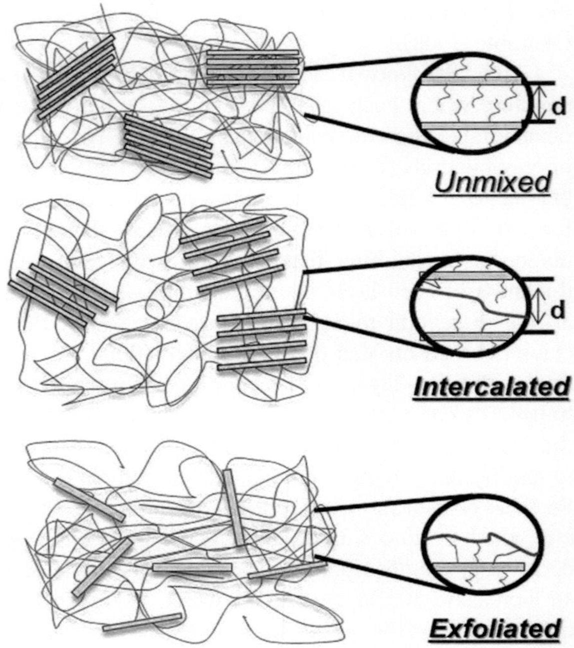

Fig. 11.5 Structures of nanocomposites based on clay as nanofillers

the clay and the uniformly dispersion of the silicate layers in the polymer matrix. To determine the nanocomposite structure, wide angle X-ray scattering (WAXS) and transmission electronic microscopy (TEM) techniques have been mainly used (Ray and Okamoto 2003; Koo 2006; Paul and Robeson 2008).

In last decades, studies based on nanocomposites have been mainly focused to develop materials based on traditional polymers. However, the use of biodegradable polymers to produce bio-nanocomposites is a new area. This is demonstrated by comparing the number of publications of nanocomposites with bio-nanocomposite topics (Fig. 11.6). Thus, the most studied bio-nanocomposites have been polylactic acid (Ozkoc and Kemaloglu 2009; Bourbigot et al. 2010; Jonoobi et al. 2010; Fortunati et al. 2012), starch (Chung et al. 2010; Heydari et al. 2012; Liu et al. 2011; Schmitt et al. 2012), polyhydroxyalkanoates (Crétois et al. 2014; Chardron et al. 2010; Sanchez-Garcia and Lagaron 2010; Yu et al. 2014; Yun et al. 2008), cellulose derivatives (Kim et al. 2010; Li et al. 2010; Quintero et al. 2013; Yang et al. 2013), and chitosan (Hsu et al. 2012; Pan et al. 2011; Pinto et al. 2012; Zhu et al. 2012).

Regarding to CA nanocomposites, different nanofillers have been used to evaluate their effect on the properties of this cellulose derivative. However, like traditional plastic materials, clays have been the nanofillers most used to produce nanocomposites. Park et al. reported the effect of a compatibilizer based on maleic anhydride-grafted cellulose acetate butyrate (CAB-g-MA) (Park et al. 2004a), and TEC plasticizer (Park et al. 2004b) on the properties of CA nanocomposites

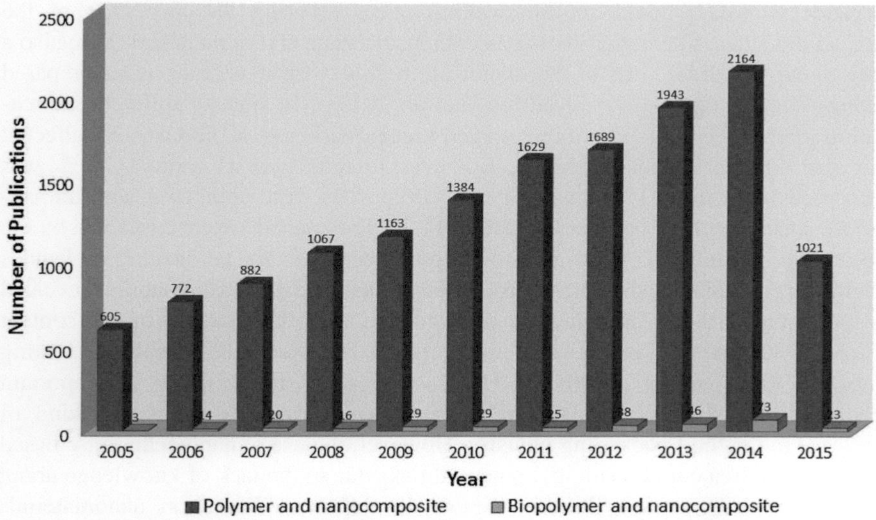

Fig. 11.6 Evolution of the number of articles per year to polymer/nanocomposite and biopolymer/nanocomposite topics. Data from ISI Web of Knowledge–Web of Science™ [v.5.17]

obtained by a melt process. In these studies, important changes on the gas barrier properties (water vapor permeability) were observed; therefore, the addition of compatibilizer favored the exfoliation of the organoclay by the cellulose acetate. Wibowo et al. (2006) reported significant changes in the mechanical properties of intercalated nanocomposites based on CA, TEC, and a commercial organoclay. In this way, nanocomposites with 5 wt% of organoclay showed an increase of around 38 % for tensile strength and 33 % for the tensile modulus regarding to CA without nanofillers. Hassan-Nejad et al. (2009) reported an unusual effect of an unmodified clay on the mechanical properties of CA films prepared by a melt process. Here, it was observed an important increase in tensile strength and Young's modulus with a 5 wt% of unmodified clay. This effect was not attributed to the structure of the nanocomposite (intercalate or exfoliate), but rather a core–shell morphology where the shell part is highly oriented which was evidenced by SEM analysis. Other study evaluated the effect of different solvents on the structure of nanocomposite based on cellulose acetate and the montmorillonite (MMT) clay which was obtained by solution casting method (Romero et al. 2009). A structural analysis of these materials evidenced a good delamination and dispersion of the MMT in the CA matrix to solvents with favorable interactions with the aluminosilicate. The formation of this structure was agreed with changes in mechanical and thermal properties of the materials. On the other hand, Lima et al. (2012) reported the elaboration of nanocomposites from CA, a commercial organoclay (Cloisite® 30B), and plasticizer by melt processing method. In this case, structures intercalated and exfoliated were evidenced and they were in concordance with interactions among the different components of the system, especially between plasticizer and organoclay which could act as swelling agent favoring the formation of the nanocomposites. Interactions between cellulose acetate and organoclays were also a key to explain differences in the intercalation grades for nanocomposites prepared using different organoclays (Rodríguez et al. 2012a). In spite of different intercalation grades, the reduction in the oxygen transmission rate (OTR) was not affected for the different nanocomposites. However, a reduction of around 50 % was observed when the OTR values for nanocomposites were compared with the cellulose acetate without organoclay (Table 11.1). These results were explained by the formation of a tortuous pathway into the polymer matrix for permeating molecules that is able to enhance the barrier properties of the material. Further studies revealed a reduction of the OTR values in concordance with the increase of the content Cloisite®30B in cellulose acetate nanocomposite films prepared by solution casting method (Rodríguez et al. 2012b). Decrease of gas permeability is an important characteristic of nanocomposite in which it has allowed to focus this kind of material to the food-packaging industry. However, the use of nanotechnology-based tools in this area could result in a potential risks due to the lack of knowledge about the novel nanomaterials. In this regard, developments based on nanomaterials require a risk assessment in order to identify and quantify potential risks that they could generate (Cushen et al. 2012). In this context, a recent study was oriented to evaluate the release of cetylpyridinium from nanocomposites based on CA and montmorillonite/cetylpyridinium bromide (MMT–CPB) organoclay to an aqueous medium by means of electrical conductivity (EC) measurements (Rodríguez et al. 2015). Results evidenced that the EC values were dependent on the content of MMT–CPB in the CA nanocomposites. Although the electrical conductivities were in agreement with MMT–CPB content in the different CA nanocomposites, the electrical conductivity changes cannot be associated exclusively with CPB migration because CA film without organoclay increased the EC of water. In addition, an important effect of the MMT–CPB on the thermal stability and opacity of the CA films was reported. The effect of the MMT–CPB content on the opacity of CA films is observed in Fig. 11.7.

On the other hand, layered double hydroxides (LDHs) have also been used as nanofillers to modify the permeability of CA films. Accordingly, Dou et al. (2014)

Table 11.1 Effect of different organoclays on oxygen transmission rate of different cellulose acetate nanocomposite films

Film	OTR (cm^3/m^2 day)
CA (Control)	662 ± 38
CA/5 wt% of Cloisite®30B	331 ± 48
CA/5 wt% of montmorillonite modified with tetradecyltrimethylammonium bromide	327 ± 35
CA/5 wt% of montmorillonite modified with hexadecyltrimethylammonium bromide	340 ± 37
CA/5 wt% of montmorillonite modified with chitosan	377 ± 33

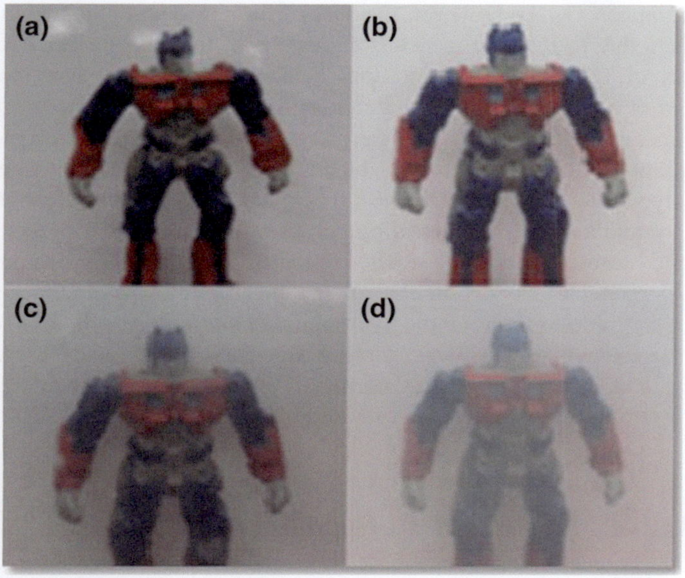

Fig. 11.7 Effect of content of montmorillonite/cetylpyridinium bromide (MMT–CPB) organoclay on the opacity of CA films. **a** CA without organoclay, **b** 2.5 wt%, **c** 5.0 wt%, and **d** 7.5 wt% of MMT–CPB

reported the reduction of oxygen permeability in flexible multilayer films based on cellulose acetate and LDH which were obtained from the spin coating of LDH nanoplatelets with different aspect ratio and cellulose acetate alternately, followed by a thermal treatment. The films were characterized by a high orientation of the LDH platelets which was favored by the formation of hydrogen-bonding network. This structure was able to change the mechanical and barrier properties of the CA films.

Considering the effect of some nanoparticles on the transport phenomena, different studies oriented to evaluate their effect on separation processes have been reported. Kim et al. (2013) studied the effect of an exfoliated flake of synthetic porous layered silicates inside CA membranes on the CO_2/CH_4 gas separation performance. The separation performance was substantially enhanced by incorporating 2, 4, or 6 wt% of nanoparticles. This study established that the CO_2 permeability increased around 54 % and CO_2/CH_4 selectivity was maintained at the same level as the pure CA membrane. A complex combination of transport mechanisms could explain the behavior of mass transfer observed. Effect of other nanoparticles as multi-walled carbon nanotubes (MWCNTs) has been evaluated on the properties of CA membranes. Ahmad et al. (2014) reported the preparation and characterization of mixed matrix membranes synthesized from cellulose acetate and MWCNTs functionalized with β-cyclodextrin using a wet phase inversion technique. These membranes showed the enhanced permeance and selectivity toward the separation of CO_2/N_2; however, the best performance was obtained at 0.1 wt% of nanofillers due to the homogeneous dispersion of this into the polymer matrix. Membranes of CA and MWCNTs functionalized have also been reported by Moghadassi et al. (2014). Here, the functionalization of MWCNTs showed better performance especially for CO_2/CH_4 gas pair separation compared with membranes based on raw MWCNTs. Therefore, blends of CA with other polymers (polyethylene glycol and styrene butadiene rubber) also affected the selectivity of the CO_2/CH_4 system and the mechanical properties of the membranes. Recently, it has been reported results about the characterization of ultrafiltration membranes based on CA and organoclay (Dehkordi et al. 2015). Incorporation of organoclay into the CA matrix was able to improve the hydrophilicity and porosity of CA membranes affecting the mass transport parameters to water and humic acid. Therefore, the mechanical and thermal properties of the CA nanocomposite membranes were modified according to the content of the organoclay.

11.3 Active Films Based on Cellulose Acetate

Bio-nanocomposites have been identified as strategic materials to the development of eco-friendly plastic materials (Darder et al. 2007). Moreover, use of these materials has risen as an important alternative to produce active materials. In this way, developing eco-materials with special functionalities such as antimicrobial or antioxidant activities could be key to the development of active food packaging.

The incorporation of antimicrobial compounds into different materials is an important strategy to provide active materials. An antimicrobial packaging is able to control the growth of spoilage or pathogenic microorganisms that are on food surfaces. Thus, the action of the antimicrobial compounds enables to increase the shelf life of packaged foods.

11.3.1 Antimicrobial Cellulose Acetate Films

The control of undesirable microorganisms in food is an important issue in the food sector since its existence can produce off-odors and changes in the aroma, color, and texture. Additionally, some microorganisms and their toxins may cause food recalls and serious foodborne outbreaks (Corrales et al. 2014). To confront this problem, the food industry has developed various technologies such as thermal processing, drying, freezing, refrigeration, irradiation, modified atmosphere, and the addition of salts and antimicrobial agents in food. However, sometimes these technologies cannot be applied in certain foods. Thus, the introduction of active agents that present antimicrobial activity in the same packaging system has been of great interest in the food industry.

Until now, a variety of substances have shown potential as antimicrobial agents; however, their use in food must be in accordance with the regulations for food contact substances (Dainelli et al. 2008).

Regarding to active cellulose acetate, different active agents have been incorporated into its structure to give antimicrobial activity. Lysozyme, an enzyme that exhibits a significant antibacterial activity which is explained by breaking the structure of peptoglycan (Corrales et al. 2014), a component of the cell wall of bacteria, has been incorporated into CA films (Gemili et al. 2009). It was evidenced that morphology of the films affected both the lysozyme release and the antimicrobial activity against *E. coli*. As a result, porous films showed higher antimicrobial activity than dense films, which is according with the release rates of lysozyme from CA films. The effect of the porosity and structure of cellulose acetate films (mono- or multilayer) was also evaluated on the potassium sorbate migration process (Uz and Altınkaya 2011). In this study, differences on the mass transfer processes were observed what was explained by alterations on the amount of the organic salt crystals and their sizes which changed according to the morphology of the films. On the other hand, Chaurasia et al. (2010) used zinc oxide nanoparticles as antimicrobial agent to produce active films. Presence of nanoparticles into the CA films was able to reduce the *E. coli* growth when the films were put in contact with this bacterium using both qualitative and quantitative methods. Bruna et al. (2014) designed active films with a montmorillonite modified with cupper (Cu^{2+}) (Fig. 11.8). Antimicrobial activity against *E. coli* increased in concordance with the content of the active agent. Therefore, these results allowed to establish that the antimicrobial activity was associated with the cupper migration from the nanocomposite films. Other CA nanocomposites have shown antibacterial effectivity. So, nanocomposites fabricated with CA, triethyl citrate, Cloisite®30B, and essential oil derivatives (thymol or cinnamaldehyde) showed activity against the growth *L. innocua* (Rodríguez et al. 2012c). Films with cinnamaldehyde showed a higher antibacterial activity than those based on thymol. In addition, CA nanocomposite films with thymol were more active than CA films without commercial organoclay. In this case, the presence of free quaternary ammonium surfactant into the CA nanocomposites would contribute to increase the antibacterial effect. Therefore, an important effect of the essential oil derivatives on the thermal properties of the CA was observed. A plasticizer effect of the essential oil derivatives was used to explain these results. Moreover, presence of essential oil derivatives affected the intercalation grade of the CA into the organoclay (Fig. 11.9) (Sepúlveda 2012). Further works was focused on the study of the specific migration of the thymol from cellulose acetate nanocomposites. This study showed

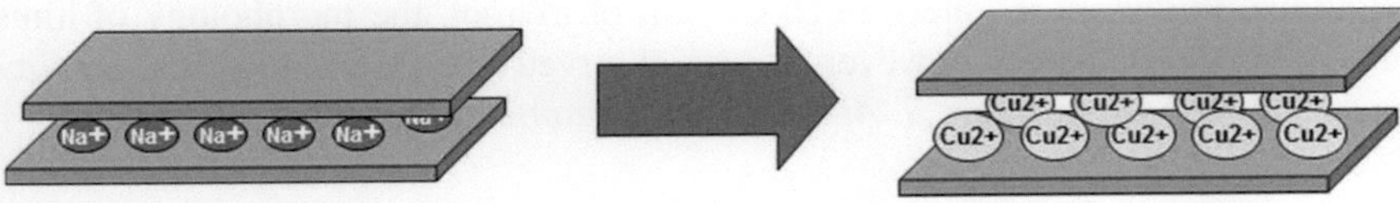

Fig. 11.8 Modification of montmorillonite to produce MMT–Cu^{2+} by a cationic interchange process (Bruna 2012)

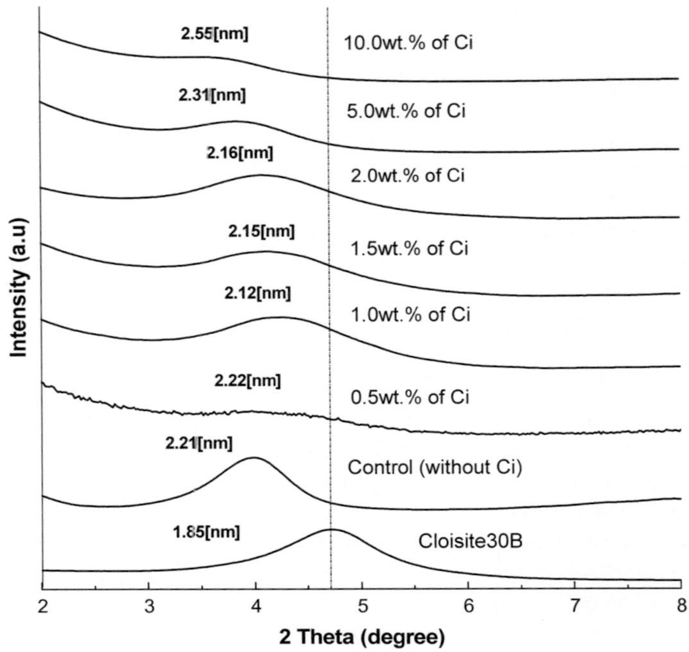

Fig. 11.9 XRD patterns of active CA nanocomposite films with different contents of cinnamaldehyde (Ci), 0–10 wt% (Sepúlveda 2012)

that presence of Cloisite®30B into the polymer matrix was not able to affect the diffusion coefficient, however, it was possible to determine that commercial organoclay affected the partition coefficient, which was highest to CA nanocomposite films (Rodríguez et al. 2014). Alternatively, active films based on CA and essential oils from oregano, cinnamon, and lemongrass have been also applied as coating on paper to produce antifungal material (Espitia et al. 2011).

Recently, it has been reported the use of supercritical fluid to include active compound into polymeric matrices. This impregnation technique is effective to the preparation of active films because it works at low or room temperature which prevents the volatilization and/or the degradation of the natural active compound (Torres et al. 2014). In this way, the impregnation of thymol into cellulose acetate films was reported (Milovanovic et al. 2015). Results showed variated yields of impregnation, which were from 4.51 to 72.26 % depending on the operating conditions. Furthermore, due to high content of thymol, the morphology of films was significantly altered; however, it was observed a high antimicrobial activity against bacteria (*E. coli* and *S. Aureus*) with microbial reduction up to 99.9 %.

11.4 Conclusion

New technologies such as nanotechnology and active packaging are strategies which can extend the use of polymeric materials from natural resources and its derivatives. Therefore, new properties or functionalities in bio-based materials can became fundamental in the availability of materials that could compete with traditional plastic from fossil origin. In this way, cellulose acetate is a promising material to applied new technologies because of its availability and easy processability through traditional industrial methods to plastic materials, offering an additional advantage compared with other bio-based materials.

References

Ahmad AL, Jawad ZA, Low SC, Zein SHS (2014) A cellulose acetate/multi-walled carbon nanotube mixed matrix membrane for CO_2/N_2 separation. J Membr Sci 451:55–66

Alexandre M, Dubois P (2000) Polymer-layered silicate nanocomposites: preparation, properties and uses of a new class of materials. Mat Sci Eng R 28:1–63

Bourbigot S, Fontaine G, Gallo A, Bellayer S (2010) Reactive extrusion of PLA and of PLA/carbon nanotubes nanocomposite: processing, characterization and flame retardancy. Polym Adv Technol 22:30–37

Bruna JE, Peñaloza A, Guarda A, Rodríguez F, Galotto MJ (2012) Development of MtCu2+/LDPE nanocomposites with antimicrobial activity for potential use in food packaging. Appl Clay Sci 58:79–87

Bruna JE, Galotto MJ, Guarda A, Rodríguez F (2014) A novel polymer based on $MtCu^{2+}$/cellulose acetate with antimicrobial activity. Carbohydr Polym 102:317–323

Corrales M, Fernández A, Han JH (2014) Antimicrobial Packaging Systems. In: Han JH (ed) Innovations in food packaging, 2nd edn. Academic Press, London

Crétois R, Follain N, Dargent E, Soulestin J, Bourbigot S, Marais S, Lebrun L (2014) Microstructure and barrier properties of PHBV/organoclays bionanocomposites. J Membr Sci 467:56–66

Cushen M, Kerry J, Morris M, Cruz-Romero M, Cummins E (2012) Nanotechnologies in the food industry—recent developments, risks and regulation. Trends Food Sci Technol 24:30–46

Chardron S, Bruzaud S, Lignot B, Elain A, Sire O (2010) Characterization of bionanocomposites based on medium chain length polyhydroxyalkanoates synthesized by *Pseudomonas oleovorans*. Polym Test 29:966–971

Chaurasia V, Chand N, Bajpai SK (2010) Water sorption properties and antimicrobial action of zinc oxide nanoparticles-loaded cellulose acetate films. J Macromol Sci Part A Pure Appl Chem 47:309–317

Chung Y-L, Ansari S, Estevez L, Hayrapetyan S, Giannelis EP, Lai H-M (2010) Preparation and properties of biodegradable starch–clay nanocomposites. Carbohydr Polym 79:391–396

Dainelli D, Gontard N, Spyropoulos D, Zondervan-van den Beuken E, Tobback P (2008) Active and intelligent food packaging: legal aspects and safety concerns. Trends Food Sci Technol 19: S103–S112

Darder M, Aranda P, Ruiz-Hitzky E (2007) Bionanocomposites: a new concept of ecological, bioinspired, and functional hybrid materials. Adv Mater 19:1309–1319

Dehkordi FS, Pakizeh M, Namvar-Mahboub M (2015) Properties and ultrafiltration efficiency of cellulose acetate/organically modified Mt (CA/OMMt) nanocomposite membrane for humic acid removal. Appl Clay Sci 105–106:178–185

Dou Y, Xu S, Liu X, Han J, Yan H, Wei M, Evans DG, Duan X (2014) Transparent, flexible films based on layered double hydroxide/cellulose acetate with excellent oxygen barrier property. Adv Funct Mater 24:514–521

Edgar KJ, Buchanan CM, Debenham JS, Rundquist PA, Seiler BD, Shelton MC, Tindall D (2001) Advances in cellulose ester performance and application. Prog Polym Sci 26:1605–1688

Espitia PJP, Soares NDFF, Botti LCM, Silva WA (2011) Effect of essential oils in the properties of cellulosic active packaging. Macromol Symp 299(300):199–205

Fischer S, Thümmler K, Volkert B, Hettrich K, Schmidt I, Fischer K (2008) Properties and applications of cellulose acetate. Macromol Symp 262:89–96

Fortunati E, Armentano I, Zhou Q, Iannoni A, Saino E, Visai L, Berglund LA, Kenny JM (2012) Multifunctional bionanocomposite films of poly(lactic acid), cellulose nanocrystals and silver nanoparticles. Carbohydr Polym 87:1596–1605

Ganster J, Fink H-P (2014) Cellulose and cellulose acetate. In: Kabasci S (ed) Bio-based plastics materials and applications. Wiley, Chichester

Gemili S, Yemenicioglu A, Altınkaya SA (2009) Development of cellulose acetate based antimicrobial food packaging materials for controlled release of lysozyme. J Food Eng 90:453–462

Gupta BS (2012) Manufactured textile fibers. In: Kent JA (ed) Handbook of industrial chemistry and biotechnology. Springer, New York

Hassan-Nejad M, Ganster J, Bohn A, Pinnow M, Volkert B (2009) Bio-based nanocomposites of cellulose acetate and nano-clay with superior mechanical properties. Macromol Symp 280:123–129

Heydari A, Alemzadeh I, Vossoughi M (2012) Functional properties of biodegradable corn starch nanocomposites for food packaging applications. Mater Des 50:954–961

Hsu S, Wang M-C, Lin J-J (2012) Biocompatibility and antimicrobial evaluation of montmorillonite/chitosan nanocomposites. Appl Clay Sci 56:53–62

Jonoobi M, Harun J, Mathew AP, Oksman K (2010) Mechanical properties of cellulose nanofiber (CNF) reinforced polylactic acid (PLA) prepared by twin screw extrusion. Compos Sci Technol 70:1742–1747

Kim D-H, Park S-Y, Kim J, Park M (2010) Preparation and properties of the single-walled carbon nanotube/cellulose nanocompo sites using N-methylmorph oline-N-oxide monohydrate. J Appl Polym Sci 117:3588–3594

Kim W, Lee JS, Bucknall DG, Koros WJ, Nair S (2013) Nanoporous layered silicate AMH-3/cellulose acetate nanocomposite membranes for gas separations. J Membr Sci 441:129–136

Klemm D, Heublein B, Fink H-P, Bohn A (2005) Cellulose: fascinating biopolymer and sustainable raw material. Angew Chem Int Ed 44:3358–3393

Koo JH (2006) Polymer nanocomposites. Processing, characterization and applications. McGraw-Hill Education, New York

Li M, Kim IH, Jeong YG (2010) Cellulose acetate/multiwalled carbon nanotube nanocomposites with improved mechanical, thermal, and electrical properties. J Appl Polym Sci 118:2475–2481

Lima JAd, Pinotti CA, Felisberti MI, Goncalves MdC (2012) Morphology and mechanical properties of nanocomposites of cellulose acetate and organic montmorillonite prepared with different plasticizers. J Appl Polym Sci 124:4628–4635

Liu Z, Zhao L, Chen M, Yu J (2011) Effect of carboxylate multi-walled carbon nanotubes on the performance of thermoplastic starch nanocomposites. Carbohydr Polym 83:447–451

Mark HF (2004) Environmentally degradable plastics. Encyclopedia of polymer science and technology, vol 6, 3rd edn. Wiley, New York

Meier MM, Kanis LA, Lima JCd, Pires ATN, Soldi V (2004) Poly(caprolactone triol) as plasticizer agent for cellulose acetate films: influence of the preparation procedure and plasticizer content on the physico-chemical properties. Polym Adv Technol 15:593–600

Milovanovic S, Stamenic M, Markovic D, Ivanovic J, Zizovic I (2015) Supercritical impregnation of cellulose acetate with thymol. J Supercrit Fluids 97:107–115

Moghadassi AR, Rajabi Z, Hosseini SM, Mohammadi M (2014) Fabrication and modification of cellulose acetate based mixed matrix membrane: gas separation and physical properties. J Ind Eng Chem 20:1050–1060

Mohanty AK, Wibowo A, Misra M, Drzal LT (2003) Development of renewable resource-based cellulose acetate bioplastic: effect of process engineering on the performance of cellulosic plastics. Polym Eng Sci 43:1151–1161

Ozkoc G, Kemaloglu S (2009) Morphology, biodegradability, mechanical, and thermal properties of nanocomposite films based on PLA and plasticized PLA. J Appl Polym Sci 114:2481–2487

Paiva LBd, Morales AR, Díaz FRV (2008) Organoclays: properties, preparation and applications. Appl Clay Sci 42:8–24

Pan Y, Wu T, Bao H, Li L (2011) Green fabrication of chitosan films reinforced with parallel aligned graphene oxide. Carbohydr Polym 83:1908–1915

Park H-M, Liang X, Mohanty AK, Misra M, Drzal LT (2004a) Effect of compatibilizer on nanostructure of the biodegradable cellulose acetate/organoclay nanocomposites. Macromolecules 37:9076–9082

Park H-M, Misra M, Drzal LT, Mohanty AK (2004b) 'Green' nanocomposites from cellulose acetate bioplastic and clay: effect of eco-friendly triethyl citrate plasticizer. Biomacromolecules 5:2281–2288

Paul DR, Robeson LM (2008) Polymer nanotechnology: nanocomposites. Polymer 49:3187–3204

Pinto RJB, Fernandes SCM, Freire CSR, Sadocco P, Causio J, Neto CP, Trindade T (2012) Antibacterial activity of optically transparent nanocomposite films based on chitosan or its derivatives and silver nanoparticles. Carbohydr Res 348:77–83

Quintero RI, Rodríguez F, Bruna J, Guarda A, Galotto MJ (2013) Cellulose acetate butyrate nanocomposites with antimicrobial properties for food packaging. Packag Technol Sci 26:249–265

Ray SS, Okamoto M (2003) Polymer/layered silicate nanocomposites: a review from preparation to processing. Prog Polym Sci 28:1539–1641

Rodríguez FJ, Galotto MJ, Guarda A, Bruna JE (2012a) Modification of cellulose acetate films using nanofillers based on organoclays. J Food Eng 110:262–268

Rodríguez FJ, Coloma A, Galotto MJ, Guarda A, Bruna JE (2012b) Effect of organoclay content and molecular weight on cellulose acetate nanocomposites properties. Polym Degrad Stab 97:1996–2001

Rodríguez F, Sepúlveda HM, Bruna J, Guarda A, Galotto MJ (2012c) Development of cellulose eco-nanocomposites with antimicrobial properties oriented for food packaging. Packag Technol Sci 26:149–160

Rodríguez FJ, Torres A, Peñaloza Á, Sepúlveda H, Galotto MJ, Guarda A, Bruna J (2014) Development of an antimicrobial material based on a nanocomposite cellulose acetate film for active food packaging. Food Addit Contam A 31:342–353

Rodríguez FJ, Cortés LA, Guarda A, Galotto MJ, Bruna JE (2015) Characterization of cetylpyridinium bromide-modified montmorillonite incorporated cellulose acetate nanocomposite films. J Mater Sci 50:3772–3780

Romero RB, Leite CAP, Gonçalves MdC (2009) The effect of the solvent on the morphology of cellulose acetate/montmorillonite nanocomposites. Polymer 50:161–170

Rudnik E (2008) Compostable polymer materials. Elsevier, Amsterdam

Sanchez-Garcia MD, Lagaron JM (2010) Novel clay-based nanobiocomposites of biopolyesters with synergistic barrier to UV light, gas, and vapour. J Appl Polym Sci 118:188–199

Schmitt H, Prashantha K, Soulestin J, Lacrampe MF, Krawczak P (2012) Preparation and properties of novel melt-blended halloysite nanotubes/wheat starch nanocomposites. Carbohydr Polym 89:920–927

Sepúlveda HM (2012) Desarrollo de eco-nanocompositos basados en acetato de celulosa-Cloisite30B con potencial actividad antimicrobiana. Department of Food Science and Technology. University of Santiago de Chile, Santiago

Torres A, Romero J, Macan A, Guarda A, Galotto MJ (2014) Near critical and supercritical impregnation and kinetic release of thymol in LLDPE films used for food packaging. J Supercrit Fluids 85:41–48

Uddin F (2008) Clays, Nanoclays, and montmorillonite minerals. Metall Mater Trans A 39:2804–2814

Uz M, Altınkaya SA (2011) Development of mono and multilayer antimicrobial food packaging materials for controlled release of potassium sorbate. LWT Food Sci Technol 44:2302–2309

Vazquez A, López M, Kortaberria G, Martín L, Mondragon I (2008) Modification of montmorillonite with cationic surfactants. Thermal and chemical analysis including CEC determination. Appl Clay Sci 41:24–36

Wertz J-L, Bédué O, Mercier JP (2010) Cellulose science and technology. EPFL Press, Florida

Wibowo AC, Misra M, Park H-M, Drzal LT, Schalek R, Mohanty AK (2006) Biodegradable nanocomposites from cellulose acetate: mechanical, morphological, and thermal properties. Comp Part A Appl Sci Manuf 37:1428–1433

Yang Z-Y, Wang W-J, Shao Z-Q, Zhu H-D, Li Y-H, Wang F-J (2013) The transparency and mechanical properties of cellulose acetate nanocomposites using cellulose nanowhiskers as fillers. Cellulose 20:159–168

Yu H-Y, Qin Z-Y, Sun B, Yang X-G, Yao J-M (2014) Reinforcement of transparent poly (3-hydroxybutyrate-co-3-hydroxyvalerate) by incorporation of functionalized carbon nanotubes as a novel bionanocomposite for food packaging. Compos Sci Technol 94:96–104

Yun SI, Gadd GE, Latella BA, Lo V, Russell RA, Holden PJ (2008) Mechanical properties of biodegradable polyhydroxyalkanoates/single wall carbon nanotube nanocomposite films. Polym Bull 61:267–275

Zhu H, Jiang R, Fu Y, Guan Y, Yao J, Xiao L, Zeng G (2012) Effective photocatalytic decolorization of methyl orange utilizing TiO_2/ZnO/chitosan nanocomposite films under simulated solar irradiation. Desalination 286:41–48

Chapter 12
Physicochemical, Morphological, and Anatomical Properties of Plant Fibers Used for Pulp and Papermaking

Kumar Anupam, Arvind Kumar Sharma, Priti Shivhare Lal and Vimlesh Bist

Abstract The plant fibers such as hardwood, softwood, grasses, annual plants and dedicated fiber crops, or those based on agricultural and industrial crop residues are the major raw materials for manufacturing different grades of pulp and paper through various processes. The processing of plant fibers into pulp and paper requires a comprehensive acquaintance of their physicochemical, morphological, and anatomical properties. Understanding the physical, chemical, morphological, and anatomical properties of plant fibers is significant in order to assess their potential toward production of pulp and paper. These properties reveal their techno-commercial suitability for pulp and paper, help in selecting the methods to be adopted for pulping and papermaking, and influence the physical as well as chemical properties of the pulp and paper produced. This chapter presents an overview of physicochemical, morphological, and anatomical properties of the plant resources that are being used as feedstocks for pulp and paper production.

Keywords Biomass · Forestry · Composition · Pulp and paper · Mechanical and strength properties

12.1 Introduction

Pulp and paper industry makes a significant contribution toward the socioeconomic prosperity of a country and hence comes under the category of high-priority industries. However, the prosperity of the pulp and paper industry depends on the availability, steady supply, efficient, and intelligent utilization of suitable raw materials. Fibers constitute the basic raw materials for pulp and paper. The pulp and

K. Anupam (✉) · A.K. Sharma · P.S. Lal · V. Bist
Physical Chemistry, Pulping and Bleaching Division, Central Pulp and Paper Research Institute, Himmat Nagar, Saharanpur, Uttarpradesh 247001, India
e-mail: kumaranupam@live.com

K.G. Ramawat and M.R. Ahuja (eds.), *Fiber Plants*, Sustainable Development and Biodiversity 13, DOI 10.1007/978-3-319-44570-0_12

paper industry is dependent on wood and certain other plant fibers. The raw material consumption pattern in the pulp and paper industry has been observed to change rapidly during the preceding few years. Earlier, the utilization of woody raw materials was higher while nonwoody raw materials had meager contributions. Thirty years back in India, the utilization of woody-, nonwoody-, and wastepaper-based fibers was 84, 7, and 9 %, respectively. However, during 2011, the share in production of paper from wood, nonwood, and recycled fibers was 31, 22, and 47 %, respectively. The nonwoody fibers used for making pulp can be classified as follows: agricultural residues, e.g., sugarcane bagasse, cotton stalks, rice straw, wheat straw etc.; natural growing plants, e.g., bamboo, reeds, sabai grass, kahi grass etc.; non wood crops grown primarily for its fiber content such as bast fiber, e.g., jute, hemp, kenaf; leaf fiber, e.g., sisal, and seed hair, e.g., cotton fiber, cotton linter. Almost all these growing plants have been tried for pulp and papermaking and various new plants are being researched upon. However, economics plays an important role here and the consideration of collection, storage, and processing sometimes put limit to the use of some materials. All these fibrous raw materials range from tiny grass plants to tall trees. Since they all are the metabolic products of living organism, it is certain that they will have infinite variation due to genetic and environment factors.

Engineering these materials of such an anatomical diversity calls for a high degree of scientific and technical understanding of their physicochemical, morphological and anatomical properties in order to obtain their correct and optimized utilization for pulp and paper. A plant fiber, to pulp and papermaker, is an elongated, tubular, cylindrical, and very small cell obtained from certain plants or parts of plants. Its diameter is quite thin and considered to be microscopic, i.e., <0.1 mm. However, its length can be significant varying from about 0.5 mm to over 120 mm. For common papermaking fibers, the length to diameter ratio lies in the range of 50–200: 1. Fibers from different sources have different physical properties, length, width, wall thickness, and cavity diameter in addition to their varying amounts of three main constituents, i.e., cellulose, hemicellulose, and lignin. On close investigation, it has been found that the fibers are built up of several different layers and have a cavity inside. The internal organization of fiber wall, the percentage chemical present in it, and its dimensions are very important features in deciding the pulp and papermaking characteristics of fibers. The fiber wall is a composite of organic materials. The organic materials are the polymers which with special arrangement make fiber a complex and highly ordered material with distinct structural and mechanical properties. This chapter presents an overview of various physical, chemical, morphological, and anatomical properties of different wood-, agro-, and grass-based fibers used for pulp and paper along with their significance and influence on the pulping and papermaking processes as well as pulp and paper properties.

12.2 Physical Characteristics of Plant Fibers

Basic density, bulk density, and moisture content are important physical characteristics of raw materials which have tremendous effects on their pulp and paper properties as well as production processes. These physical characteristics are general financial gauge of raw materials quality for papermaking process. Basic density determines the shipment expenditures, pulp manufacturing capacity, chemical quantity, and energy consumption. Bulk density or packing density estimates the quantity of plant fibers on the oven-dried basis that can be packed or fed into a pulping digester. Density is one of the most significant characteristics of raw materials from the point of view of production of good quality pulp and paper because it has an impressive effect on the strength and various other physical properties of pulp and paper. The quantity of raw materials required to manufacture one ton of air-dried pulp can be estimated from the density and pulp yield (Magaton et al. 2009). Moisture content of the raw materials affects their dimensional stability and those experience brisk moisture changes are unattractive due to prominent effects on their physical and mechanical properties (Saravanan et al. 2013). These physical properties differ from plant to plant due to disparity in plant variety, within species to species of same plant and age of the plant. The basic density and bulk density of wood increases while moisture content decreases with increase in age of the tree. The density of wood species is strongly associated with the thickness of cell walls and its percentage, the number of vessels, fibers, and wood rays wherein increase of fibers and rays in wood tissue, increase in thickness of cell walls, and decrease in the number of vessels lead to increase in density (Samariha 2011). Sometimes wood density is independent of annual ring width as well as early to late wood ratios wherein increase in annual ring width increases the density of hardwoods but decreases the density of softwood (Samariha 2011).

12.3 Proximate Chemical Properties of Plant Fibers

Chemical composition of wood is important for the preliminary characterization of lignocellulosic raw materials for evaluating its pulp and papermaking potential. It gives an assessment of fibrous and nonfibrous materials. Estimation of chemical components is also relevant from the point of view of raw material storage. The raw materials to be utilized for pulp and paper are generally characterized in terms of the amount of ash, water solubility, alcohol–benzene extractives, sodium hydroxide solubility, lignin, holocellulose, and pentosans. Ash content of any fibrous lignocellulosic mass suggests the quantity of inorganics such as Si, Ca, and Mg found in them. The ash content of the fibrous raw materials indicates the presence of inorganics. The inorganic content varies from 0.5 to 1.0 % in softwoods and hardwoods, 1.0–2.5 % in bamboo and bagasse, 5.0–10.0 % in wheat straw and other grasses while more than 15.0 % in rice straw. The higher inorganic content in the nonwoody fibrous raw materials is due to the presence of silica. The presence of silica causes severe problems, especially in the chemical recovery process. The lignocellulosic material in the plant fibers degrades during storage but ash content which has inorganics increases with storage period. The water solubles of wood include gum, low molecular weight phenolics, pectins, low molecular weight polysaccharides, and other polar extractives such as free amino acids and alkaloids (Anupam et al. 2014). It gives an idea about the anticipated pulp yield in case of mechanical pulp. When the fibrous raw materials is subjected to storage, above constituents reduce with time. The increase in water solubles signals the degradation of the lignocellulosic precursor. With an increase in storage period, cellulosic components present in raw materials decay due to fungus invasion which results in the rise of their water solubility.

Alcohol–benzene extractives are generally higher in soft woods as compared to hardwoods and grasses. The low molecular compounds of various types are extractable from the fibrous raw materials with water or organic solvents excluding lignin and hemicelluloses. The extractives, which are extraneous components, include aliphatic and aromatic hydrocarbons, terpenes and their derivatives, alcohols, aldehydes, phenols, and quinones. Some woods contain essential oils, resin acids, and sterols whereas others yield tannins and coloring matter. The color, odor, taste, or unusual flammability can be attributed to extractives. They may interfere with the pulping process, cause foaming, and sometimes cause corrosion to the equipments. Juvenile wood contains higher extractives than matured one. During the storage, alcohol–benzene solubility decreases as some of the organic substances like essential oils present in eucalyptus, pine etc. get evaporated. It is good to store the freshly harvested raw materials for a period of 2–3 months in order to decrease the extractives. Sodium hydroxide solubility, perhaps the single super most information parameter, indicates about the physical and chemical characteristics of lignocellulosic material. It also dissolves all wood components extractable by means of hot water. In addition to water soluble, it also dissolves almost all the polar components in wood via their sodium salt formation. A tiny proportion of lignin, hemicelluloses, and cellulose having small molecular mass are also extractable with sodium hydroxide. Sodium hydroxide solubility can be employed as an analytic parameter to gauge the microbial degradation of a particular raw material and the relative chemical pulp yield of raw material to be obtained (Anupam et al. 2014).

Klason lignin is an indication of acid insoluble lignin present in wood. More or less, it represents the substantial amount of lignin in lignocellulosic masses as the amount of acid soluble lignin is relatively almost insignificant. The digestion of active pulping chemicals, the duration of total pulping process, stiffness of fiber, and bleachability are directly proportional to the amount of lignin present in plant materials (Dutt et al. 2012). Holocellulose is the major component, i.e., 60–80 % of wood cell wall and represents the collective amount of cellulose and hemicelluloses. Pulp and paper quality is very much dependent on physical and chemical characteristics of holocellulose. Cellulose molecule that includes proper organized $C_6H_{12}O_6$ particles is known as alpha cellulose and has degree of polymerization

600–2000 while beta and gamma cellulose have degree of polymerisation around 400–600 and 25–400, respectively. Alpha cellulose is insoluble even in 17.5 % NaOH while beta and gamma cellulose are soluble and go into solution phase. Cellulose and hemicelluloses can be precipitated out by absolute alcohol. Hemicelluloses are low molecular weight polysaccharides having degree of polymerisation 100–250. The total cellulosic portion, α-cellulose, and hemicelluloses play key role for production of good quality pulp and paper. Nieschlag et al. (1960) developed a grading system according to which plant materials having alpha cellulose >34 % are regarded as the best suited pulp and paper precursor. The amount, chemical construction, distribution, and degree of polymerization of hemicelluloses have a cutting edge effect on mechanical strength properties of paper and it has been found that the higher the hemicelluloses percentage, the larger the swelling performance of pulp which leads to augment in tensile index, burst index, and double folds of paper sheet while decrease in electrical power consumption of a beater or refiner (Tyagi et al. 2004). As raw materials are stored in open climate, the reduction in alkali and water solubilities affects total mass of the wood. This results in increment of holocellulose content. Pentosans include low molecular weight carbohydrate sugar units principally xylose and arabinose. These get hydrolyzed and converted to furfural with strong hydrochloric acid. Pentosans quantity provides a rough estimate of total hemicelluloses present in raw materials particularly hemicelluloses having pentosan as building unit. During storage, hemicelluloses are degraded to some extent due to degradation of low molecular weight short chain pentose sugars (Anupam et al. 2014). Figure 12.1 presents proximate chemical properties of some of the plant fibers used for pulp and papermaking.

12.4 Morphological Features of Plant Fibers

Fibers are the most useful cellulose material in the pulp. These are normally long, flexible, and form the basic network in the paper. Fibers contribute to the basic strength of the paper. Fibers are long narrow cells with tapering ends and central canals known as lumen. The fibers depending upon origin differ significantly. Weight proportion of fibers in pulp varies from 95 % in softwoods, 65–75 % in hardwoods, and 55–65 % in agricultural residues. Morphological characteristics of fibers such as length, diameter, cell wall thickness, lumen size, and their derived morphological factors such as Runkel ratio, flexibility coefficient, slenderness ratio, and felting factor are significant parameters to evaluate the pulp quality of fibers in raw materials (Oluwafemi and Sotannde 2007). Fiber length, fiber cell wall thickness, fiber width, and fiber lumen of some plant fibers utilized for pulp and papermaking are shown in Fig. 12.2. The comparative study of cross-sectional dimensions of various pulp and papermaking plant fibers is demonstrated in Fig. 12.3.

The papermaking pulps have fiber population with varying lengths. The heterogeneity of the fiber population influences the papermaking, and the

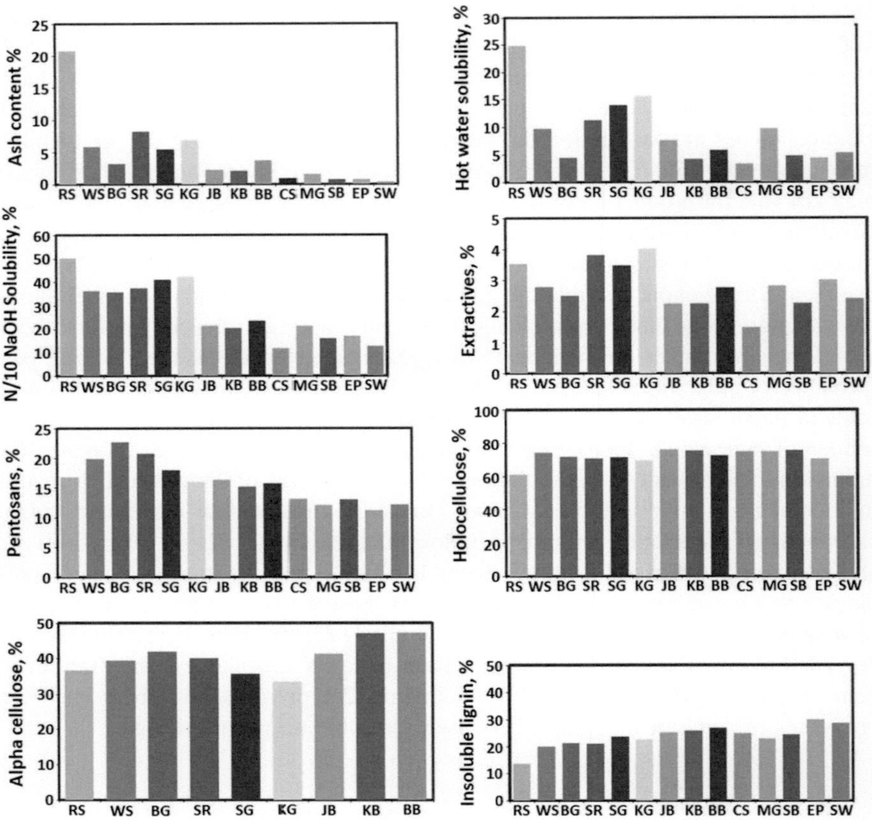

Fig. 12.1 Proximate chemical composition of different fibrous raw materials (*RS* rice straw, *WS* wheat straw, *BG* bagasse, *SR* sarkanda grass, *SG* sabai grass, *KG* Kans grass, *JB* Jati bamboo, *KB* Kako bamboo, *BB* Bhulka bamboo, *CS* Casuraina, *MG* mango, *SB* subabul, *EP* eucalyptus, *SW* softwood). http://www.dcpulppaper.org/gifs/report24.pdf

knowledge and understanding of the fiber length distribution is highly essential in predicting the behavior of a raw material in the papermaking process. Generally, the fiber length is taken as average for a source, which is relatively an easy expression and gives a broad idea for comparison purpose. The average fiber length varies from 3.5 mm in softwoods, 0.8–1.2 mm in hardwoods, 2–5 mm in bast and leafy fibers, and 1.0–2.5 mm in straws, bagasse, and bamboo. Usually, it is a belief that a paper made from a long fiber will give paper more strength than that made from a shorter fiber. However, with usage of different kinds of raw materials, it is possible to produce good quality papers from even short fibers. Fiber length also influences the general structure and surface properties of paper sheet. The fibers in a paper web are randomly distributed, and depending on the length of the fiber, the number of fiber crossings increases or decreases. If the number of fiber crossings increases due

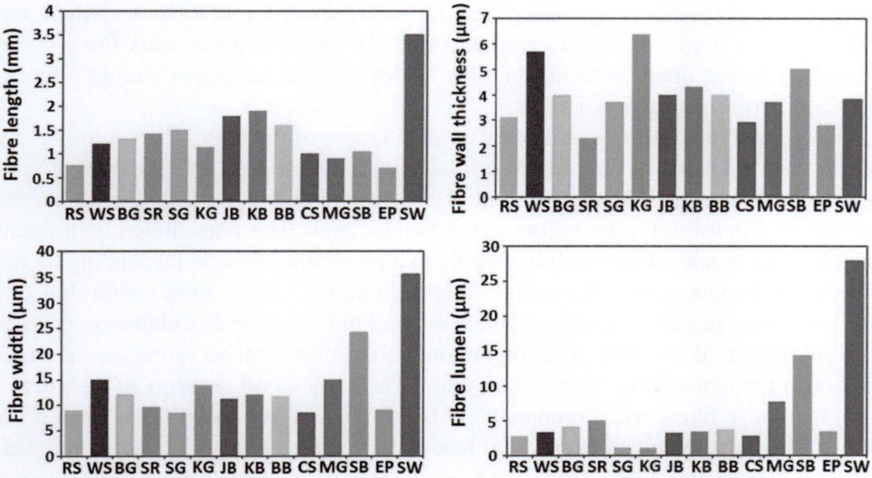

Fig. 12.2 Fiber length, wall thickness, width, and lumen of different fibrous raw materials (*RS* rice straw, *WS* wheat straw, *BG* bagasse, *SR* sarkanda grass, *SG* sabai grass, *KG* Kans grass, *JB* Jati bamboo, *KB* Kako bamboo, *BB* Bhulka bamboo, *CS* Casuraina, *MG* mango, *SB* subabul, *EP* eucalyptus, *SW* softwood). http://www.dcpulppaper.org/gifs/report24.pdf

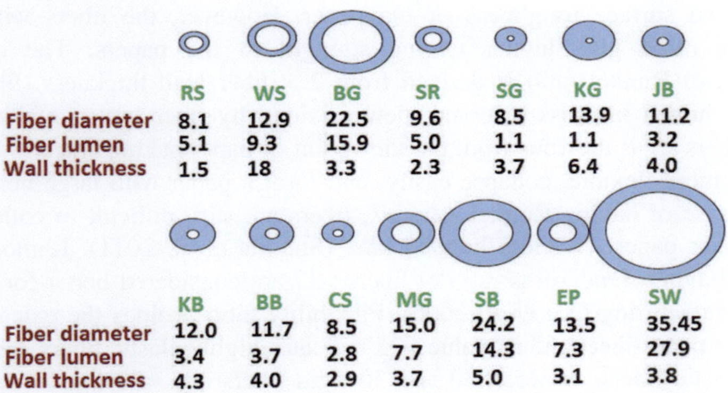

Fig. 12.3 Comparative cross-sectional dimensions of different fibrous raw materials (*RS* rice straw, *WS* wheat straw, *BG* bagasse, *SR* sarkanda grass, *SG* sabai grass, *KG* Kans grass, *JB* Jati bamboo, *KB* Kako bamboo, *BB* Bhulka bamboo, *CS* Casuraina, *MG* mango, *SB* subabul, *EP* eucalyptus, *SW* softwood). http://www.dcpulppaper.org/gifs/report24.pdf

to longer fibers, the web becomes stronger and the paper sheet has higher wet web strength as well as the dry strength. The wet web strength is very critical in high-speed machines. The terms width and diameter of the fibers are normally used for all the practical purposes for the same dimension. The unrefined fibers are normally tubular in structures which become flattened on refining. In the paper web when the fibers cross over randomly, the area of fiber crossover is influenced by fiber width. If the fibers are wide, then the area per crossover increases where the fibers are held together which contribute to the strength of paper web. For a given fiber length, the fibers with higher fiber width give higher paper strength due to increased crossover area per fiber.

The length to width ratio is known as slenderness of the fiber. The central cavity in the fiber is known as fiber lumen which is void. Depending on the extent of void space, the fiber may flatten to different extents, as the fiber is refined. The higher the extent of collapsibility, the higher is the bonded area. The fiber lumen is different for different species. Fiber wall is specific to a given fiber source. Depending on the fiber wall thickness, the response of fibers to refining varies. Fibers with thin cell walls collapse readily. Increase in fiber cell wall thickness leads to increase in mass per unit length of the fiber. The importance of the fiber wall on properties of paper had been recognized for a number of years. The early wood portions of the growth ring may have fibers with comparatively thin wall, and late wood fibers have thick walls. The pulps obtained from wood having thin-walled fibers give dense and well bonded sheets and those from thick wall give bulky sheets with high tearing resistance. It is apparent that thin-walled cells collapse and confirm to other fibers easily to give a dense bonded sheet of paper due to their high flexibility. Fibers with higher coarseness are called coarser fibers and with lower coarseness as finer fibers. Fibers of higher coarseness are stiff and difficult to collapse which lead to poor bonding due to less bonding area and lower strength. This will result in increased porosity and surface roughness of the paper. However, the fibers with higher coarseness often give higher tearing strength to the papers. The derivative expression of Runkel ratio is derived from 2 × fiber wall thickness (fiber width excluding lumen in cross-sectional view) divided by fiber lumen width. Runkel ratio of fibers <1 is desirous from the viewpoint of papermaking since in such case fibers are more flexible, collapse easily, and form a paper with large bonded area while in case of having Runkel ratio >1, fibers are stiff, difficult to collapse, and form bulkier paper with less bonded area (Sharma et al. 2011). Lignocellulosic materials having slenderness ratio of fibers >33 are considered better for pulp and paper manufacturing (Xu et al. 2006). Flexibility ratio decides the extent of fiber bonding in paper sheet and its value >75 indicates highly elastic fibers, between 50 and 75 elastic fibers, between 30 and 50 rigid fibers and <30 highly rigid fibers (Ekhuemelo and Tor 2013). Figure 12.4 shows the photomicrographs of fibers of indigenous woody and agro biomass used for pulp and paper production.

The submicroscopic parameters of the fibers such as microfibrillar orientation in various layers of fibers have also significant influence on the papermaking. The length of microfibril is not well defined but their diameters are above 10–20 nm. Microfibril occurs in small bundles or macrofibril. These, in turn, can be organized into thin sheets or lamellae which give the wall a layered architecture. At the fiber surface, the microfibrils form a thin net-like surface covering the primary wall. However, in the bulk of fiber wall or secondary wall, the microfibrils occur in parallel arrays, or sheets of preferred orientation which is spiral about the fiber, producing layered construction. The orientation of different wall lamellae from the fiber axis is termed as microfibril angle. Thin primary wall (0.1 μm) has

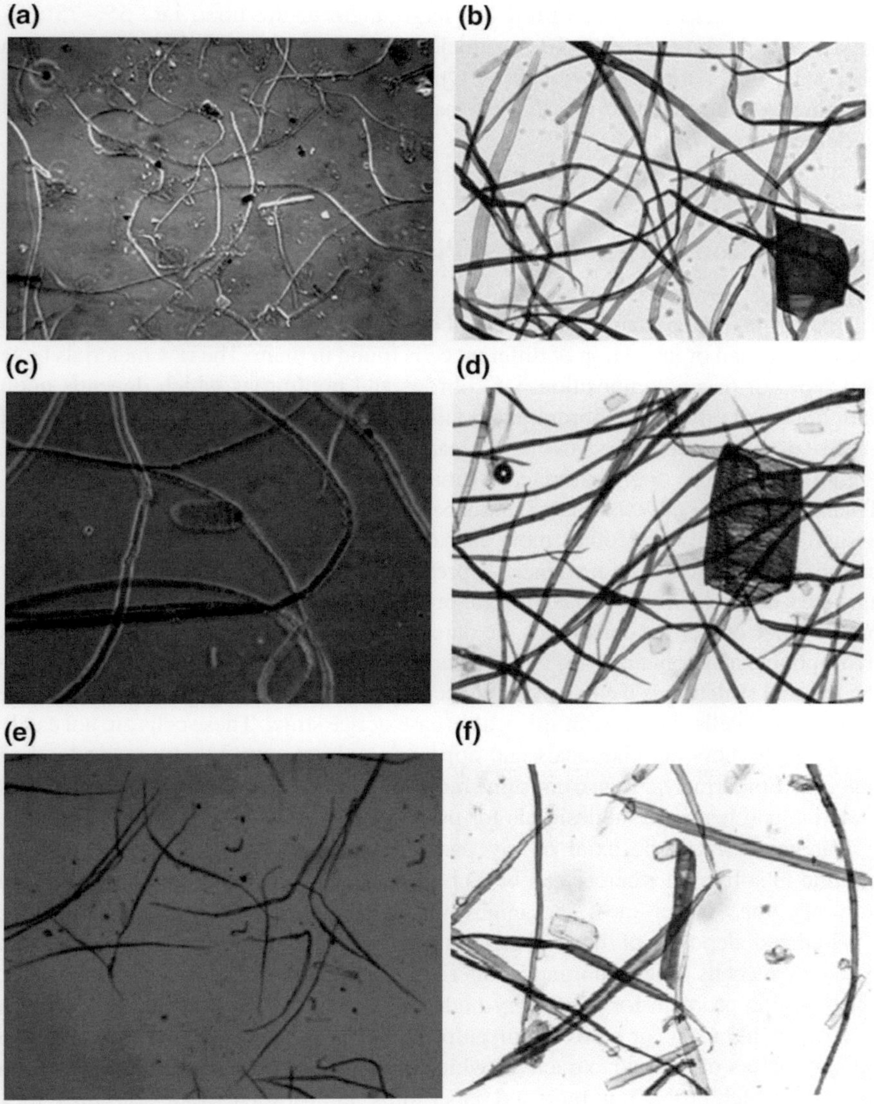

Fig. 12.4 Photomicrograph of fibers of **a** rice straw pulp, **b** acacia pulp, **c** bamboo pulp, **d** eucalyptus pulp, **e** subabul pulp, **f** wheat straw pulp. http://www.dcpulppaper.org/gifs/report24.pdf

microfibrils forming a more or less irregular interwoven pattern. Cellulose content of primary wall is very less, about 10.0 %, and is embedded in a matrix of pectic material. In the inner most part of primary wall, the cellulose microfibril begins to exhibit an ordered arrangement. They exhibit three different layers: the first and outermost layer (S_1), the secondary layer (S_2) and the innermost layer (S_3). In S_1

(0.1–0.2 μm thick), the microfibril has large angle to the fiber, i.e., 55°–75°, S_2 makes a thick layer about 2–10 μm thick and has microfibril orientation close to fiber axis. In $S_{2,}$ the fibrillar angle is typically 5°–20°. S_3 layer has similar construction to that of S_1, but it is thinner than S_1 (0.07–0.08 μm) and has microfibril angle between 60° and 90°.

12.5 Anatomical Features of Plant Fibers

The papermaking eminence and potential of fibrous raw materials highly depend on constitution and organization of different cells found in them. The raw materials have two types of tissue composition, i.e., fibrous and nonfibrous which depends upon their botanical origin. The fibrous and nonfibrous cells usually present in the different grades of pulps are fibers, tracheids, parenchyma, vessels, and epidermal cells. Figure 12.5 shows the anatomical features of some papermaking plant fibers. Tracheids are generally found in gymnosperm species. On the other hand, nonfibrous tissue, i.e., vessels are found more in case of hardwood compared to other raw materials. Epidermal cells are generally prevalent in wheat straw, rice straw, and other agro-based sources. Epidermal cells are present in abundance in monocotyledonous pulps. Since epidermis constitutes the outer most covering of nonwoody plants, it is hydrophobic and contains wax-like substance called cutin. In nonwoody plants since the internal structure is weak, epidermal cells give strength and protection to internal tissue. These cells also contain silica, as in case of rice straw. These cells are not easily separated by cooking chemicals and appear in pulp as masses with sharp-toothed margins. Epidermal cells also consume more amounts of chemicals, contain siliceous material, and hence are undesirable for pulping.

Parenchyma can be axial or ray parenchyma. It is present about <10 % by volume in softwood whereas can be as high as 50 % in nonwoods. These cells are normally the site of most inorganics which can be K, Mg, Mn, Ca, and Si. Crystallized deposits of these materials and amorphous of silica are present in parenchyma cells. These inorganic materials contribute to ash in wood or nonwood. They are the problem for some very high purity pulp and promote scale formation in equipments used for recovery of pulping chemicals. Parenchyma cells are also the originators of organic extractives which are undesirable in dissolving grade pulp and cause pitch problem in pulp and paper mills. In absorbent grades of pulps, they render the fibers less wettable reducing liquid absorption rate. Parenchyma cells are sometimes as small as 10–100 μm and thus are responsible for giving fines content to pulp causing higher drainage rate of stock. These cells also contain gums, oils, resins, latex tannins, starch, silica, or calcium oxalate and thus consume more chemicals during pulping. They are very thin-walled cells of poor strength so they are not desirable in paper sheet and are generally preferred to be screened out, e.g., during depithing process.

Vessel elements present in all the angiosperm plants whether woody or nonwoody hardwoods contain vessel volumes ranging from 30 to 50 % by volume.

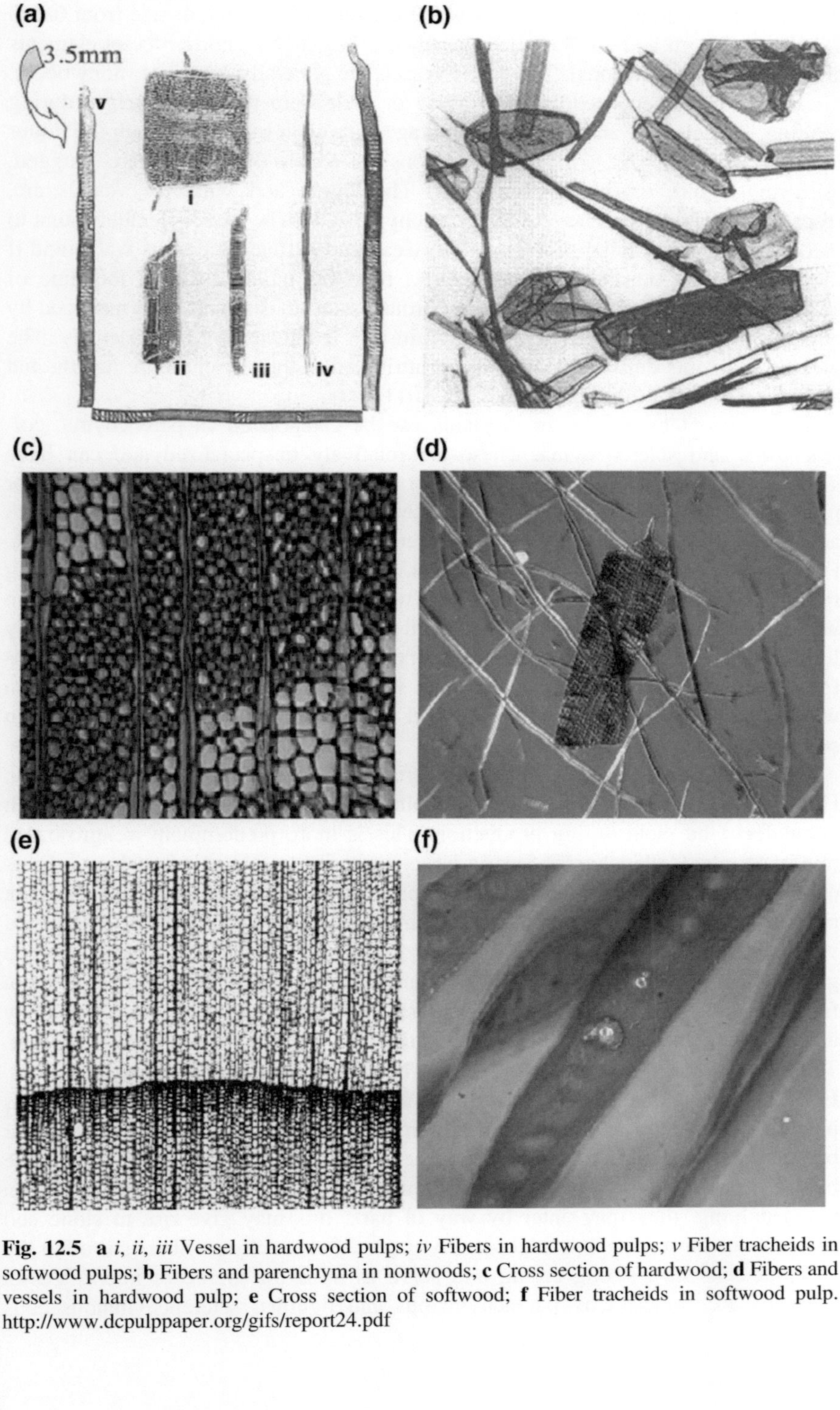

Fig. 12.5 **a** *i*, *ii*, *iii* Vessel in hardwood pulps; *iv* Fibers in hardwood pulps; *v* Fiber tracheids in softwood pulps; **b** Fibers and parenchyma in nonwoods; **c** Cross section of hardwood; **d** Fibers and vessels in hardwood pulp; **e** Cross section of softwood; **f** Fiber tracheids in softwood pulp. http://www.dcpulppaper.org/gifs/report24.pdf

A high vessel volume tends to lower wood density. They range in size from 0.2 to 1.3 mm in length and 20–300 μm in diameter. In general, shorter vessel elements are also wider (hardwoods). The larger vessels are generally narrower (nonwoods). These vessel elements tend to pull away or pick from the sheet surface during printing. Vessels play an important advantageous role in transport of pulping liquor in wood chips. There are some situations in which the vessels are plugged, impairing liquid, or liquor conduction. The length and width of vessels are, respectively, shorter and larger than parenchyma cells. The vessel elements burst to pieces during the pulping process due to weak and extremely porous walls; and if they are loosely attached to fiber matrix, they get pulled away at the time of printing. The capacity of fibrous or nonfibrous tissues to disintegrate is managed by quotient of lumen and fiber width which in turn is affected by basic density. The picking problems during printing can be attributed to the vessel width, length, and numbers per unit weight (Sharma et al. 2011).

Fundamental tissue system in plants can be categorized as parenchyma, collenchyma, and sclerenchyma. These are relatively unspecialized and can have various physiological functions. These are commonly referred as ground tissue or pith tissue or nonfibrous tissue and also often act as a sort of filler tissue. The length and width of parenchyma are dependent on the variety of raw materials. Parenchyma, in softwoods and hardwoods, are small living cells often functioning for photosynthesis, storage, secretion, and wound healing. They are commonly thin walled, but in central regions of stems and roots may be thick walled and heavily lignified. They may also accumulate crystals, e.g., calcium oxalates, tannins (polyphenols), or food materials such as starch, fats, and oil. In monocots (non wood), parenchyma are much more varied in size and can be as large as fibers in some plants.

The strength of paper is not much affected by the presence of parenchyma. However, it creates problems during washing due to poor drainage property which also affects the yield during production. This is more predominant in agro-based raw materials. Collenchymas have a large similarity to that of parenchyma; however, these cells are bit longer and have special wall thickening. The cell walls are rich in pectin. These are often located near the surface of young stems and in leaf veins. They are not of any relevance to wood pulping but are present in some grasses in large volumes along with parenchyma and it is desirable to separate them from fiber cells, e.g., bagasse. Sclerenchyma are usually dead cells at maturity, they are thick walled and heavily lignified functioning primarily for mechanical support. These are two types of cells which can be distinguished as fiber and sclereids. Fibers are similar to xylem, sclerenchyma fibers are extra xylary in location such as phloem of dicots or bast fibers (jute, hemp, etc.) present in vicinity of vascular bundles. Sclereids are variably shaped cells often branched as well. They give hardness and rigidity to tissue such as phloem, leaves, seed coats, and shells. In wood pulping, they may enter by way of bark, this may give rise to stone cell problems or dirt or speck in high quality papers. Sclerenchyma fibers surround the vascular cells in a strong casing and serve as an extra source of fibers. Vessel elements are enclosed by parenchymatous and lignified sclerenchymatous cells

(Sharma et al. 2011). Lignified parenchymatous ground tissue comprises of fibers having wide cell wall, i.e., sclerenchyma cells which prove to be a valuable material for pulp and papermaking fibers.

12.6 Conclusion

This article presents an overview of different types of lignocellulosic raw materials utilized for pulp and paper production. There are various sources for lignocellulosic feedstocks and include trees, grasses, agricultural wastes, plant residues, and dedicated fiber crops. The quality of paper products made from pulp fiber is very much dependent upon the physical, chemical, morphological, and anatomical properties of plant fibers. The age of wood is also one of the most crucial parameters because it decides the amount of extractives, cellulose, hemicelluloses, lignin, fiber length, fiber density, cell wall thickness, vessels, parenchyma, etc. which ultimately affects the pulping process as well as pulp and paper properties. Plant fibers grown in different ecological and environmental conditions affect the morphology, chemical composition, physical properties, and anatomical features due to soil and water conditions. Further, when considerable expansion of agro sylviculture estates along with reforestation ventures developed on congenitally modified species of trees are taken seriously these days by pulp and paper industry understanding the morphological and anatomical features of plant fibers become inevitable. More research is also needed in the areas of state-of-the-art modern raw material handling, storage, and pre-treatment technologies because these technologies also alter the fiber quality and properties. A structured strategic plan should be adopted to address the challenges related to fiber material requirement, quality, and supply to ensure the demand and supply of quality pulp and paper products.

References

Anupam K, Lal PS, Bist V, Sharma AK, Swaroop V (2014) Raw material selection for pulping and papermaking using TOPSIS multiple criteria decision making design. Environ Prog Sustain Energy 33(3):1034–1041

Dutt D, Sharma AK, Agnihotri S, Gautam A (2012) Characterization of dogs tooth grass and its delignification by soda pulping process. J Sci Technol 1(8):434–447

Ekhuemelo DO, Tor K (2013) Assessment of fibre characteristics and suitability of maize husk and stalk for pulp and paper production. J Res For Wildl Environ 5(1):41–49

Magaton ADS, Colodette JL, Gouvêa ADFG, Gomide JL, Muguet MCDS, Pedrazzi C (2009) Eucalyptus wood quality and its impact on kraft pulp production and use. TAPPI J 8:32–39

Nieschlag HJ, Nelson GH, Wolff JA, Perdue RE (1960) A search for new fiber crops. Tappi J 43 (3):193–201

Oluwafemi OA, Sotannde OA (2007) The relationship between fibre characteristics and pulp-sheet properties of *Leucaena leucocephala* (Lam.) de wit. Middle-East J Sci Res 2(2):63–68

Samariha A (2011) The Influence of Trees's Age on the Physical Properties and Fiber Length of *Eucalyptus camaldulensis* in the Zabol Region at Iran. Middle-East J Sci Res 8(5):851–854

Saravanan V, Parthiban KT, Kumar P, Marimuthu P (2013) Wood characterization studies on *Melia dubia* cav. for pulp and paper industry at different age gradation. Res J Recent Sci 2:183–188

Sharma AK, Dutt D, Upadhyaya JS, Roy TK (2011) Anatomical, morphological and chemical characterization of *Bambusa tulda, Dendrocalamus hamiltonii, Bambusa balcooa, Malocana baccifera, Bambusa arundinacea and Eucalyptus tereticornis*. BioResources 6:5062–5073

Tyagi CH, Dutt D, Pokharel D, Malik RS (2004) Studies on soda and soda AQ pulping of *Eulaiopsis binata*. Indian J Chem Technol 11(1):127–134

Xu F, Zhong XC, Sun RC, Lu Q (2006) Anatomical, structure and lignin distribution in cell wall of *Caragana korshinskii*. Ind Crops Prod 24:186–193

Epilogue

Fiber or clothes are the basic need of human beings and essential after food and shelter. Ever-increasing population particularly in Southern Hemisphere and China presents a difficult challenge to feed and provide clothes to all. Besides clothing, plant fibers are used for several daily needs and households such as bags, ropes, curtains, and in papermaking. Thus, there is need to explore and exploit new resources to meet the demand by studying their biology, fiber properties, and applications. Search for new plant-based fiber is continuous and new resources are put to test for various domestic appliances, cloth manufacture, and composite formation.

Cotton is one of the oldest cultivated plants and categorized as the most economically important non-food crop. Cotton cultivation provides livelihood to over 100 million farmers. However, cotton cultivation is neither highly productive nor self-sustainable. Sustainable cultivation requires better use of land, water, transport, and labor with benefits to environment and farmers (community). Use of synthetic fiber is declining, and demand for natural fibers such as cotton and linen is increasing day by day. With increasing middle class in Asian, Latin American, and African countries, increase in cotton demand is envisaged being a preferred fabric. Growing organic cotton is increasing in several countries because of environmental issues and consumer preferences, and the demand is growing at 20 % per annum. Currently, ~20 countries are growing organic cotton including USA, Turkey, Syria, China, and India.

This book was aimed to present a current scenario of state of technology for fiber resources, new technologies being developed for improvement of fiber and plant traits as well as new applications of fiber. Fiber being a renewable resource finds its applications in automobile industry, medical appliances, and food industry besides its traditional use as fabric and papermaking. Useful information is given in the chapters on global scenario of cotton cultivation, development of regeneration methods for cotton and bamboo using plant biotechnological tools, cotton fiber biotechnology toward improvement of cotton fiber and genes involved in the process, agronomic traits for transgenic cotton, genomics of jute toward better understanding of fiber formation, linen fiber and its processing, and applications of natural fiber in cellulose acetate film and paper industry. New resources of plant fiber are also described.

K.G. Ramawat and M.R. Ahuja (eds.), *Fiber Plants*, Sustainable Development and Biodiversity 13, DOI 10.1007/978-3-319-44570-0

How natural resources of fiber plants are domesticated, cultivated, and improved is summarized in Fig. 1. Efforts are being made with new tools of molecular biology to develop designer crops with diverse but specific use. Understanding the underlying mechanism of cellulose formation and deposition will provide insight to

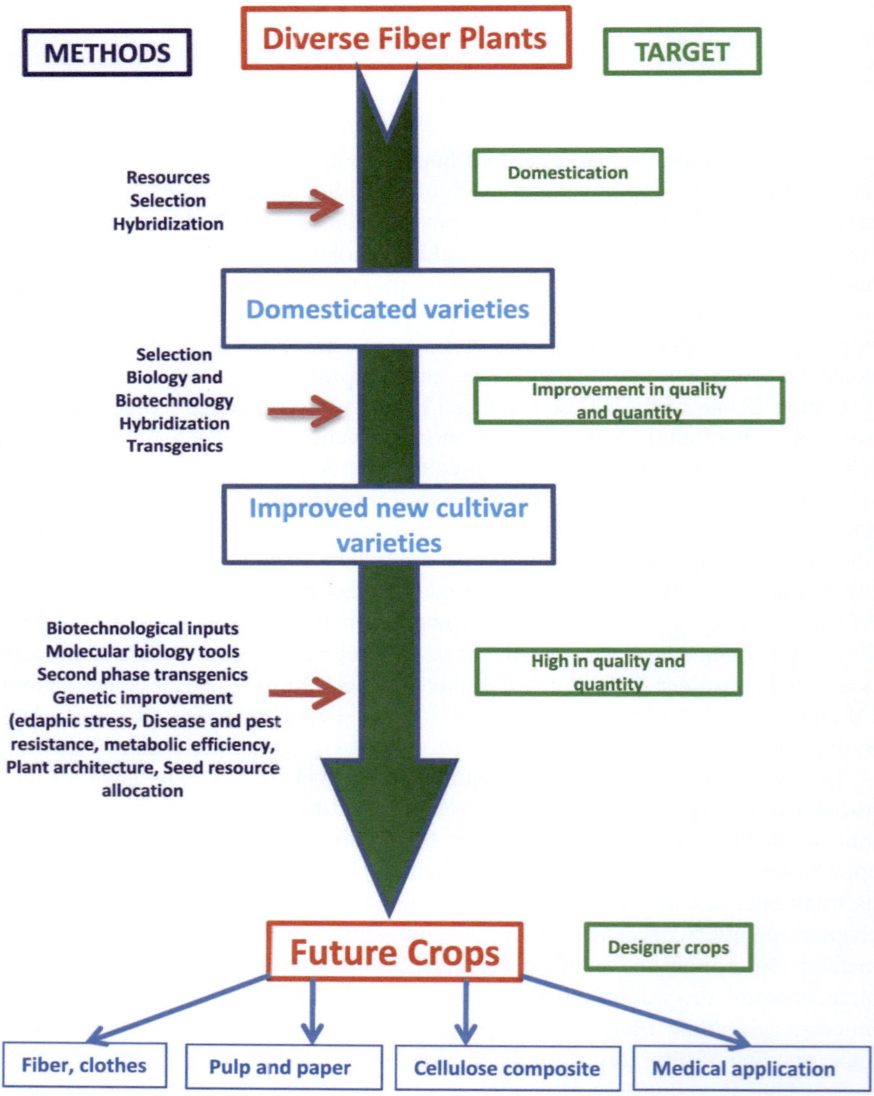

Fig. 1 Schematic presentation of improvement of plants and cultivar varieties to obtain plants with desirable traits over the time using modern tools and techniques of plant breeding, molecular biology, and biotechnology

develop new varieties with desirable characteristic for specific application. Cotton fiber is produced as elongation of outer cell layer of ovules (seeds), making it a difficult trait to modify by conventional methods. Multigene resistance using trangenes and understanding the cellulose biosynthesis and its deposition in the cell walls may lead to the development of plants with better fiber or less woody biomass. Such plants will be useful in developing technology for plant-based biofuels production as an alternative to fossil fuel, which are available in limited amounts and causing pollutions. In this direction, proteins responsible for the cellulose deposition have been identified which will help in developing crops with desired cellulose deposition (Zhang et al. 2016).

The consumer preference for natural dye coloring of fiber leads to the development of new technologies. Declining/banned use of synthetic dyes leads to greater use of natural dyes. Coloring cotton for cloth manufacture is a technology in itself. This is associated with several technical and biological problems such as resources, availability, fastness of the color, and color itself. The idea of developing colored cotton is almost two decades old, and attempts are being made to develop new colored and durable fabric using new technologies such as genetics, genetic engineering, genomics, nanotechnology (nanotech coating), air cleaning, no iron, feel and stretch with simultaneously curbing water wastage (Semizer-Cuming et al. 2015; Reddy and Yang 2015; Ma et al. 2016).

Loss of crop due to insects and pests is another major problem. Bt cotton imparting resistance to bollworm developed by Monsanto is grown in several countries for the last two decades after in its introduction in 1996 and in 2006. Monsanto introduced a stacked second generation product, Bollgard® with Roundup® ready flex cotton (genes for Bt and Roundup resistance); yet scanty information is available on its impact on other species, environment, labor economics and poverty, use of pesticides, and nontarget insects (Zhao et al. 2016).

Though cotton is one of the oldest crops, still production yield and quality can be increased by studying biology of pollinators (it is an outcrossing crop and pollen grains are sticky, thus wind pollination is not possible), breeding long lint cultivars with local varieties imparting traits like disease and insect resistance such as bt cotton, drought resistance, genes (enzymes) involved in cellulose biosynthesis and deposition and so on (Watanabe et al. 2015). Since male sterile lines are available, they ensure good hybrid seed production. Some of these issues are addressed in this book.

Composites made with natural fibers including cotton are gaining importance as well as research support because they are lightweighted, save fuel, and have sufficient strength. Parts made of natural composites are not only used for automobile industries but are also used in the most advanced jet like Boeing 787 dreamliner (Akampumuza et al. 2016; Koniuszewska and Kaczmar 2016). In conclusion, we have been able to address some of the issues in various chapters about improvement of fiber and its quality, agronomic traits such as plant structure, seed resources allocation, and metabolic efficiency to improve the plant and its products, and molecular tools and techniques to understand the process of fiber formation. We hope this will be useful book for all those interested in plant fiber.

References

Akampumuza O, Wambua P, Ahmed A, Li W, Qin X (2016) Review of the applications of biocomposites in the automotive industry. Polym Compos. doi:10.1002/pc.23847

Koniuszewska AG, Kaczmar JW (2016) Application of polymer based composite materials in transportation. Prog Rubber Plast Recycl Technol 32(1):1

Ma M, Luo S, Hu Z, Tang Z, Zhou W (2016) Antioxidant properties of naturally brown-colored cotton fibers. Text Res J 86(3):256–263

Reddy N, Yang Y (2015) Colored cottons. In: Innovative biofibers from renewable resources. Springer, Berlin, pp 331–345

Semizer-Cuming D, Altan F, Akdemir H, Tosun M, Gurel A and Tanyolac B (2015) QTL analysis of fiber color and fiber quality in natural green colored cotton (*Gossypium hirsutum* L.). Turk J Field Crops 20(1)

Watanabe Y, Meents MJ, McDonnell LM, Barkwill S, Sampathkumar A, Cartwright HN, et al (2015) Visualization of cellulose synthases in *Arabidopsis* secondary cell walls. Science 350 (6257):198. doi:10.1126/science.aac7446

Zhang Y, Nikolovski N, Sorieul M, Vellosillo T, McFarlane HE, Dupree R, Kesten C, et al (2016) Golgi-localized STELLO proteins regulate the assembly and trafficking of cellulose synthase complexes in *Arabidopsis*. Nat Commun 7:11656. doi:10.1038/ncomms11656

Zhao Y, Zhang S, Luo JY, Wang CY, Lv LM, Wang XP, Cui JJ, Lei CL (2016) Bt proteins Cry1Ah and Cry2Ab do not affect cotton aphid *Aphis gossypii* and ladybeetle *Propylea japonica*. Sci Rep 6:20368. doi:10.1038/srep20368

Index

K.G. Ramawat and M.R. Ahuja (eds.), *Fiber Plants*, Sustainable Development and Biodiversity 13, DOI 10.1007/978-3-319-44570-0

MIX
Papier aus verantwortungsvollen Quellen
Paper from responsible sources
FSC® C105338

If you have any concerns about our products,
you can contact us on
ProductSafety@springernature.com

In case Publisher is established outside the EU,
the EU authorized representative is:
Springer Nature Customer Service Center GmbH
Europaplatz 3, 69115 Heidelberg, Germany

Printed by Libri Plureos GmbH
in Hamburg, Germany